Handbook of

Fiber Finish

Technology

Handbook of

Fiber Finish

Technology

PHILIP E. SLADE

PHILIP E. SLADE

Gulf Research Services
Pensacola, Florida

CRC Press
Taylor & Francis Group
Boca Raton   London   New York

CRC Press is an imprint of the

CRC Press is an imprint of the
Taylor & Francis Group, an **informa** business

CRC Press
Taylor & Francis Group
6000 Broken Sound Parkway NW, Suite 300
Boca Raton, FL 33487-2742

First issued in paperback 2019

ISBN-13: 978-0-8247-0048-5 (hbk)
ISBN-13: 978-0-367-40086-6 (pbk)

Visit the Taylor & Francis Web site at
http://www.taylorandfrancis.com

and the CRC Press Web site at
http://www.crcpress.com

Preface

The textile industry, including yarn, fabric, and end–product manufacturing, is one of the largest employers in the world. In the United States alone, more than 320,000 people are engaged in working in the overall textile industry. According to the American Textile Manufacturers Institute, the textile contribution to the Gross Domestic Product of 61 billion dollars is second only to that of the automobile industry, at 67.5 billion dollars. The area surrounding Dalton, Georgia, has about 30,000 people who work in carpet manufacturing. Of course, in other parts of the world many more people comprise the work force. Sources in India estimate that about 15 million individuals are directly employed by textile industries. These large numbers call attention to the need for concise information about fibers and their conversion into a final product.

A majority of textile products contain natural fibers that nature has supplied with their own lubricant coating to aid in further processing. Man–made fibers, however, must be treated during spinning with a formulation of various chemical substances to coat those fibers for assistance in end–use manufacturing. The purpose of this book is to provide the reader with information about the components in these finishes, their chemical and physical properties, and their effects on fibers and yarn. When the author first became involved in the field of fiber finishes, there was no one source of that information, and data had to be obtained from each supplier. This handbook is designed to assemble that information into a single, comprehensive treatise.

For many years, most fiber and fabric finishes were compounded by trying the material on test yarn to see if it would be satisfactory. Gradually, a more scientific approach was developed, and a formulation could be appraised by physical and chemical testing. This approach is also discussed in the book. It is hoped that it will be beneficial to those in the textile industry who need to know more about finishes.

There are many people without whose help this endeavor would have never been completed. The first is my wife, Nancy, who encouraged me to attempt and complete this project. Her support is greatly appreciated. Other people who helped in gathering the many papers and patents necessary for a comprehensive treatise are the librarians at the Burlington Textiles Library at North Carolina State University, Georgia Tech, Duke University Chemistry Library, and The University of Southern Mississippi, and especially Janice LaMotte at the Monsanto Technical Center Library. I also want to acknowledge the generous help of my daughter, Kara, who spent hours away from her own graduate studies to assist in the library research. Many other individuals at a number of component suppliers sent data sheets for their products, especially Ethox, Henkel, Hoechst–Celanese, ICI, Stepan, Inolex, and several other companies. The kind assistance of Yash Kamath at the Textile Research Institute is gratefully acknowledged.

Several other people helped in reading the manuscript to insure technical accuracy and finding my typographical mistakes: Dr. Debra Hild, Dr. Robert Peoples, John Foryt, David Floyd, Dr. Gordon Berkstresser, and Dr. Kaye Wolf Angus, all my former co–workers at Monsanto. The advice and helpful suggestions by my editors, Linda Schonberg and Anita Lekhwani, will certainly make the book more readable and increase its value as a contribution to the technical literature.

Philip E. Slade

Science is simply common sense at its best, that is, rigidly accurate in observation, and merciless to fallacy in logic.

Thomas Henry Huxley
(1825-1895)

Contents

CHAPTER 8. OTHER FINISH ADDITIVES: ANTIOXIDANTS, DEFOAMERS, ANTIMICROBIALS, AND WETTING AGENTS

CHAPTER 1

INTRODUCTION

Although many valuable products are fabricated each day from synthetic fibers, these items could never exist unless a lubricating finish had been applied to the fibers during the extrusion or spinning process. A package, bobbin, or bale cannot be made without application of a spin finish. No drawing or texturing could take place and additional processing of the yarn could not be accomplished, either through internal manufacturing or at a customer's plant, without the presence of this finish. The fibers would be a useless, tangled mass of extruded polymer unless the creativity of the finish chemists results in the formulation of a lubricating mixture that is applied early in the manufacturing process. These may appear to be strong statements to a casual observer, but they are not without validity since *all* natural fibers are created with a lubricating finish on the surface of those products.

Wool is coated with a product known as wool fat, which in its purified form is called lanolin. One curious observation is that the lubricating fat content of wool varies depending upon the country of origin [1]. This is shown in Table 1-1. The composition of wool fat has been investigated and it consists mostly (73%) of the esters of cholesterol and other fatty alcohols, some fatty acid esters and free alcohols (25%), and small amounts of free fatty acids and hydrocarbons [1].

Cotton also has a coating on the surface of the fiber, called cotton wax. This wax is composed of about 43% free monohydroxy aliphatic alcohols, about 13% phytosterols, both in the free form and as their esters, about 22% esters of wax acids, and about 8% hydrocarbons and some smaller amounts

1

Table 1–1

Fat Content of Wool from Various Sources

Wool	Wool Fat (%)
Australian	35 to 46
African	28 to 41
Peruvian	9 to 14
English	9

(Ref. [1]. Reprinted by permission of Van Nostrand Reinhold)

of other products [1].

As a silkworm spins its cocoon, the silk fiber, or fibroin, is held together by a substance called sericin, or silk gum. A cocoon consists of about 80% fibroin and 20% sericin. The sericin, a protein, is the natural silk finish that is necessary for future processing of that fiber. Some of the sericin is removed by boiling water during processing so that the cocoon can be unwound, but the remainder is left to prevent damage to those delicate fibers [2].

It would be virtually impossible to duplicate either wool fat, cotton wax, or sericin as a spin finish formulation since most of the components of these compositions are not commercially available. The finish chemist must, therefore, formulate a mixture of available chemical components that will allow synthetic fibers to be processed on textile machinery into a final product that the consumer will buy.

In this handbook of finish science and technology, we will explore the chemistry, physics, and engineering of finishes and their components. The testing of those products to predict how the materials will process on synthetic fiber manufacturing equipment, and how these fibers are processed into a final product, will also be examined.

1.1 ART AND TECHNOLOGY

The synthetic fiber industry began in the middle of the 19th century with the introduction of nitrocellulose fibers, but full commercial production of this type of fiber did not commence until after 1890. The nitrocellulose process was followed by the cuprammonium process and then the viscose process, for which commercial production began in about 1905. Some spin finishes were necessary for fibers made using these processes, but little is known about their compositions. With the production of rayon, then known as artificial silk, during the First World War, mineral–oil–based finishes emulsified with a soap began to be produced. Their formulation was based mainly on trial and error and the compositions were safeguarded as secret mixtures. Until about twenty years ago, spin finish development was almost entirely an art, and one gained experience through many failures and only with an instinct that a formulation would be satisfactory. Since then, a substantial amount of time and energy have gone into the transformation of finish development from art into technology. This transition was accomplished in both the synthetic fiber producers' research laboratories and in the laboratories of finish formulators or component suppliers.

There are many difficulties when finish formulation is perceived as an art, without logical scientific reasoning. Some people never seem to develop the knack for putting together a group of components and arriving at a finish composition that will solve a problem. There are major disadvantages, however, for remaining in the art mentality:

1. Individual chemists tend to become highly specialized in their jobs, which might limit their chances for advancement or in their professional development.
2. For the fiber company, a large amount of critical technology accumulates in the skills of a few people who might not be inclined to pass this technology along to others.
3. It takes a very long incubation time for a technologist to become skilled in finish development. In rapidly changing fiber markets, that time may be excessive if a person who is skilled in finishes should leave the company or retire. Despite being apprenticed to some excellent tutors, this author did not achieve a comfortable feeling about finish formulation for at least five years after entering the field.

Although one can never completely change finish development from an art into an exact science, it is undeniably worth the effort to attempt that approach.

This manual is not intended to be the last word in finish technology. In the fiber manufacturing business, change is ever present. Each time that a technique is perfected or a new component evaluated, another improvement thrusts itself upon the horizon. It is important to broaden the knowledge base and to determine that the latest formulation is certainly not the last. One must always work toward making his or her latest creation obsolete, and efforts *must be directed toward something better*, for if they are not, a competitor will achieve that goal.

1.2 REQUIREMENTS FOR SPIN FINISHES

A spin finish must satisfy many requirements [3] to allow the yarn on which it has been applied to be processed satisfactorily. The finish chemists must keep these in mind when compounding their formulations. The considerations may be summarized as:

1. **Safety.** The finish must not contain any materials that are toxic or cause injury to persons in contact with the finish, either in the liquid state or on the yarn. One area that must be addressed specifically is that of waste disposal. Many finish components are removed from the yarn during dyeing or some other processing step, and these must not be hazardous to the environment.

2. **Yarn Lubrication.** Perhaps one of the most important characteristics of a spin finish is to provide the proper lubricity for the yarn or fiber. This lubricity includes both fiber to metal and fiber to fiber lubrication. It is most important to note the use of the phrase *proper lubricity*, because many people who are not skilled in the science of fiber lubrication assume that the lower the friction between the fiber and another object the better the fiber will process, but that is not always the case. Often one must adjust the lubrication of the yarn to higher frictional values to yield enhanced yarn performance.

3. **Antistatic Properties.** The finish must be able to dispel the static electrical charges that build up on the yarn. While static dissipation is attributed chiefly to the conductance of a water layer, the use of an antistatic agent to enhance this conductance is very important.

4. **Microorganism Growth.** The finish must not allow the growth of bacteria, fungi, or yeast since these organisms could seriously interfere with the processing of the yarn.

5. **Wetting.** If the finish does not wet the yarn properly, the application will be uneven and this will present problems in processing.

6. **Emulsion Quality.** Since most finishes in the United States are made as an oil–in–water emulsion, the emulsion must be a stable suspension with small particle size and a large negative zeta potential.

7. **Chemical Effects.** The finish should neither absorb too quickly into the fiber, nor react with the fiber in any way.

8. **Oxidation.** During prolonged storage of the fiber, finishes should not undergo any oxidative degradation.

9. **Viscosity.** The viscosity of the components and the final finish formulation should be homogeneous and be relatively free of temperature effects.

10. **Removal.** Spin finishes should be easily removed during the scouring process prior to dyeing.

11. **Cost.** The components should be available at a reasonable price.
12. **Biodegradability**. The finish must be biodegradable in the public waste treatment facilities that are available to the plants involved.
13. **Thermal Properties**. The finish should have the correct thermal properties to allow the proper processing on heated equipment.

Although it is probably impossible to formulate a finish that meets all these requirements, it is possible to compromise and thus achieve as nearly as possible the ideal composition.

1.3 COMPONENTS IN SPIN FINISHES

Every spin finish consists of one or more components [4]–[6], with the choice of materials dependent upon the end use of the yarn being prepared. Generally, a finish will contain the following materials:

1. **Lubricants.** These components may be water insoluble materials such as esters, alkanolamides, mineral oils, long chain fatty acids or alcohols, fluorocarbons, or silicones. Water soluble materials as ethylene oxide–propylene oxide copolymers are very valuable as components. At times, emulsifiers, cohesive agents, or antistats may provide lubrication properties.
2. **Emulsifiers.** If the lubricant is water insoluble and if the finish is to be applied as an oil–in–water emulsion, an emulsifier is necessary to obtain the proper suspension. Some typical emulsifiers are ethoxylated glycerides, ethoxylated fatty acids or fatty alcohols, and polyglycol esters. Anionic emulsifiers, such as sodium lauryl sulfate normally used in dishwashing detergents, find limited use in spin finishes.
3. **Antistatic Agents**. Anionic, cationic, nonionic, and some amphoteric components are used in spin finishes as antistatic agents. The anionic components are mostly phosphated alcohols while the cationic components are generally quaternary amines. Some of the nonionic materials include betaines and amine oxides.
4. **Antimicrobial Compounds.** The growth of bacteria, yeast, and fungi can be controlled by using one or more of many antimicrobial compounds that are available for use in aqueous systems.
5. **Other Components.** Other materials such as cohesive agents, humectants, wetting agents, and similar materials can be and are frequently used in spin finishes.

Thus, as it can easily be observed, a spin finish is a complex mixture of a wide range of chemical components and a mixture that is required to assume many roles in fiber processing. Nevertheless, no matter how good a finish may be, it can never completely cover up poor polymer quality or a spinning system that is mechanically unsatisfactory.

REFERENCES

1. Warth, A. H. (1947). *The Chemistry and Technology of Waxes*, Van Nostrand Reinhold, New York, pp. 75, 79, 126
2. Potter, M. D. and Corbman, B. P. (1967). *Textile Fiber to Fabric*, Gregg/McGraw–Hill, New York, p. 264
3. Vaidya, A. A. (1982). The Chemistry and Technology of Spin Finishes, *Synthetic Fibres*, October/December, p. 24
4. Redston, J. P., Bernholz, W. F., and Schlatter, C. (1973). Chemicals Used as Spin Finishes for Man–Made Fibers, *Textile. Res. J.*, **43**: 325
5. Redston, J. P., Bernholz, W. F., and Nahta, R. C. (1971). Emulsifier Choice in Design of Finishes for Man–Made Fibers, *J. Amer. Oil Chem Soc.*, **48**: 344
6. Bernholz, W. F., Redston, J. P., and Schlatter, C. (1984). Spin Finish Usage and Compounding for Man–Made Fibers in *Emulsions and Emulsion Technology, Part III*, K. J. Lissant, Ed., Marcel Dekker, New York, pp. 215–239

CHAPTER 2

TENSION, FRICTION, AND LUBRICATION

During both the manufacturing and subsequent processing of synthetic fibers, the proper control of yarn tension is extremely important. Mechanical equipment must be designed so that the yarn tension needed for each step in the process is optimized. Rolls, guides, drive motors, texturing units, tangle chambers, and similar devices are carefully chosen so that the yarn will acquire the tension essential for that process. There are two forces involved in this control of tension. One is a mechanical *driving* force such as a rotating roll, a loom shuttle forcing yarn through the warp by some mechanism, an air or steam jet propelling the yarn in a forward direction, or some similar action. The second force is a *retarding*, or frictional, force caused by the yarn passing over surfaces that are fixed or moving at a forward speed slower than the yarn speed. The clutch that links the forward driving forces to the retarding force is the spin finish.

2.1 GENERAL MECHANISM OF FRICTION

When one solid body slides over another under the influence of a load, W, a finite force resists that motion. This force is called the frictional force and is designated F_s for static and F_k for kinetic operating conditions [1].

7

Leonardo da Vinci (1452–1519) proposed two of the basic laws of sliding friction [2] of solids:

1. Friction is independent of the area of the solids.
2. Friction is directly proportional to the load that is normal to the sliding direction.

A simple experiment can easily demonstrate the viability of these laws. A brick pulled along a surface displays the same frictional force if the brick is either lying flat or standing on its edge. If the load is doubled by placing a second brick on top of the first, the force required to move the bricks is twice as large. The addition of a third brick results in a threefold increase in friction. This means that for any particular pair of solids, the ratio of:

$$\frac{Friction(F)}{Load(W)} \qquad (2\text{--}1)$$

is constant and this value is called the coefficient of friction, μ.

The French engineer Amontons redefined these laws in 1699 and added his own mathematical interpretation. His observations were verified in 1785 by C. A. Coulomb [4], and Equation (2–1) is commonly referred to as Coulomb's law. The two latter investigators suggested two additional basic laws to those proposed by da Vinci, namely:

3. The static coefficient of friction is greater than the kinetic coefficient.
4. Coefficients of friction are independent of sliding speed.

A common example of friction is one of two metals sliding against each other. It is generally agreed that unlubricated metals produce a frictional force arising from two separate components. The first, which is usually the main one, is the adhesion of the surfaces at the real areas of contact. Since no surface is truly smooth, these real areas of contact are sections of the solid that protrude from the main surface and are usually much smaller than the total surface area of that solid. These protrusions are called asperities. The main component of friction between these sliding surfaces is the force necessary to shear the junctions at these protrusions. Very simply stated, if A is the total area of shear and s the average shear strength of the weaker material, the adhesion component of friction can be stated as:

$$F_{\text{adhesion}} = A\,s \qquad (2\text{--}2)$$

A metal surface exhibits a nearly normal or Gaussian distribution of asperity heights [5] according to the relationship:

$$\phi(s) = [\frac{1}{\sqrt{(2\pi)}}]\exp(-\tfrac{1}{2}s^2) \qquad (2\text{--}3)$$

in which $\phi(s)ds$ expresses the probability that a particular asperity on the surface has a height between s and $(s + ds)$ above some reference plane. If this height distribution is assumed, it can be shown that the actual area of contact is directly proportional to the load:

$$A = K_1 W \qquad (2\text{--}4)$$

where K_1 is a constant. The mean actual pressure P_a of those systems is defined by the relationship:

$$W = A P_a \qquad (2\text{--}5)$$

and it follows from Equations (2–4) and (2–5) that P_a is constant and independent of load.

The second frictional component is deformation, such as plowing or grooving, of the softer material by the harder one. For two similar metals, this component is much smaller than the adhesion one, but for a soft metal sliding against a harder one, or a polymer against a hard surface, it can become important. This is the deformation term, and the total frictional force is the sum of the two:

$$F = F_{\text{adhesion}} + F_{\text{deformation}} \qquad (2\text{--}6)$$

Equation (2–6) may be combined with Equation (2–2) and the result is:

$$F = A s + F_{\text{deformation}} \tag{2-7}$$

Since for metals the deformation term is insignificant, Equation (2-2) can be viewed as contributing the entire resistance F to motion. If we now divide both sides of Equation (2-2) by W, an expression for the coefficient of sliding friction, μ, is defined:

$$\mu = \frac{F}{W} = \frac{A s}{A p_a} = \frac{s}{p_a} \tag{2-8}$$

This relationship thus describes the simple theory of unlubricated friction for metals [5]. Although this might be a possible oversimplification of unlubricated friction, the general principles still apply.

2.2 POLYMER FRICTION

The common concepts of the adhesion theory of metallic friction have been extended to include polymer friction. For example, it has been confirmed experimentally that during sliding of one polymer against another, strong adhesion does occur, and that fragments sheared off from one polymer have been observed adhering to the other polymer [1][6]. As a result, for most polymers, reasonable agreement has been found among shear strength, yield pressure, and the coefficient of friction, as shown in Equation (2-8). Bahadur [7] experimentally confirmed this agreement. Adams [8] simultaneously measured the friction and apparent contact area of nylon on glass and concluded that the tangential force of friction had little effect on contact area. He also interpreted the data from the preceding paper in terms of the adhesion theory of friction with the additional hypothesis that the shear strength of the contact area between nylon and glass increases with pressure [9]. One exception to the generalization of the adhesion theory of polymer friction is poly(tetrafluoroethylene). The very low surface energy of this material causes almost no adhesion to occur [10]. It would thus seem that there is reasonable agreement between the adhesion theory of metal friction and polymer friction. In contrast, there is a considerable difference in the nature of deformation between these two materials. Whereas metals generally exhibit only plastic deformation, with polymers one must also consider elastic deformation over certain ranges of loads [11]. As a result of this viscoelastic deformation there is a hysteresis loss with polymeric substances. Similarly, the viscoelastic deformation also influences the total frictional force in Equation (2–6) and the classical laws of da Vinci, Amontons, and Coulomb no longer hold [9].

Before considering the concepts of fiber lubrication, some experiments by Bowers, Clinton, and Zisman [12] on the lubrication of nylon film should be summarized. This paper has an excellent description of stick–slip and also some data on lubricants for nylon. These authors state:

Sliding of dry nylon on dry nylon always produced intermittent motion and seizure or stick–slip. When this occurs, the static coefficient of friction, μ_s, can be calculated from the maximum frictional displacement of the slider at the "stick", while a close approximation to the kinetic coefficient of friction, μ_k, is calculated from the frictional displacement value halfway between the maximum displacement during the "stick" and the minimum value during the "slip".

The effect of various lubricants on nylon friction is listed in Table 2–1. The authors caution that the coefficient of friction as listed may contain some component of the hydrodynamic, or high speed, friction because of the velocities of the slider. The data in Table 2–1 show that effective boundary, or slow speed, lubricants must be able to absorb either chemically or physically at the surface and thus decrease the real area of contact, and therefore the adhesion, of the two rubbing surfaces. All three of the most effective lubricants in their solutions appear to be better lubricants than the solvents. Stearic acid gave a significantly lower μ_s than octadecylamine even though these chemicals contain the same number of carbons in their aliphatic chain. This suggests that the acid groups bond more strongly to the nylon surface than do the amine groups. Perfluorolauric acid lubrication resulted in higher values than did stearic acid although some evidence has been reported [13] that fluorinated material adsorbs more effectively on the surface of nylon by hydrogen bonding than does stearic acid. The difference in the coefficients of friction may result from weaker cohesion forces between the $-CF_2-$ groups of neighboring molecules than between the $-CH_2-$ groups. This difference in cohesion between adjacent molecules may be even more important than the length of the carbon chain. It may also explain why pelargonic acid is a better lubricant than perfluorolauric acid even though there are three fewer carbon atoms on the former molecule than the latter. It is also feasible that pelargonic acid is a better lubricant than nonyl alcohol since the acid group should adhere to the nylon better than the alcohol group. The adsorption theory of polymer lubrication has been confirmed

Table 2–1

Boundary Coefficients of Friction for Nylon

Lubricant	Nylon on Nylon		Steel on Nylon	
	Static	Kinetic	Static	Kinetic
Dry	0.46	0.37	0.37	0.34
Water	0.52	0.33	0.23	0.19
Ethylene glycol	0.58	0.19	0.20	0.16
Glycerol	0.36	0.19	0.23	0.18
n–Decane	0.37	0.27	0.34	0.30
n–Hexadecane	0.34	0.18	0.30	0.26
Perfluorolube oil	0.58	0.24	0.30	0.19
Perfluorotrihexyl amine	0.43	0.20	0.31	0.14
n-Nonyl alcohol	0.35	0.16	0.23	0.14
Pelargonic acid	0.31	0.14	0.17	0.12
Oleic acid	0.29	0.13	0.15	0.08
Stearic acid in hexadecane (0.1%)	0.22	0.13	0.13	0.08
pri–n–octadecylamine in hexadecane (0.1%)	0.30	0.20	0.17	0.14
Perfluorolauric acid in decane	0.33	0.26	0.30	0.28
Polymethylsiloxane	0.43	0.17	0.19	0.12

(*Reference [13]. Reprinted with permission of Ind. and Eng. Chem.*
Copyright 1954 American Chemical Society.)

by Cohen and Tabor [14] and Fort [15]. These experiments, and other research reported in later sections, suggests that an effective boundary lubricant for nylon should be attracted to the surface and have good cohesive energy between the molecules.

2.3.1 Introduction

Textile yarns pass over many different surfaces during their manufacturing and processing steps. Guides of differing surface roughness are contacted, rolls drive the yarn toward the end of a machine, and fibers rub against each other in processes such as texturing, drafting, weaving, and knitting. In all these processes fiber lubricants are necessary to control friction, wear, and static electricity. The friction must not only be controlled but must be kept at a uniform level to make a quality product. In order to attain consistent friction and tension, the finish must be applied uniformly so that an even distribution of lubricant along the threadline can be achieved.

The critical elements in the control of friction are (1) the pressure of the yarn against the surface, which is largely a function of the pre-tension of the yarn before the contact point; (2) the lap angle that the yarn makes around the surface with which it is in contact; and (3) the effective coefficient of friction between the yarn and the surface at a given speed. The yarn speed at the contact point is also important, but mostly because of secondary effects. These effects at slow speeds are due to interactions between yarn motion, yarn elasticity, and tension. At high speeds the effects on friction are due to changes in contact, air entrainment, heating at the contact point, the material of construction of the contacting surface, and the viscosity and uniformity of the lubricant.

A technique commonly used for measuring the friction of yarns is to determine the tension before (T_1) and after (T_2) a friction point and to relate these measurements by the ratio of T_1 to T_2, by the difference between the tension values, or by calculating a numerical coefficient of friction. The latter number is much more useful and values are easily compared. The friction coefficient is calculated by the capstan formula for coil friction [16]–[18]. This equation is:

$$\frac{T_2}{T_1} = e^{\mu\theta} \tag{2-9}$$

where:

T_1	=	tension before the friction point,
T_2	=	tension after the friction point,
e	=	base of the Napierian logarithm (2.718),
μ	=	coefficient of friction, and
θ	=	lap angle in radians.

Equation (2–9) can be rearranged and converted to base 10 logarithms and we find:

$$\mu = \frac{\log_{10}(T_2/T_1)}{0.4343\,\theta} \tag{2–10}$$

Schlatter has suggested [19] that the equations for coil friction were derived from the basic laws of metallic friction and they would not hold for fibers or textiles. Bowden and Young [20] developed an equation that has been shown to apply to textile materials:

$$F = aW^n \tag{2–11}$$

Where F and W are identical to the terms in Equation (2–1), a is a constant that is equal to μ when $n = 1$, and n a measure of the elasticity or plasticity of the materials in frictional contact. For two perfectly plastic substances $n = 1$, and $n = 0.67$ for two perfectly elastic ones.

Using this equation, Howell [21] and Lincoln [22] proposed a revision to the capstan equation:

$$\frac{T_2}{T_1} = \left[1 + (1-n)\,a\,\theta\,\left(\frac{r}{T_1}\right)^{1-n}\right]^{\frac{1}{(1-n)}} \tag{2–12}$$

in which r is the radius of the cylinder employed as the friction contact surface. As $n \to 1$, Equation (2–12) becomes:

$$\frac{T_2}{T_1} = e^{a\theta\,(r/T_1)^{1-n}} \tag{2–13}$$

which corresponds to Equation (2–9) in the special case when:

$$\mu = a\left(\frac{r}{T_1}\right)^{1-n} \tag{2–14}$$

or when $n = 1$. Thus T_2 / T_1 is not constant but depends on the normal force as determined by T_1, on r, and on n. This has been substantiated experimentally and theoretically [23]–[25].

These equations do not apply for yarn or fiber to which a finish has been added. The concept of n presupposes a free and random interaction of one surface with the asperities or points of adhesion of the other surface in frictional contact. A finish present on the surface will hinder or completely prevent this interaction [19]. In studies of liquid lubricated yarn, Equations (2–9) and (2–10) are normally the formulae of choice to calculate the coefficient of friction. The generally accepted experimental procedure is to plot μ against yarn speed or some similar parameter.

Figure 2–1 represents the idealized general frictional behavior of a spin finish lubricated yarn, showing the probable relationships among the coefficient of friction, the lubricant viscosity, the threadline speed, and the pressure between the yarn and the surface with which the yarn is in contact [1][18][26]. The dashed line symbolizes the friction contribution from the boundary lubrication, the dotted line represents the contribution from the hydrodynamic lubrication, and the solid line is the composite of the two. The solid line is typical of friction measurements observed in practice. At very low speeds the boundary friction component of the total friction is at a high value, while at high speeds the hydrodynamic friction is quite high. The semi–boundary region is the transition between the two other regions and represents the lowest friction for the entire system. The speed ranges for boundary friction are assumed to be from 10^{-4} to 10^{-1} meters/minute (m/min), the semi–boundary speeds range from 10^{-1} to 5 m/min, and the hydrodynamic speeds above 5 m/min [26]. The friction at the slowest speed range is called *boundary* because the mechanical, physical, and chemical forces existing at the boundary or interface between the two surfaces are dominant in determining friction. The *hydrodynamic* friction at the higher speeds is given that name because the friction characteristics are primarily dependent upon hydrodynamics, or the flow and shear properties of the lubricant, that is, the viscosity, chemical structure, wetting, and affinity for the surfaces. The intermediate *semi–boundary* region, of course, is a combination of all these parameters. These regions of frictional behavior will be reviewed in greater detail in the next few sections, and the various processing factors that influence friction will be discussed.

2.3.2 Boundary Friction

The primary characteristic of boundary friction is the tendency to stick and slip rather than to move smoothly at uniform speed. This is not a unique property of fibers and yarns but is a fundamental characteristic of all elastic

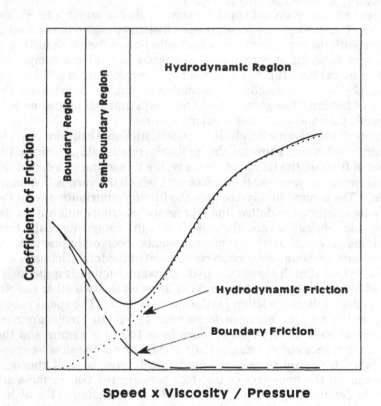

Figure 2–1. General Frictional Behavior of Liquid Lubricated Yarn.
(Reference [18]. Reprinted by permission of the Textile Res. J.)

surfaces in contact with another surface. In this handbook, however, the discussion is limited to yarn friction.

The boundary friction is greatest when the two surfaces that are in contact are clean. At rest, these two surfaces under pressure tend to adhere by intermolecular attraction and the deformation of the fiber surface by the asperities on a harder solid contact. When tension develops in a yarn causing it to move, the tension must build until the force generated is large enough to overcome this adhesion and deformation. When this happens, the movement is sudden and continues only until the tension is relieved, at which time the yarn comes to rest and the adhesion and deformation forces are again more prominent. This process is repeated, if the yarn speed and tensions are constant, until the surfaces are permanently changed. The common term for this behavior is stick–slip. The friction terms used to characterize stick–slip are: (1) the *static coefficient*, a high value prevailing at the point just before the initiation of movement (stick); and (2) a lower *dynamic or kinetic coefficient*, which is typical of the condition just at the point before movement stops (slip). The difference between the two is often quite large, with the static coefficient sometimes twice the dynamic value. Yarn speed, pressure, and contact surfaces have a profound impact on stick–slip behavior. High pressure, high friction, and a plastic yarn tightly conforming to a mirror surface can combine to yield such a large static friction coefficient that the yarn may reach a breaking tension before the static friction forces can be overcome.

One must not complacently assume that boundary friction is limited to a few special situations, such as rope twisting or tire cord production, but it is a very basic phenomenon that is a part of any two surfaces which may be in contact and that have molecular interaction and molecular elasticity. Synthetic yarns, being both plastic and elastic, are highly susceptible to large ranges of deformation under conditions of contact pressure and longitudinal tension. While stick–slip characteristics are more obvious at slow speeds, they certainly exist at higher speeds. Speed of movement is the most common feature in initiating stick–slip, but anything that can cause temporary molecular interaction between the surfaces can start the process.

Let us consider two examples in which boundary friction can play an important role:

1. *Package formation.* Although yarn might be traveling at high rates of longitudinal speed, as the traverse mechanism moves the yarn back and forth across the package, it will come to the edge of the bobbin before it reverses. At the point of the reversal, the yarn speed, in relation to the movement *across* the package, becomes zero. At this reversal, the most important frictional process is boundary friction. In many instances, the addition of a good cohesive agent to the spin finish can improve package formation.

2. *Roll surfaces.* As yarn moves around a drive or feed roll, the yarn speed when *compared to the roll speed* approaches the range for boundary friction. On a draw roll, the yarn is accelerating but still moving slower than the roll surface until, at some point on the surface, it is in equilibrium. Although the yarn is accelerating to the draw roll speed, its speed compared to the roll surface is

decelerating to zero. Again the boundary friction characteristics predominate.

This is not to imply that stick–slip behavior is bad, because it is essential to most yarn processing steps. Without this mechanism, yarn would be even more difficult to control than it is now. Like all fundamental aspects, stick–slip exists, it is real, and we must learn to cope with it and use the mechanism to our advantage.

Several experimental techniques have been proposed for the measurement of boundary friction and these can be divided into two general principles [27]. The first is the *point contact* method, in which one fiber is rubbed against another fiber at right angles to it [28]. A variation of this is the inclined plane method, in which one fiber forms an inclined place and the sliding fiber is loaded and looped around the inclined fiber. The second general technique is the *extended line contact* method, in which a fiber is always in contact with other fibers by twisting together or on a cylindrical surface, either as the bulk material or covered with fibers. A procedure describing the yarn twisting technique is described by Prevorsek and Sharma [29] and is used in some commercial equipment. Röder [30] depicted an apparatus in which a single fiber slides on a layer of the same fibers arranged on the surface of a cylinder. An improved type of this equipment used to determine boundary friction was proposed by Fort and Olsen [31]. The apparatus consists of a small cylindrical roll wound with several layers of the yarn to be tested. A slow speed drive rotates this roll at linear yarn speeds of 0.0001 to 0.5000 meters per minute. Another section of the same yarn is placed over this roll so that the lap angle is 180°. A fixed weight is attached to one end and a tension measuring device with an electrical output determines the force on the other end of the yarn. Typical stick–slip behavior may be recorded by a computer or on a strip chart recorder. A chart representing data that may be obtained from this apparatus is shown by Figure 2–2. The stick part of the friction is represented by the static curve while the slip coefficient of friction is given by the dynamic curve. Fort and Olsen [31] discussed the effects of various lubricants on fiber lubrication and found results similar to those reported by Bowers, Clinton, and Zisman [12]. Since mineral oil has no polar groups, it cannot chemically bond to the fiber surface and that lubricant did not appreciably reduce the boundary coefficient of friction. However, stearic acid and *n*–octylamine were very effective in the reduction of the friction values. It is interesting that oleic acid, which contained the same number of carbon atoms in the aliphatic chain as stearic acid, is not nearly as effective as stearic in the reduction of friction. The authors attribute that difference to the unsaturation in the oleic chain which reduces the cohesion between the adsorbed molecules. It is apparent that effective boundary lubrication can be achieved only from

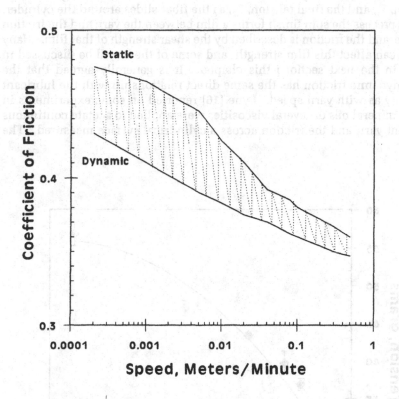

Figure 2–2. Typical Fiber to Fiber Friction at Boundary Speeds.

molecules which are polar so that they are attracted to the nylon surface and ones that tend to form a cohesive layer on that surface.

The point contact procedures are considered to be the most suitable for studying yarn to yarn friction in its purest form but the extended line contact methods probably yield the most reproducible results.

2.3.3 Hydrodynamic Friction

A widespread concept associated with friction is the sliding of one surface against another at high speeds. This concept is true for fiber friction as well as metallic friction. In studying the frictional behavior of fibers in the yarn over a cylindrical surface, the most widely employed technique consists of passing the fiber or yarn over that surface and then measuring the initial tension, T_1, and the final tension, T_2, as the fiber slides around the cylinder. In this process the spin finish forms a film between the yarn and the friction surface and the friction is described by the shear strength of that film. Many things can affect this film strength and some of these will be discussed in detail in the next sections this chapter. It is generally agreed that the hydrodynamic friction has the same direct relationship with the lubricant viscosity as with yarn speed. Lyne [16] reported on some experiments in which mineral oils of several viscosities were applied to acetate continuous filament yarn and the friction across a cylindrical surface measured. The

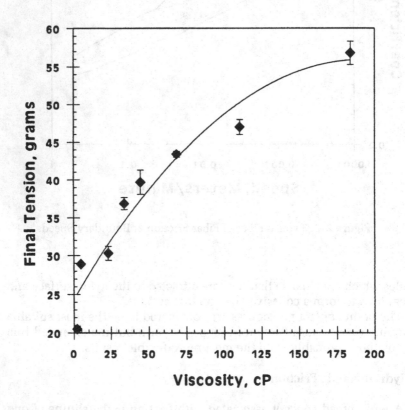

Figure 2–3. Friction of Acetate Yarn versus Lubricant Viscosity.
(Reference [16]. Reprinted by permission of the J. Textile Inst.)

data is presented by Figure 2–3. This plot shows the dramatic increase of the final tension as the oil viscosity increases, especially at viscosities ranging up to about 100 centipoise (cPs). Since speed also contributes to friction, Hansen and Tabor [17] used the data of Lyne and calculated a factor of the viscosity, Z, and the yarn speed, V, to give a better relationship. They propose that it is well known in metal lubrication that velocity has the same effect as the viscosity of the lubricant: that is, if the velocity is increased by a factor of two, the frictional behavior is the same as if the velocity had been kept constant and the viscosity increased by a factor of two. Thus the friction should depend only on the product of V times Z. Figure 2–4 shows a plot of the final tension of the yarn against VZ for which in one case the viscosity was held constant at 44.7 cPs and in the other the speed was constant at 100

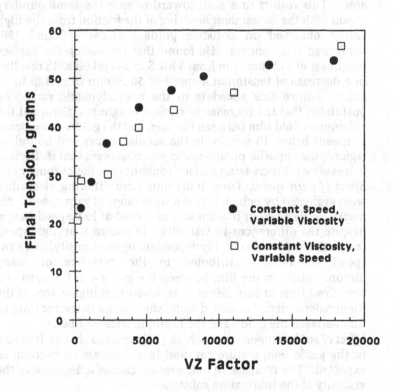

Figure 2–4. The Hansen–Tabor VZ Factor versus Yarn Tension. *(Reference [17]. Reprinted by permission of the Textile Res. J.)*

meter per minute. Since the two curves are almost identical, it can be assumed that yarn friction behaves similarly to metal friction and thus fits the hydrodynamic model.

2.3.4 Influence of Yarn Attributes, Finishing Agents, and Contact Surfaces on Friction

Several authors have published papers on the various features of yarn, finishes, and contact surfaces which can affect friction. These aspects are detailed below.

1. *Effect of guide surface roughness.* According to Olsen [18], an increase in the roughness of a guide surface can be considered analogous to an increase in pressure between the fiber and the guide as a result of the decrease in the fiber–guide surface contact area. This results in a shift toward or into the semiboundary region with the consequent lowering of the friction from the high values observed on polished guide surfaces. Schick [26] confirmed this concept. He found that increasing the surface roughness of a chrome pin from 4 RMS to 20 and 60 RMS resulted in a decrease of tension at a speed of 50 m/min from 135 to 80 grams. Since this speed is in the hydrodynamic range, he postulated that the increase in surface roughness disrupted the continuous fluid film between the yarn and the guide. In contrast, at speeds below 10 m/min (in the semiboundary and boundary regions) the opposite phenomenon was observed, that is, friction increases with increasing surface roughness of the chrome pin.

2. *Effect of yarn speed.* Three lubricants with differing viscosities were evaluated by Schick [32] at a wide range of yarn speeds. No marked difference in friction was observed at low speed ranges despite the differences in viscosity. In contrast, at high speed ranges the friction was highly dependent upon viscosity. The slow speed effects are attributed to the presence of many discontinuities in the film between the guide and the yarn and also direct fiber to chrome contacts, whereas at higher speeds the friction dependence is related to the shear stress in the continuous film between the guide and the liquid lubricated fiber.

3. *Effect of guide temperature.* Schick [33] reported that an increase in the guide temperature resulted in a decrease in friction, as expected. The increase in temperature causes a decrease in the viscosity of the lubricating substance.

4. *Effect of contact angle.* The effects of the contact angle that the yarn makes around a cylindrical pin have also been studied [33]. In this work it was found that the change of contact angle causes

a proportional change in the contact area and that in the boundary region an increase in contact area leads to an increase in fiber wear with a consequent increase in stick–slip. In the hydrodynamic region the friction follows the equation:

$$F = \eta(A \frac{V}{d})$$
(2–15)

where F is the frictional force, A the area of contact between the fiber and the guide, V the yarn speed, d the thickness of the lubricant film, and η the viscosity of the lubricant [34].

5. *Effect of pretension.* Another feature which appears to have a major influence on the frictional properties of yarn is the pretension [33]. It has been demonstrated by several sources [26] [27][33] that stick–slip is magnified with increasing pretension at the fiber to fiber (F/F) interface, which was attributed to the elevated pressure at the areas of contact. Additional work [33] at high speeds showed that at the fiber to metal (F/M) interface the friction is also raised with increasing pretension, and frictional values obtained with the smooth pin exceed those with the rough pin. In line with the explanation given for the boundary region, the increase in friction with higher pretension in the hydrodynamic regions is attributed to the increased pressure at the F/M interface with a consequently enlarged contact area. An increase in the latter leads to greater shearing of the continuous fluid film at the F/M interface.

6. *Effect of denier.* Schick [33] also found, as expected, that the friction increases significantly with increasing total yarn denier, especially in the hydrodynamic region. This is again attributed to the increase in the area of contact at the F/M interface.

7. *Effect of moisture regain.* A study of the effects of moisture on hydrophilic fibers, such as rayon, was carried out [33]. A marked increase in friction was observed in the hydrodynamic region on increasing the RH from 12 to 69%. This increase in friction is attributed again to an increase in area of contact at the F/M interface by the swollen rayon yarn. Röder [30] found that the coefficient of friction of wool fibers increased dramatically at relative humidities over 70 %. Moisture could also affect certain finish compositions because of the formation of certain non–lubricating liquid crystals in the emulsifier components.

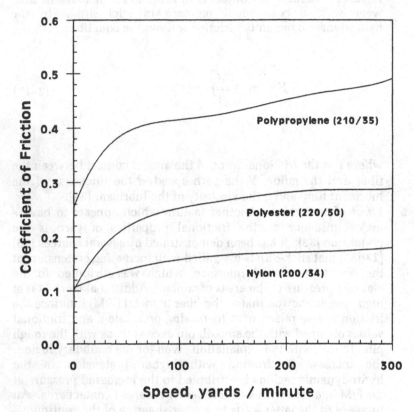

Figure 2–5. Effect of Fiber Material at Constant Denier on Friction.
(Reference [35]. Reprinted by permission of the Textile Res. J.)

8. *Effect of fiber material.* It is a well established fact that different
   fiber materials require specific lubricants. A comparative study
   [35] of the frictional properties of three hydrophobic fiber
   materials of comparable 200–220 total denier and pretension of
   0.05 g/den was carried out and the pertinent data are displayed in
   Figure 2–5. The lubricant used in these experiments was
   Nopcostat® 1296, which has a viscosity of 22.4 cPs at room
   temperature. As this plot shows, the friction increases in the
   following order: nylon 6,6, polyester, and polypropylene. In order
   to establish a link between frictional and wetting properties of
   these materials, the contact angles of the lubricants were

measured on films cast from the same materials from which the fibers were made. The results from this test are shown by Table 2–2. It follows from these data that the excellent wetting of nylon 6,6 by the lubricant is reflected in lower friction. The poorer wetting of the other polymers is also noted with the increase in friction [35]. We must assume that good wetting of the fiber surface by the lubricant is necessary for low friction.

Table 2–2

Contact Angle of Nopcostat® 1296 on Polymer Sheets

Polymer	Critical Surface Tension Dynes/cm	Contact Angle in Degrees
Nylon 6,6	42	4
Polyester	43	6
Polypropylene	24	8

(Reference [35]. Reprinted by permission of the Textile Res. J.)

9. *Effect of lubricant viscosity.* Experiments by Schick [35] have shown that under conditions of good wetting that friction is independent of lubricant viscosity in the boundary region, but highly dependent on lubricant viscosity in the hydrodynamic region. The independence of lubricant viscosity contribution to friction in the boundary region is attributed to the existence of many discontinuities in the film between the guide and the fiber, whereas the dependence in the hydrodynamic region is caused by an increase in the shear stress in the continuous film between the guide and the fiber with increasing viscosity. These results confirm the earlier investigations of Lyne [16].

10. *Effect of the amount of lubricant on the fiber.* In another group of experiments, Schick [26] varied the amount of the finish applied to the yarn from 0.25 to 2.0%. In the measurement of friction versus speed, the curves leveled off to a plateau at about 0.5% weight on fiber. In other work Olsen [18] found a minimum in the finish level versus friction plot at 0.15% weight on fiber that he attributed to the amount required to form a mono–molecular

film on the fiber. This value of the minimum friction agrees very closely with the 0.20% finish found by Röder [30].

11. *Effect of fiber luster on friction.* Luster is an important surface property of synthetic fibers, and its intensity is dependent on the amount of TiO_2 added during polymerization. Three general types of fiber luster are normally encountered: (1) bright fibers which may contain up to 0.1% TiO_2; (2) semidull fibers con–taining about 0.5% TiO_2; and (3) dull fibers containing about 2.0% TiO_2. According to Scheier and Lyons [36], variations in TiO_2 content also affect the surface geometry of the fiber, that is, bright fibers have a smooth surface while dull fibers have a rough surface. These differences in fiber luster have now been related to the frictional properties of these fibers [37]. These experiments showed that the friction of nylon yarns in the boundary and semiboundary regions, as the yarn was being passed over a smooth chrome pin, increases in the following order: bright, semidull, and dull. This increase in the friction in these regions is attributed to the increased fiber wear with increasing surface roughness. The corresponding frictional properties in the hydrodynamic region display the reverse order, that is, friction decreases with increasing fiber roughness. It is postulated that increased fiber roughness disrupts the continuous fluid film that exists between the fiber and the guide, resulting in a shift towards conditions simulating a semi–boundary region with consequent lowering of friction.

12. *Effect of molecular groups on fiber friction.* Very little difference between the effects of –COOH and the –OH groups (*i.e.,* stearic acid and stearyl alcohol) in a molecular structure on fiber friction was noted, but there was a marked decrease in friction upon the introduction of an amide group. Increases in the number of ethylene oxide groups results in an increase in friction since the component viscosity also rises [30].

13. *Effect of crystallinity on friction.* Steinbuch [38] found that as the crystallinity of nylon 6 is increased, the density, stiffness, yield stress, hardness, and abrasion resistance increase. At low levels of crystallinity, however, an increase in crystallinity results in a decrease in the dry coefficient of friction.

14. *Effect of cross section on friction.* Experiments performed on viscose tire cord [39] indicated that the friction coefficients were not related to the shape of the cross–section. These differences in cross–section were caused only by varying manufacturing parameters or raw materials and not by changes in spinneret hole design.

2.3.5 Relationships Among Friction and Other Yarn and Fabric Properties

1. *Relationships among friction, sonic modulus, abrasion resistance, and fatigue parameters.* Studies were made using mercerized and cross linked cotton yarn in which yarn to yarn friction, sonic modulus, abrasion resistance, and fatigue properties were measured [40]. These authors found a good relationship among all of the properties for mercerized cotton, but the correlations when cross linked cotton was the base fiber were not as apparent.

2. *Relationship between friction and yarn and fabric mechanics.* An investigation of woven cotton and aramid fabric revealed that friction is very important in the deformation characteristics of the weaves [41]. Yarn pull–out experiments revealed that interfacial friction directly influences yarn stiffness and therefore weave stiffness. The external filament to filament contacts, as contrasted to internal yarn mechanics, also directly influence the forces required to remove yarns from fabrics. A modification of the frictional characteristics at the smallest scale in a weave structure thus directly changes the behavior of the weave on a larger scale.

3. *Influence of fiber to fiber friction on yarn strength.* Experiments in measuring the breaking strength of polyester yarn have shown that yarn to yarn friction can be a very valuable prediction of yarn strength [27]. Unlubricated polyester yarn was treated with various experimental spin finishes and the yarn strength was determined. The correlation between yarn friction and strength was quite good.

4. *Relationship between friction and dynamic viscoelastic properties of yarn.* Prevorsek and Sharma [29] reached two conclusions during their experiments examining fiber to fiber friction and dynamic viscoelasticity of nylon, polyester, and polypropylene. The first is that the adhesive effect in friction is negligible for these fibers. The second is that changes in fiber morphology resulting from heat treatments, draw ratio, and degree of crystallinity will result in only minor effects on the coefficient of friction. These conclusions agree with data obtained by Ono and Komatsu [42] in which the frictional properties of tire cords were found to have minor effects on viscoelasticity.

5. *Influence of fiber friction on yarn bending recovery.* Fabric wrinkling is reported [43] to be largely dependent on the yarn deficiency angle, $\Delta\beta_y$, which is a function of the denier and the number of filaments. The yarn deficiency angle was found to be directly proportional to the fiber–fiber coefficient of friction and inversely proportional to the fiber modulus.

6. *Influence of fiber friction on sliver, roving, and yarn.* The cotton sliver formed on a card is apparently more dependent upon fiber torsional and bending rigidities than fiber to fiber friction [44]. In roving drafting force, however, fiber friction plays a more dominant role than torsion and bending rigidities. In yarn, fiber tenacity and elongation contribute most to yarn tenacity and elongation, with fiber friction and rigidities secondary in importance.

REFERENCES

1. Schick, M. J. (1975). Friction and Lubrication of Synthetic Fibers, in *Surface Characteristics of Fibers and Textiles*, (M. J. Schick, ed.), Marcel Dekker, New York, pp. 2, 3, 5

2. Richter, I. A., ed. (1952). *Selections from the Notebooks of Leonardo da Vinci*, Oxford University Press, London, England, p. 79

3. Amontons, G. (1699). *Histoire de l'Académie Royale des Sciences avec les Mémoires de Mathématique et de Physique*, Paris, France, p. 206

4. Coulomb, C. A. (1785). Théorie des Machines Simples, *Mémoires de Mathématique et de Physique, présentés a l'Académie Royale des Sciences, Paris*

5. Moore, D. F. (1975). *Principles and Applications of Tribology*, Pergamon Press, Oxford, UK, p. 41

6. Chapman, J. A., Pascoe, M. W., and Tabor, D. (1954). The Friction and Wear of Fibers, *Conference on Fiber Friction*, Ghent, Belgium, September, 1954, pp. P3–P19

7. Bahadur, S. (1974). Dependence of Polymer Sliding Friction on Normal Load and Contact Pressure, *Wear*, **29**: 323

8. Adams, N. (1963). Friction and Deformation of Nylon. I. Experimental, *J. Appl. Poly. Sci.*, **7**: 2075

9. Adams, N. (1963). Friction and Deformation of Nylon. II. Theoretical., *J. Appl. Poly. Sci.*, **7**: 2105

10. Allen, A. J. G. (1957). Wettability and Friction of Polytetrafluoroethylene Film: Effect of Prebonding Treatment, *J. Poly. Sci*, **24**: 461

11. Meland, T. (1962). Frictional Properties and Wear of Plastic Materials, *Tek. Ukeblad*, Oslo, Norway, **109**: 531

12. Bowers, R. C., Clinton, W. C., and Zisman, W.A. (1954). Friction and Lubrication of Nylon, *Ind. and Engineering Chem.*, **46**: 2416

13. Ellison, A. H. and Zisman, W. A. (1954). Wettability Studies of Nylon, Poly(ethylene terephthalate), and Polystyrene, *J. Phys. Chem.*, **58**: 503

14. Cohen, S. C. and Tabor, D. (1966). The Friction and Lubrication of Polymers, *Proc. Roy. Soc. (London)*, **Ser. A 291**: 186

15. Fort, T., Jr. (1962). Adsorption and Boundary Friction on Polymer Surfaces, *J. Phys. Chem.*, **66**: 1136

16. Lyne, D. G. (1955). The Dynamic Friction Between Cellulose Acetate Yarn and a Cylindrical Metal Surface, *J. Textile Inst.*, **46**: P112

17. Hansen, W. W. and Tabor, D. (1957). Hydrodynamic Factors in the Friction of Fibers and Yarns, *Textile Res. J.*, **27** : 300

18. Olsen, J. S. (1969). Frictional Behavior of Textile Yarns, *Textile Res. J.*, **39**: 31

19. Schlatter, C., Olney, R. A., and Baer, B. N. (1959), Concerning the Mechanism of Fiber and Yarn Lubrication, *Textile Res. J.*, **29**: 200

20. Bowden, F. P. and Young, J. E. (1951). Friction of Clean Metals and the Influence of Adsorbed Films, *Proc. Roy. Soc. (London)*, **A208**: 444

21. Howell, H. G. (1953). The General Case of Friction of a String Round a Cylinder, *J. Textile Inst.*, **44**: T359

22. Lincoln, B. (1954). The Frictional Properties of the Wool Fibre, *J. Textile Inst.*, **45**: T92

23. Howell, H. G., Mieszkis, K. W., and Tabor, D. (1959). *Friction in Textiles*, Butterworths, London, England, pp. 33–42

24. Rubenstein, C. (1958). The Friction and Lubrication of Yarns, *J. Textile Inst.*, **49**: T13

25. Rubinstein, C. (1956). General Theory of the Surface Friction of Solids, *Proc. Phys. Soc. London*, **B69**: 921

26. Schick, M. J. (1973). Friction and Lubrication of Synthetic Fibers. Part I: Effect of Guide Surface Roughness and Speed on Fiber Friction, *Textile Res. J.*, **43**: 103

27. de Jong, H. G. (1993). Yarn–to–Yarn Friction in Relation to Some Properties of Fiber Materials, *Textile Res. J.*, **63**: 14

28. Howell, H. G. (1951). Inter–Fiber Friction, *J. Textile Inst.*, **42**: T521

29. Prevorsek, D. C. and Sharma, R. K. (1979). Fiber–Fiber Coefficient of Friction: Effects of Modulus and Tan δ, *J. Appl. Poly Sci.*, **23**:173

30. Röder, H. L. (1953). Measurement of the Influence of Finishing Agents on the Friction of Fibers, *J. Textile Inst.*, **44**: T247

31. Fort, T. and Olsen, J. S. (1961). Boundary Friction of Textile Yarns, *Textile Res. J.*, **31**: 1007

32. Schick, M. J. (1973). Friction and Lubrication of Synthetic Fibers. Part II: Two Component Systems, *Textile Res. J.*, **43**: 198

33. Schick, M. J. (1973). Friction and Lubrication of Synthetic Fibers. Part III: Effect of Guide Temperature, Loop Size, Pretension, Denier, and Moisture Regain on Fiber Friction, *Textile Res. J.*, **43**: 254

34. Schlatter, C. and Demas, H. J. (1962). Friction Studies on Caprolan® Yarn, *Textile Res. J.*, **32**: 87

35. Schick, M. J. (1973). Friction and Lubrication of Synthetic Fibers. Part IV. Effect of Fiber Material and Lubricant Viscosity and Concentration, *Textile Res. J.*, **43**: 342

36. Scheier, S. C. and Lyons, W. J. (1964). Studies of the Surface Geometry of Fibers. Part II: Improved Instrumentation and Representative Results on Camel Hair and Dacron®, *Textile Res. J.*, **34**: 410

37.   Schick, M. J. (1974). Friction and Lubrication of Synthetic Fibers. Part V. Effect of Fiber Luster, Guide Material, Charge, and Critical Surface Tension on Fiber Friction., *Textile Res. J.*, **44**: 758

38.   Steinbuch, R. Th. (1964). How the Crystallization of Nylon Affects Processing and Properties, *Mod. Plastics*, **42**: 137

39.   Kittelman, W. (1964). Surface Characteristics of Elementary Viscose Cord Filaments., *Faserforsch. Textiltech.*, Berlin, Germany, **15**: 406

40.   Subramaniam, V., Muthukrishnan, P., and Naik, S. V. (1991). Sonic Modulus of Cotton Yarn and its Relationship to Abrasion Resistance, Friction, and Fatigue Parameters, *Textile Res. J.*, **61**: 58

41.   Briscoe, B. J. and Motamedi, F. (1990). Role of Interfacial Friction and Lubrication in Yarn and Fabric Mechanics, *Textile Res. J.*, **60**: 697

42.   Ono, R. and Komatsu, N. (1962). Dynamic Viscoelastic Properties of Tire Cords. I. Behavior at Low Amplitude, *Sen–i Gakkaishi*, Tokyo, Japan, **18**: 15

43.   Butler, R. H., Lamb, G. E. R., and Prevorsek, D. C. (1975). Influence of Fiber Properties on Wrinkling Behavior of Fabrics. Part III. Bending Recovery of Yarns, *Textile Res. J.*, **45**: 285

44.   DeLuca, L. B. and Thibodeaux, D. P. (1992). The Relative Importance of Fiber Friction and Torsional and Bending Rigidities in Cotton Sliver, Roving, and Yarn, *Textile Res. J.*, **62**: 192

CHAPTER 3

LABORATORY TESTING AND COMPONENT PROPERTIES

A spin finish formulation consists of many different types of chemical components. Included in these finishes are lubricants, emulsifiers, antistats, wetting agents, and several additional materials added for specific reasons. These are discussed briefly in Chapter 1, but are described in greater detail in subsequent sections. Finish component suppliers will, upon request, submit a sample of a new substance for evaluation. When this sample of a new chemical is received, the normal procedure is to subject it to several laboratory tests to obtain some basic data about its behavior before compounding the material into a finish. This chapter contains some procedures used to test these components and some information about the chemical structure that causes the component to act as it does. In addition, some procedures to analyze the yarn prepared with new compositions are discussed.

3.1 PRELIMINARY COMPONENT EVALUATION

3.1.1 Thermal Stability

Since textile fibers and yarns are frequently subjected to contact with hot surfaces during drawing, texturing, drying or similar processes, the tendency of finish components to change during heating must be tested. Several components will vaporize, leaving little lubricant on the yarn. Some materials will not volatilize but will leave a sticky residue on the yarn. One of the more troublesome aspects, however, is the tendency for the material to form a varnish or char of organic residues on the hot surface. Often these varnishes are very difficult to remove from the surface and may require the use of caustic oven cleaner type chemicals to renew the clean surface. In addition to lost production while the equipment is being cleaned, there is a significant safety hazard to plant personnel during the application of the materials. The best way to prevent the formation of undesirable residues is to select finish components that reduce their formation. It is necessary to devise some type of test for estimating component heat stability early in the evaluation procedure. Several types of tests are used in practice and are described below.

1. *Thermogravimetry*. Thermogravimetric analysis (TGA) is probably the best technique that can be utilized to study the thermal properties of finish components [1]-[3]. Materials may be tested either in air for examining their behavior in an oxidative atmosphere or in an inert gas to study only their volatility. In the procedure most often used for TGA, the sample is heated at a fixed rate of temperature increase, *i.e.*, 10°C per minute, and a plot of weight loss versus temperature is recorded. In addition to a visual observation of the chart, several features of the TGA data trace for the decomposition of the material being analyzed can be noted.

A single–feature criterion of the decomposition, such as the initial decomposition temperature, is very difficult to apply consistently. On some gently sloping curves, that initial temperature is almost impossible to define. Other single areas of the decomposition curve, such as the point at which the rate of decomposition is at a maximum or the final temperature of the weight change, are neither unique nor consistent for a wide range of materials. Cumulative weight change data, however, are more reliable. One of these is the half–volatilization point, which is defined as the temperature, T_s, at which half of the ultimate weight has occurred. Doyle [4] has extensively discussed the calculations involved in thermal analysis, including the kinetics of decomposition. The thermograms of two finish components, sorbitan monolaurate and sorbitan monolaurate, with 20 moles of ethylene oxide, are shown in Figure 3–1. These analyses were performed using a nitrogen purge and a heating rate of 10°C per minute [5].

2. *Other Thermal Analysis Techniques*. Several other techniques are available for laboratory studies of the thermal stability of finish components and finish oil blends that do not necessitate the purchase of expensive equipment. One of these consists of heating a small amount of the sample in a dish on a hot plate at different temperatures. For example, one might heat the substance at temperatures that range from 200°C to 350°C at fixed

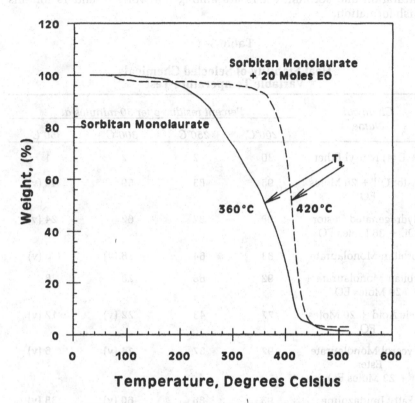

Figure 3–1. TGA Thermograms of Sorbitan Monolaurate and Sorbitan Monolaurate with 20 Moles Ethylene Oxide.

temperature intervals and for specific times. If the samples are weighed, heating at each temperature data may be obtained that can give a relative indication of thermal stability. Possible shortcomings in this technique are: contact between the sample container and the hot plate may not be consistent, temperature measurement and control may vary with position on the hot plate, and weighing errors might occur. The procedure does have the advantage of the investigator being able to examine the sample at each step, and the ease of determining the nature of the residues [5].

Examples of data obtained by this method are presented in Table 3–1. The presence of certain organic functional groups in a component can result in the formation of a very hard varnish on heating at elevated temperatures. Unsaturation and sorbitan esters are among the worst offenders for this varnish formation.

Table 3–1

Heat Stability of Selected Chemicals
Variable Temperature Test

Chemical Name	Percent residue after 30 minutes at			
	200°C	250°C	300°C	350°C
POE (4) Lauryl Ether	30	2	2	1
Castor Oil + 26 Moles EO	95	83	59	46 (v)
Hydrogenated Castor Oil + 26 Moles EO	98	87	62	24 (v)
Sorbitan Monolaurate	83	64	38 (v)	6 (v)
Sorbitan Monolaurate + 20 Moles EO	92	68	28	8
Oleic Acid + 20 Moles EO	77	43	22 (v)	17 (v)
Glycerol Monolaurate Ester + 23 Moles EO	97	57	14 (v)	6 (v)
Fatty Imidazoline Quaternary Amine	93	86	60 (v)	38 (v)

Note: The symbol (v) shows the formation of a varnish.

Other heat stability procedures that are in common use are isothermal heating tests. In one technique about 1 gram of the sample is weighed into a small dish and placed in an oven at a selected constant temperature for 24 hours. The conditions of the residue and the weight loss are recorded. The ease of deposit removal may be subjectively rated by washing with warm tap water. This test is especially useful in predicting how a finish will perform on hot surfaces, such as draw rolls or pins, and as a prediction of the effectiveness of antioxidants since the oven atmosphere is normally air. Another thermal stability/fuming test has also been described [6]. A 5–gram sample of the product to be tested is weighed into a small aluminum weighing dish. Place the dish containing the sample on a hot plate preheated to 265°C. The time in seconds for initial smoking to begin is recorded and heating continued for 15 minutes. After the dish containing the sample is removed and cooled to room temperature, it is reweighed and the percent weight loss is recorded. The thermal stability and fuming data obtained with this method for the same materials listed in Table 3–1 is presented by Table 3–2.

Table 3–2

Heat Stability From the Thermal Stability/Fuming Test

Chemical Name	Heat Stability	
	Fuming, seconds	Percent Varnish
POE (4) Lauryl Ether	29.0	0
Castor Oil + 26 Moles EO	4.0	40
Hydrogenated Castor Oil + 26 Moles EO	2.5	15
Sorbitan Monolaurate	8.0	35
Sorbitan Monolaurate + 20 Moles EO	5.0	8
Oleic Acid + 20 Moles EO	3.0	10
Glycerol Monolaurate Ester + 23 Moles EO	3.0	5
Fatty Imidazoline Quaternary Amine	10	70

(Reference [6]. Reprinted with permission of ICI Americas)

The volatility of finish components cannot be predicted with any degree of certainty, but several generalizations can be made. Mineral oils and low molecular weight esters will easily volatilize. Poly(ethylene glycol) esters of fatty acids (PEG esters), poly(ethylene oxide)–alcohol ethers (POE alcohols), and ethylene oxide–propylene oxide (EO–PO) copolymers decompose by a process called *autoxidation* [7]–[9]. In autoxidation, oxygen attacks the carbon next to the ether linkage and forms a hydroperoxide. This unstable molecular segment results in chain scission and subsequent volatilization of the short chain fragments. Frequently the use of components containing EO and/or PO is desirable since volatility in certain processing steps is necessary.

As one would expect, varnish formation is dependent upon bond cleavage and subsequent cross–linking into intractable polymers. Unsaturated aliphatic fatty acids or alcohols form these varnishes readily, as do sorbitan esters. The differences in varnish formation between castor oil (consisting of about 5% oleic acid, 4% linoleic acid, and 88% ricinoleic acid) and the hydrogenated version are readily apparent in Table 3–2. Sorbitan monolaurate will develop a higher level of varnish than ethoxylated sorbitan monolaurate since the ethylene oxide part of the molecule decomposes leaving almost no residue.

3.1.2 Viscosity

Since Lyne [10] has shown that the hydrodynamic friction of yarn to which a spin finish has been applied is directly proportional to the viscosity of the lubricant, the determination of viscosity is very important in the initial screening of components. Although dilute solution viscosity as measured with an Ostwald, Cannon–Fenske or Ubbelohde capillary viscometer will yield a very accurate indication of viscosity [11] , a simple rotational method is appropriate for finish materials [12]. A typical instrument for measuring rotational viscosity is manufactured by Brookfield Engineering [13].

A small quantity of component, about 75 milliliters, is placed in a container at a temperature as close as possible to 25°C. Immerse the selected spindle into the liquid and tilt the spindle to eliminate air bubbles. Raise the sample until the indentation on the spindle coincides with the surface of the liquid component [14]. After rotation begins and the viscosity indication dial has stabilized, the viscosity can be read. The Brookfield Viscometer measures the viscosity in centipoise (cP), or 0.01 poise, the cgs unit for absolute viscosity. For most materials that are liquid at room temperature, viscosity measurements are made at 25°C. Sometimes where the components are viscous liquids or semi–solids under ambient conditions, viscosity measurements are made at elevated temperatures. These are usually 40°C, 100°F, or 212°F.

Frequently chemical data sheets report the viscosity in other units, especially as centistokes (cSt) or as Saybolt Universal Seconds (SUS). A centistoke is 0.01 stokes, the cgs unit of kinematic viscosity. The relationship between cP and cSt is:

$$cSt = \frac{cP}{density} \tag{3-1}$$

Usually the viscosity should be reported as the kinematic viscosity in cSt, but some SUS data can be found in past literature. Unfortunately, the SUS viscosity equivalent to cSt viscosity varies with the temperature at which the measurements are made. The basic conversion values for 100°F are given in ASTM Method D 2161–87 [15]. If the kinematic viscosity is between 1.81 cSt and 500 cSt at 100°F, the tables in the ASTM method referenced can be used. Equations for viscosity conversion at other temperatures are also given in the ASTM method, but these are rarely necessary for finish components.

3.2 LABORATORY TESTING OF YARN WITH FINISH

After the preliminary evaluation of the thermal properties and viscosity of a component, the investigator may decide to apply that component to some fiber, or to formulate a complete finish for application. Although one might have some idea about the component behavior from the initial studies, its effectiveness in a finish cannot be determined exactly until applied to the yarn for which it is intended. Finish application could be accomplished by several routes, but one of the most successful is to pump a dilute solution of emulsion onto dry (unfinished) yarn with a laboratory applicator or on a pilot plant spinning position. Over application of the new formulation onto yarn already coated with a finish should be avoided since yarn test results will be confusing. It is possible to remove most of the original finish from the yarn by scouring with hot water or a mixture of water and isopropanol. This can be very successful with an in–line scour/finish application system. After the new finish has been applied, further testing is possible.

3.2.1 Very Slow Speed Fiber to Fiber Friction

Very slow speed fiber to fiber friction is a little understood and even less appreciated phenomenon in yarn processing; however, it is a fundamental parameter and very important in all phases of yarn manufacturing. When the term *very slow speed friction* is used it commonly refers to the friction in the boundary region. The boundary friction mechanism is dominated by surface conformity, molecular attraction between the surfaces in contact, and the elastic nature of one or both surfaces in the contact region. From an operational standpoint, this means contact speeds ranging from 10^{-4} meters/minute to 10^{-1} meters/minute and perhaps extending into the semi–boundary or transition speed range up to about 5 meters/minute. Primary areas of interest in continuous filament apparel, tire and industrial yarns are in the yarn to yarn contact as the packages are being formed. Carding and sliver cohesion are steps in staple yarn processing in which slow speed fiber to fiber friction is very important. Frequently (in a carpet body, for example), a fully manufactured product may depend on slow speed friction for characterization of performance.

One important feature of slow speed friction is the characteristic intermittent motion that results from the interaction of an elastic yarn contacting another surface. This motion is commonly called stick–slip, which is precisely what happens. The yarn alternately sticks and slips on the friction surface until the contact speed becomes high enough to maintain the yarn in continual extension and motion. The characteristics of this stick–slip amplitude, frequency, uniformity, and speed at which it ceases, all suggest how the yarn might behave in a similar end–use situation.

Two different types of equipment can be used to measure very slow speed fiber to fiber friction. One of these is the TRI/SCAN™ Surface Force Analyzer manufactured and sold by the Textile Research Institute. This instrument uses the *point contact* system [16] for friction measurement and some results obtained with the equipment have been reported by Kamath [17]. Another instrument using the *extended line contact* technique, or capstan method, was first described by Röder [18] and then modified by Fort and Olsen [19]. The apparatus consists of a slow speed motor fitted with a gear reduction system to allow rotational speeds that can be varied from about 0.0001 to about 0.50 meters per minute. A cylinder is wrapped with a complete cover of the test yarn. The wrapping should be at a low helix angle so that the yarn will be in almost parallel rows. The traverse mechanism from a fishing reel is quite satisfactory to obtain proper yarn confirmation on the cylinder.

A strand of the same test yarn is draped over the cylinder, with one end attached to some strain measuring device, such as a strain gauge or through a Rothschild tension head. The other free end is loaded with a fixed weight. The cylinder is rotated so that tension is applied to the fixed end and the electrical output from the strain device transmitted to a strip chart recorder or to a computer input system. Other details for the capstan method can be found in ASTM Method D 3412–89 [20]. The coefficient of friction is calculated using the capstan or belt equation:

$$\frac{T_2}{T_1} = e^{\mu \theta} \tag{3-2}$$

where:

T_1 = input tension,
T_2 = output tension,
μ = coefficient of friction, and
θ = wrap angle in radians.

Personnel at ICI Americas [6] have measured the fiber to fiber friction of several components manufactured by that company and have provided a list of those materials in their literature. The data is listed in Tables 3–3, 3–4, and 3–5. The data presented in these tables are only some examples of components that can be selected to alter the fiber to fiber friction of synthetic fibers.

Table 3–3

Low Friction Materials

Low Stick–Slip	Medium Stick–Slip
Sorbitan ester of POE glyceride	POE (20) sorbitan monopalmitate
Most esters and/or softeners	POE (20) sorbitan monostearate
POE (200) castor oil or hydrogenated castor oil	POE (20) sorbitan monooleate
POE (25) hydrogenated castor oil	Diethyl sulfate quaternary of ethoxylated fatty amine

(Reference [6]. Reprinted with permission of ICI Americas)

Table 3–4

Medium Friction Materials

Low Stick–Slip	Medium Stick–Slip	High Stick–Slip
POE (4) sorbitan monolaurate	Sorbitan monolaurate	POE (10) oleyl alcohol
Sorbitan monostearate	Sorbitan monopalmitate	POE (20) oleyl alcohol
POE (4) sorbitan monostearate	Sorbitan monooleate	POE fatty amine
	Sorbitan trioleate	POE (9) laurate
	POE (20) sorbitan monolaurate	POE (8) stearate
	POE (20) sorbitan tristearate	POE (7) POP (2) lauric acid, methyl capped
	POE (2) sorbitan trioleate	POE (9) POP (2) lauric acid, methyl capped
	POE (40) sorbitol septaoleate	POE (9) POP (2) oleic acid, methyl capped

Table 3–4, Continued

Low Stick–Slip	Medium Stick–Slip	High Stick–Slip
	POE (40) sorbitol hexaloeate	POE (12) POP (5) oleic acid, decyl capped
	POE (20) stearyl alcohol	POE (23) glycerol monolaurate
	POE (12) tridecyl alcohol	
	POE (6) tridecyl alcohol	
	POE (9) lauric acid methyl capped	
	POE (8) oleic acid, butyl capped	
	POE (9) oleic acid, methyl capped	
	Hexyl alcohol phosphate, K^+ salt	

(Reference [6]. Reprinted with permission of ICI Americas)

Table 3–5

High Friction Materials

Low Stick–Slip	Medium Stick–Slip	High Stick–Slip
		POE (4) lauryl alcohol
		POE (6) C_{12}–C_{13} alcohol, methyl capped

(Reference [6]. Reprinted with permission of ICI Americas)

3.2.2 High Speed Friction

High speed friction, used primarily for fiber to hard surface measurements, is a somewhat simpler idea to understand than boundary fiber to fiber friction. Fiber to metal friction is directly related to the viscosity of the lubricant [10][21], with a few exceptions. If the structure of the molecule is either rod–like or ball–like, the viscosity relationship is excellent, but the exceptions are due to emulsifiers that may be highly branched molecules and may have comb or similar molecular structures. An early instrument designed to measure high speed friction was the device described by Zaukelies [22]. Although it is relatively easy for one to build their own instrument for high speed yarn friction determination, commercial equipment is available. One such instrument is the Rothschild F–Meter, Model R–1188 [23], sold by Lawson–Hemphill Sales, Inc., Spartanburg, S.C. Yarn is withdrawn from a cone or bobbin and passes through an adjustable input tension device where the friction coefficient is calculated by measuring the yarn tension before and after the friction pin. Both yarn to surface (metal, ceramic, or other) and yarn to yarn friction at speeds in the hydrodynamic region (5 meters per minute or greater) can be measured with the Rothschild instrument. A diagram of the yarn path through the Rothschild F–meter is shown in Figure 3–2. Mayer [24] described the use of the F–Meter in studying the influence of softeners and finishes on cotton yarn. He found

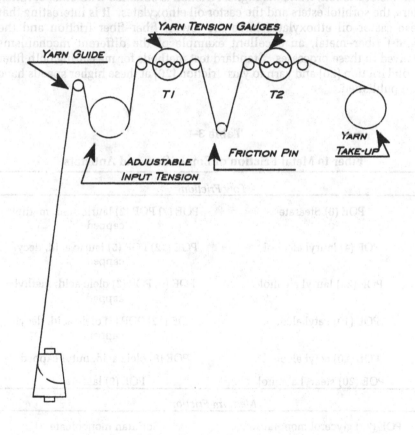

Figure 3–2. Layout of Yarn Path Through Rothschild F–Meter.
(Ref. [23]. Reprinted with permission of Lawson–Hemphill Sales, Inc.)

that the coefficient of friction of these yarns against steel decreased by a factor of two after 0.5% of a cationic softener was added. The variability of the yarn to yarn friction also decreased significantly at that same level.

A similar device is the ATLAB® friction tester [25]. This instrument may be obtained from Custom Scientific Instruments, Whippany, N.J. ICI has reported fiber to metal friction data using the ATLAB® instrument for a number of emulsifiers and the effects on friction due to the structure of those emulsifiers can be seen in Table 3–6.

It is readily noted in reading Table 3–6 that the lowest friction is shown by the straight chain poly(ethylene glycol) esters and ethers, especially with hydrophobe chain lengths of 12 carbons or longer. The shorter hydrophobe chain lengths and some sorbitan esters have higher fiber–metal friction, while the highest friction is obtained with other sorbitan esters, the sorbitol esters and the castor oil ethoxylates. It is interesting that these castor oil ethoxylates have the lowest fiber–fiber friction and the highest fiber–metal, an excellent example of the different mechanisms involved in these processes. Standard test methods for measuring both fiber to solid friction [26] and yarn to yarn friction [20] at these higher speeds have been published.

Table 3–6

Fiber to Metal Friction of Emulsifiers and Antistats

Low Friction	
POE (8) Stearate	POE (7) POP (2) lauric acid, methyl capped
POE (4) lauryl alcohol	POE (12) POP (5) lauric acid, decyl capped
POE (23) lauryl alcohol	POE (9) POP (2) oleic acid, methyl capped
POE (10) cetyl alcohol	POE (12) POP (5) oleic acid, decyl capped
POE (20) cetyl alcohol	POE (8) oleic acid, butyl capped
POE (20) stearyl alcohol	POE (9) laurate

Medium Friction	
POE (23) glycerol monolaurate	Sorbitan monooleate
POE (6) tridecyl alcohol	Sorbitan trioleate
POE (9) oleic acid, methyl capped	POE (4) sorbitan monostearate

Table 3–6, Continued

Medium Friction	
Potassium salt of hexyl alcohol phosphate	POE (20) sorbitan tristearate
Sorbitan monolaurate	POE (20) stearyl alcohol
Sorbitan monopalmitate	POE (12) tridecyl alcohol
Sorbitan monostearate	POE (20) hydrogenated tallow amine
Sorbitan tristearate	POE (4) sorbitan monolaurate
POE (20) sorbitan monolaurate	

High Friction	
POE (20) sorbitan monostearate	POE (25) hydrogenated castor oil
POE (20) sorbitan monooleate	POE sorbitol oleate–laurate
POE (2) sorbitan trioleate	POE (40) sorbitol septaoleate
	POE (50) sorbitol hexaoleate

(Reference [25]. Reprinted with permission of ICI Americas)

3.2.3 Static Electricity and Finish Level on Fiber

A series of experiments devised to examine the relationship between the static electrical charge developed on the fiber and the uniformity of spin finish were performed. In the first part of these experiments, polyester yarn was made with various levels of spin finish and the static measured with a Rothschild Static Voltmeter, Model R–4021 [27], over a four week period. The data are shown in Table 3–7. The finish on yarn was calculated from flow rates and previously established relationships of flow rates to finish on yarn. These data are shown graphically on Figure 3–3 [5].

The plot, and the data in the table, clearly show that the static charge developed on the fiber is inversely related to the finish on yarn, that is, lower finish induces a higher charge on the yarn. If the finish on yarn is at a level of about 0.5% or greater, the static charge is changed very little with an increase in the finish amount.

Because of this experiment, a second part of the test was performed. A special test stand was set up with a metered finish applicator. Yarn prepared with very low finish on yarn was passed through the applicator and the motor driving the pump was stopped and started at regular intervals. The period of time that the finish pump was turned off varied from 1850 meters to 60 meters.

Table 3–7

Static Charge on Polyester Yarn With Time

Finish on Yarn	Static in Volts				
	Day 1	Day 4	Day 7	Day 14	Day 28
0.2	124.5	142.5	172.5	255.0	250.0
0.3	84.8	100.5	142.5	185.0	180.0
0.4	–	45.0	64.5	110.0	117.0
0.5	33.2	33.0	43.5	65.0	81.0
0.65	–	10.5	36.0	65.0	60.0
0.70	19.5	25.5	30.0	60.0	57.0

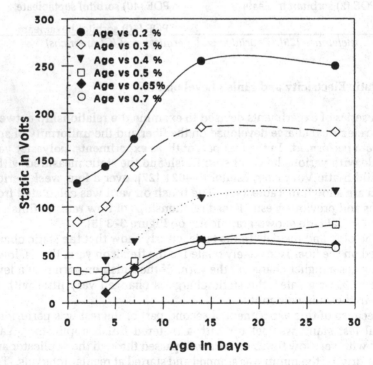

Figure 3–3. Effect of Static Charge versus Time for Polyester Yarn Made with Different Finish Levels.

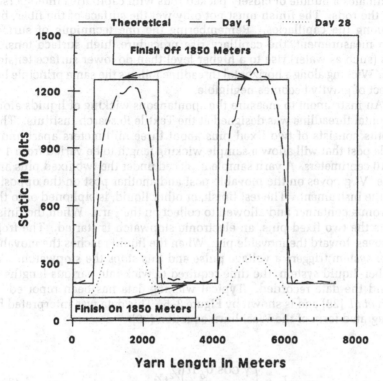

Figure 3–4. Static Electricity on Polyester Yarn Made with Intermittent Finish Application.

The results from this test showed that actual finish on yarn, as measured by the static generated after one day aging, almost duplicated the theoretical off–on curve. The same relationship held after 28 days, but at a lower static level since much of the finish had absorbed into the fiber. Figure 3–4 shows the data for the 1850 meter experiment [5].

3.2.4 Wicking

During the application of a finish to yarn in the spinning process, the spreading of the finish along that moving threadline is more complex than one might initially assume. The surface tension of the finish emulsion should be low enough to initially wet the polymer surface, below about 46 milli–Newtons/meter (the critical surface energy) for nylon; however, if it is too low the wicking along the threadline will be inhibited. This seeming contradiction is not as strange as it might seem. As yarn is being spun, it approximates a bundle of closely packed rods with capillaries interspersed among the rods. The finish must not only wet the surface of the fiber, but wick along the capillaries. Remembering the one technique of surface tension measurement, the capillary rise procedure, high surface tension liquids (such as water) rise to a higher level than do lower surface tension liquids. Wicking along a horizontal threadline follows the same principle but the effect of gravity becomes negligible.

An instrument to measure the spontaneous wicking of liquids along a horizontal threadline was designed at the Textile Research Institute. This apparatus consists of two fixed pins about three millimeters apart and a movable post that will allow a sample wicking length to be varied from 1 to about 10 centimeters. A yarn sample is placed under the two fixed pins and into the "V" grooves on the movable post and another post on the opposite end of the instrument. The test finish, or other liquid, is siphoned onto the yarn from a container and allowed to collect on the yarn. When the finish contacts the two fixed pins, an electronic stopwatch is started. The front then moves toward the movable pin. When the liquid reaches the movable pin, the system triggers a voltage pulse and this stops the stopwatch. For each fiber–liquid system, the time required to wick into various lengths of yarn and the data recorded. Typical wicking data has been reported by Kamath et al. [28] and is shown by Figure 3–5. This data was interpreted by the integrated form of the Washburn equation [29]:

$$L = \left(\frac{\gamma \cos \theta_a R}{2\eta}\right)^{1/2} t^{1/2} \tag{3-3}$$

where:

L	=	wicking length,
R	=	average effective pore radius,
γ	=	surface tension of the liquid,
θ_a	=	advancing contact angle of the liquid with the solid surface,
η	=	viscosity of the liquid, and
t	=	time.

According to Equation (3–3), a plot of L against $t^{1/2}$ should be linear with a slope given by $(R \gamma \cos \theta_a / 2 \eta)^{1/2}$, which is the wicking constant for a particular yarn/finish system. It is obvious from this equation that the wicking rate is directly related to the adhesion tension, $\gamma \cos \theta_a$, so that as the surface tension is increased, the wicking length is also increased.

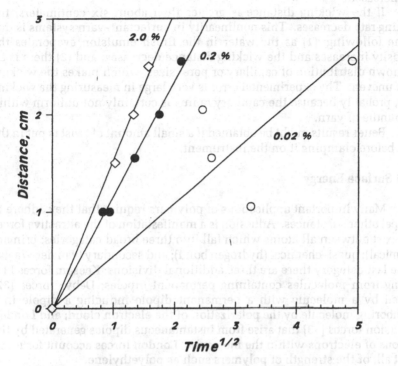

Figure 3–5. Wicking of Tergitol® NP–10 into Poly(ethylene
terephthalate) Yarn: Effect of Surfactant Concentration.
(Reference [28]. Reprinted by permission of the Textile Res. J.)

In a two component water–surfactant system, as presented by Figure
3–5, adsorption of the surfactant molecules on the solid surface at the
wicking front results in an increase in the surface tension of the liquid with
a corresponding decrease in cos θ_a. Equilibrium conditions are reestablished
when surfactant molecules diffuse from the more concentrated regions into
the leading edge of the meniscus. These effects, sometimes identified as
transient effects, of depletion and renewal of surfactants at the liquid
surface are discussed in a paper by Hirt *et al.* [30] on measuring the dynamic
surface tension of solutions. The overall results of adsorption and resultant
surface tension gradient contractions at the surface, known as the Marangoni
effect, are erratic wicking behavior and a lower wicking rate. These effects

are more pronounced in dilute solutions and decrease as the concentrations are increased.

If the wicking distance is greater than about six centimeters, the wicking rate decreases. This nonlinearity in surfactant–yarn systems is due to the following: (1) as the water in the finish emulsion evaporates the viscosity increases and the wicking distance decreases; and (2) there is an unknown distribution of capillary or pore sizes, which makes the wicking front uneven. The experimental error is very large in measuring the wicking time, probably because the capillary radius is certainly not uniform within the bundle of yarn.

Better results might be obtained if a small amount of twist is put in the yarn before clamping it on the instrument.

3.2.5 Surface Energy

Many important applications of polymers require that they adhere to or repel other substances. Adhesion is a manifestation of the attractive forces that exist between all atoms which fall into three broad categories: primary (chemical); quasi–chemical (hydrogen bond); and secondary (van der Waals). In the last category there are three additional divisions: Keesom forces [31] arising from molecules containing permanent dipoles; Debye forces [32] caused by a molecule with a permanent dipole inducing a dipole in a neighboring molecule by the polarization of the electron cloud; and London dispersion forces [33] that arise from instantaneous dipoles generated by the motions of electrons within the molecule. London forces account for most, if not all, of the strength of polymers such as polyethylene.

Several years ago the Textile Research Institute developed a method to measure the contact angle of liquids on single filaments [34]. During the past few years, scientists at that institution have used this method to study spin finish uniformity on various types of fiber materials [35]. In some other experiments with contact angles, a technique was devised which allows the calculation of polar, dispersion, and total surface energies of polymers [36]. The total surface energy of a solid is theoretically similar to the surface tension of a liquid. The TRI contact angle device, TRI/SCAN™, was adapted for use in surface energy measurements and these instruments have been widely used by both research and development personnel. This experimental technique was implemented by Slade and Hild [37] to obtain contact angles and to calculate the surface energies of nylon. Table 3–8 compares this data with other values from the literature. The agreement among the different methods is quite good. Practical examples of the effects of surface energy are the waterproofing of textile fabric, treatment with Scotch–gard™ fabric protector to repel spills and stains and the use of fluorocarbons on carpet fibers to reduce soiling. The treated fibers have a lower surface energy than the soil or water so these are not attracted to the surface.

Table 3–8

Comparison of Nylon 6,6 Surface Energy Values

Source	Surface Energy, ergs/cm²		
	Polar, γ^p	Dispersion, γ^d	Total, γ^t
Slade [5], equilibrium contact angle on nylon film	18.4	34.4	52.8
Slade and Hild [37]	15.4	32.4	47.8
Wu [36]	12.9	35.0	47.9
Owens and Wendt [38]	6.2	40.8	47.0
Zisman [39]	–	–	46
Dalal [40]	13.4	35.8	49.2

(Reference [37]. Reprinted with permission of Textile Res. J.)

3.2.6 Wettability Scanning and Finish Distribution

The same equipment used for surface energy measurements can also be utilized to examine the distribution of spin finish both along the fiber and the amount of finish around the perimeter of the fiber at any one point. A small loop of wire, about 5 mm in diameter, is formed and placed in a special holder. The fiber sample is passed through the loop and a weight is attached to provide a constant tension. A liquid membrane is formed in the loop by carefully allowing the loop to contact the liquid surface. The fiber is then scanned with the liquid membrane and the force measured with a recording microbalance. This technique was also developed at TRI and has been reported by personnel from that institution [41]. Vector analyses of the forces at the solid–liquid interface of the two menisci are given by:

$$F = P \gamma_{LV} (\cos \theta_a - \theta_r) \qquad (3–4)$$

where:

F	=	force recorded by the microbalance,
P	=	fiber perimeter,
γ_{LV}	=	membrane liquid surface tension,
$\cos \theta_a$	=	cosine of the advancing contact angle, and
$\cos \theta_r$	=	cosine of the receding contact angle.

Cos θ, which can be considered a constant for a given solid–liquid system, can be determined separately by the short fiber immersion method as used for surface energy measurements. A special computer software program will allow the degree of finish coverage to be calculated.

3.2.7 Finish Absorption Into Yarn

As Figure 3–4 shows, the static charges on yarn with spin finish applied to the surface definitely changes with time. While it might be possible for this change to be attributed to a loss of charge through an electrical discharge into a moist atmosphere, another explanation might be the loss of antistatic agent through the absorption of the finish into the yarn.

Polyester polymer and undrawn polyester yarn have very low crystallinity. The crystal structure can be developed by drawing, heat treatment, or by solvents. The effects of non–aqueous solvent absorption on polyester yarn are well known [42][43]. These solvents, such as methylene chloride or dimethylformamide, induce crystallization in the outer regions of crystallinity. The diffusion of dyes into solvent–crystallized polyester is enhanced, probably by a disorientation of the ordered noncrystalline regions resulting in increased segmental mobility that occurs during the solvent treatment. Since the crystallinity of spun nylon is essentially equivalent to that of the drawn fiber, solvent crystallization of that polymer is of little importance.

Several studies have been conducted to determine the effects of the absorption of spin finish components into several types of yarn. In a study of fabrics made from cotton, nylon, and polyester [44], treatment with anionic, nonionic, or cationic surfactants or with dimethyl formamide resulted in very small changes to the physical properties or dyeing characteristics of these fabrics. The most significant changes were in wetting or wicking properties. These results are not unexpected since the yarns used to make the fabrics have already reached their normal orientation and crystallinity. Needles *et al.* [45] examined both polyester and nylon heat–treated fabric in the presence and absence of a nonionic surfactant, Triton® X–100, manufactured by Rohm and Haas. Triton® X–100 is α–[4–(1,1,3,3,–tetramethylbutyl)phenyl]–ω–hydroxypoly(oxy–1,2–ethane-diyl) with a structure of:

$$(CH_3)_3C-CH_2-\underset{\underset{CH_3}{|}}{\overset{\overset{CH_3}{|}}{C}}-\underset{}{\bigcirc}-O(CH_2CH_2O)_9H$$

The dyeing properties of untreated and treated polyester were studied using three related 1,4–substituted anthraquinone dyes. In general, the heat–treated and surfactant/heat–treated polyester dyed to a deeper shade than the untreated fabric. Surfactant/heat–treated polyester had an overall higher uptake of dyes than polyester annealed without surfactant. This data suggests that the absorption of the surfactant disrupted the orientation in the amorphous areas of the polymer allowing greater dye penetration. In the studies of nylon, the authors found that Triton® X–100 reduced dye uptake at most heat–treatment temperatures except for 1,4–diaminoanthraquinone at 150°C and 175°C where it slightly increased dye uptake. These studies also do not truly represent the absorption of finish components into freshly spun fiber since the experiments were performed on fabric.

The specific interactions between partially oriented (POY) polyester yarn and polyethylene glycols (PEG) were described by Selivansky et al. [46]. The yarn was commercial 116/34 denier poly(ethylene terephthalate) (PET) POY spun at 295°C spinning block temperature and wound up at 3300 meters/minute. The finishes included 15% aqueous solutions of PEG of different number average molecular weights (400, 600, 1000, 1450, 1540, 2000, 6000), and of diethylene glycol, ethylene glycol, and deionized water. About 0.4% by weight of the non–aqueous portion of the finish was added. These yarns were annealed at temperatures between 20°C and 90°C for one hour in a forced air oven and dyed with CI disperse blue 27. A pronounced dependance of the PEG dye blocking efficiency and PEG molecular weight was found. PEG in a molecular weight range of 500 ± 100 was significantly more effective in blocking the uptake of dye than those of either lower or higher molecular weight. Apparently the absorption of PEG–500 ± 100 enhances the intermolecular interactions through plasticization. This component, located in a limited volume at the fiber surface, activates the interactions that bring about the higher degree of cohesion and order and the resulting low temperature dyeing kinetics. In contrast, PEG 2000 and PEG 6000 enhance dye uptake to levels greater than that of non–finish treated yarn.

Not only do the surfactant components absorb into the fiber and affect yarn properties, but other components can cause equally unwanted results. Low molecular weight lubricants, such as low viscosity mineral oil or butyl stearate, will absorb into yarn and thus can significantly increase the friction across guide surfaces. Another serious problem is the absorption of the antistatic agent and the subsequent dramatic increase in static generated in yarn processing. Low molecular weight antistats, such as hexyl alcohol phosphate, might be acceptable on polyester but have been found to be quite unsatisfactory on nylon. This particular component will disrupt the polymer molecular order in nylon to such as a degree that other components are absorbed more rapidly than finishes without the antistat.

3.2.8 Amount of Finish on Yarn

Spin finishes are applied to fibers and yarn during spinning or in subsequent processing. The finishes are applied either as a neat oil mixture, as an emulsion, or with the oil dissolved in a solvent. If the oil mixture is emulsified or dissolved, the water or solvent has either evaporated or absorbed into the yarn before it is wound up on the package or prior to crimping and cutting of staple fiber. In all examples, this results in an oil mixture remaining on the fiber surface. There are three general techniques for determining the amount of these finish oils that have been applied to the yarns and fibers. These are:

A. The extraction of the oils with a suitable solvent and determination of the finish oils by evaporating the solvent and weighing the oil residues.

B. A solvent extraction of the oils and then measuring the oil content using an instrumental procedure.

C. Measuring the oil content on the yarn without extraction by some instrumental technique.

The selection of the procedure to be used depends upon the situation for which the data are needed. Often in research and development laboratories the simple gravimetric technique is adequate, but in fiber manufacturing plants this procedure is too slow to maintain control of the spinning process. An instrumental method utilizing robotic sampling, measurement, and data reporting is essential for monitoring finish–on–yarn (FOY) during fiber production.

1. *Extraction and gravimetric determination of oils.* Prior to about 1980, finish oils were normally extracted with carbon tetrachloride, chloroform, or some other chlorinated hydrocarbon. Since that date, these solvents have been found to be extremely hazardous; they can cause extensive liver damage and in many cases are carcinogenic. The American Association of Textile Chemists and Colorists (AATCC) developed a method [47] for determining the extractable content of textile materials using 1,1,1–trichloroethane, CAS# 71–55–6, as the solvent. The structure of this compound is:

$$\text{Cl}-\underset{\underset{\text{Cl}}{|}}{\overset{\overset{\text{Cl}}{|}}{\text{C}}}-\text{CH}_3$$

and although not as toxic as some of the other chlorinated hydrocarbons, it should be used with caution. This compound is moderately toxic by ingestion, inhalation, and skin contact [48]. In this method the oils, fats and waxes are removed by Soxhlet extraction and the solvent is allowed to evaporate in a hood. The amount of extractables is calculated by the loss in weight of the sample. Another procedure for establishing the amount of extractable matter in textiles has been published by ASTM [49]. Although the ASTM method also employs a Soxhlet extraction approach for removing the oily contaminates on the fiber, yarn, or fabrics, this technique uses a different solvent. The solvent recommended is called HH, or 1,1,2–trichloro–1,2,2–trifluoroethane, CAS# 76–13–1. Other synonyms are fluorocarbon 113 or Freon® 113. The structure of this compound is:

$$\text{Cl}-\overset{\displaystyle\overset{F}{|}}{\underset{\displaystyle\underset{F}{|}}{C}}-\overset{\displaystyle\overset{Cl}{|}}{\underset{\displaystyle\underset{Cl}{|}}{C}}-F$$

HH solvent is considered only mildly toxic by ingestion and inhalation [48] and it is believed to have little effect on the depletion of the ozone layer in the atmosphere. The principal barriers to the use of HH solvent are its limited availability and its increasing cost. Other than the solvent, this ASTM procedure differs from the AATCC method in the drying step of the textile material. After air drying, the sample is placed in an oven at $105\pm3\,°C$ until successive weighing result in a change of less than 0.001 grams.

An innovative new technique has recently been introduced to replace Soxhlet extraction in the gravimetric determination of extractable substances from textiles. This procedure is supercritical fluid extraction. Drews et al. [50] have shown that extraction with unmodified supercritical CO_2 (SFE) could successfully duplicate the results obtained using fluorocarbon 113 Soxhlet extraction. The authors used SFE to extract commercial finishes from polyester, nylon, and polypropylene fibers. Instrumentation for the extractions included a Dionex 703M, a Hewlett–Packard 7680 A, a Suprex PrepMaster equipped with an AccuTrap and automated variable restrictor, and an ISCO SFX 2–10 with a Model 100 DX pump. Based upon the data obtained during these experiments, it appears that SFE can be successfully used to replace Soxhlet extractions for a wide range of textile and fiber applications. In nearly all cases of the SFE versus Soxhlet, as well as in inter–instrument studies, the correlations observed were excellent.

2. *Extraction with instrumental analysis of oil content.* The most conventional procedure for instrumentally measuring the oil content of solutions extracted from textile materials is by infrared spectroscopy. Almost all finish components, with the exception of silicones, consist of molecules that have carbon–hydrogen bonds. If one can extract the fibers with a solvent

that does not contain that particular bond, it is possible to measure the absorbance at some particular wavelength and thus to quantitatively determine the amount of oil present. Fluorocarbon 113 is a good choice for a solvent since its toxicity is relatively low; however, if used in large quantities the cost can become exorbitant.

Quantitative studies in the infrared region are based on the Beer–Lambert law, $A = \varepsilon bc$, where ε and b are constants. For a particular absorption band a calibration curve is constructed by plotting A against c for solutions of known concentration [51]. A standard amount of fiber, for example 10.000 grams, is weighed into a sealed container and some fixed amount of solvent is pipetted into the container. A simple room temperature extraction is made and the absorbance of the carbon–hydrogen stretching vibration at a wavenumber of about 2900–3000 cm^{-1} is determined. This value is then compared to that of the standard and the FOY can be calculated. Some errors might be seen if there is a large amount of monomer and/or dimer extracted with the finish. To correct this error, a blank solution of an extract of the same yarn that has been spun without finish must be made.

3. *Determination of finish on yarn without extraction*. Perhaps one of the most important new techniques for measuring the finish on yarn is by pulsed NMR. This method uses the solid yarn, fiber, or textile sample without any extraction and FOY results can be obtained in less than 30 seconds. One instrument that is commercially available is the Oxford[20+] Pulse NMR [52]. By using a 20 Mhz proton permanent magnet and radio frequently pulses this instrument can differentiate between hydrogens in either the liquid or solid phases of the sample. Applications include oil or fat content in cosmetics, oilseeds, foodstuffs, and oils on or in polymers and fibers. In addition, it can be used to measure water in seeds, powdered milk, and pharmaceuticals.

A series of fiber samples were analyzed for finish on yarn by packing them into an 18 mm diameter probe and the mass maximized by packing with a Teflon plug [53]. After establishing a calibration curve, the data were obtained and are presented in Table 3–9. The fourth sample in Table 3–9, standard FOY of 0.395, was analyzed 10 times to obtain the precision of data obtained with the QP[20+]. The results were 0.39, 0.39, 0.39, 0.40, 0.39, 0.39, 0.39, 0.39, 0.39, 0.39 with a mean of 0.39 and a standard deviation of 0.006. Since this technique yields accurate and precise results in a short time and does not require the use of hazardous or expensive solvents, it appears to be an exceptional tool for measuring finish on yarn.

3.3 SPECIFICATION ANALYSES

When a finish component has been selected for commercial use in a mixture which will be applied to fiber, certain laboratory analyses must be performed to insure consistent quality. Most of these tests, usually called

Table 3-9

Spin Finish on Fiber by Pulsed NMR

Standard FOY	FOY by NMR			Mean	Standard Deviation
	Run 1	Run 2	Run 3		
0.146	0.14	0.14	0.13	0.14	0.005
0.202	0.19	0.20	0.20	0.20	0.008
0.299	0.31	0.30	0.30	0.30	0.006
0.395	0.38	0.40	0.39	0.39	0.008
0.498	0.49	0.49	0.48	0.49	0.005
0.541	0.54	0.53	0.53	0.53	0.008

(Ref. [53]. Reprinted with permission of Oxford Instruments)

specifications, are negotiated with mutual agreement between the purchaser and supplier. These tests usually measure the functional groups in the component so that the chemical identity remains consistent for each lot purchased. Unfortunately, even the best specifications do not accurately predict how the material will perform in its end use. The transition from laboratory evaluation to pilot plant application and eventually to manufacturing is often replete with snares that trap even the most careful investigator.

Quantitative analytical methods for the determination of the functional groups [54] of fats and oils came into general use in the last century and have been a great help in making possible the systematic characterization and comparison of these types of materials. Some of the most important of these fat characteristics are saponification value, free acid content, hydroxyl value, and iodine value. Other properties that may be used to establish purchasing specifications include viscosity, color, water content, titer, specific gravity, pH, cloud point, and turbidity at room temperature. In addition, information obtained from the gas chromatographic separation of the esters of the fatty acids can give very valuable information about their distribution in the product. The amount of reacted poly(ethylene glycol) in nonionic surfactants, as well as the level of free PEG, can also be critical in preparing good quality emulsions. These procedures are discussed in the following sections.

Although the standard procedures described in this chapter are generally those proposed by the American Society for Testing and Materials (ASTM), they are essentially equivalent to other published methods. These additional techniques are recommended by the American Oil Chemists' Society (AOCS) [55], the International Union of Pure and Applied Chemistry

(IUPAC) [56], and in books by Hamilton and Rossell [57] and Mehlenbacher [58].

3.3.1 Saponification Value (Number)

Many lubricants used as components in spin finishes are esters that may be split by water (hydrolysis) in the presence of either acidic or basic catalysts [59]. Alkaline hydrolysis frequently is called *saponification*, because it is the type of reaction used in the preparation of soaps. This base–catalyzed hydrolysis goes to completion and requires one equivalent of alkali for each equivalent of ester, because the acid formed in the base–catalyzed equilibrium reacts irreversibly with the catalyst to form a salt and water. Because this reaction goes to completion, alkaline saponification can be used as a quantitative procedure for the measurement of ester content. In practice the saponification value, also known as saponification number (or "sap" number), is obtained by refluxing the sample with alcoholic potassium hydroxide. The excess KOH is then back–titrated with standardized hydrochloric acid to a phenolphthalein end–point. The saponification number is expressed as the number of milligrams of KOH required to saponify 1 gram of sample. In all these analyses, a reagent blank with the sample omitted is run concurrently. The equation used to calculate the saponification number is:

$$Saponification\ number = \frac{(B - A)(N_{HCl}) \times (56.1)}{C} \tag{3-5}$$

where:

B	=	ml of HCl for titration of the blank,
A	=	ml of HCl for titration of the sample,
N_{HCl}	=	normality of HCl solution,
C	=	grams of sample used, and
56.1	=	formula weight of KOH (mg/meq).

Complete instructions for determination of saponification number for natural and synthetic waxes may be found in ASTM Method D 1387–89 [60] and for drying oils, and fatty acids in ASTM Method D 1962–85 [61].

3.3.2 Acid Value (Number)

Many finish components are either fatty acids or derivatives of those fatty acids. Often it becomes necessary to determine the amount of free acid present in the sample. The number obtained from this determination is called either the acid value or acid number. It is defined as the number of milligrams of potassium hydroxide required to neutralize the acids present in 1 gram of the sample. In this technique the sample is dissolved in the appropriate solvent; xylene for waxes, ethanol for fatty acids and their reaction products, and a 50:50 mixture of toluene and ethanol for polyols. The free acids are then titrated with either an aqueous or ethanolic potassium hydroxide solution to a phenolphthalein end–point. The acid value is calculated as:

$$Acid\ Value = \frac{(V N \times 56.1)}{W} \tag{3-6}$$

where:

V = ml of KOH solution required for titration of the sample,
N = normality of the KOH solution,
56.1 = formula weight of KOH, and
W = sample weight in grams.

For waxes and polyols, 0.1 N KOH is recommended as the titrating solvent and 0.5 N KOH for fatty acids. Detailed directions for acid number can be found in ASTM Methods for waxes [62], fatty acids [63], and polyols [64].

3.3.3 Hydroxyl Value (Number)

The hydroxyl value, or hydroxyl number, is, by definition, the number of milligrams of potassium hydroxide equivalent to the hydroxyl content of 1 gram of the sample. There are several procedures for determining the hydroxyl content of finish components, but in most methods the hydroxyl groups in the sample are reacted with either acetic anhydride or phthalic anhydride solutions. The sample–anhydride solution is then reacted under reflux conditions or in a pressure bottle. These two conditions will give essentially equivalent results. The excess anhydride is hydrolyzed with water and the resulting acid is titrated with standard potassium hydroxide to a phenolphthalein end–point. For most finish materials, acetic anhydride [65][66], is the anhydride of choice, but for some polyols phthalic anhydride is recommended [67]. Aldehydes, primary and secondary amines, and other compounds that react with acetic anhydride, interfere with the results. Any free acid present in the sample will also be titrated causing a low hydroxyl number; however, the hydroxyl value may be corrected by determining the acid value of the sample. The hydroxyl number is calculated as:

$$Hydroxyl\ Value = \frac{B + (SA/C) - V}{S} N \times 56.1 \tag{3-7}$$

where:

A = ml of KOH used for titration of the acid value,
B = ml of KOH used for the reagent blank,
C = grams of sample used for the acid value,
V = ml of KOH for titration of the acetylated sample,
N = normality of the KOH solution,
56.1 = formula weight of KOH, and
S = sample weight of hydroxyl value sample.

If a sample is supposed to consist of a pure monoglyceride, that is, with two unreacted hydroxyl groups, it should have a hydroxyl value twice the saponification value; the hydroxyl value of a pure diglyceride is half of its saponification value.

3.3.4 Iodine Value

One factor that the spin finish formulator must consider in the selection of components for that finish is the degree of unsaturation in the carbon chain. Although unsaturation usually reduces the viscosity, and subsequently the hydrodynamic friction, it can also result in a large amount of varnish and char on hot surfaces. The double bonds are easily oxidized to a complex mixture that can be cross-linked to a typical varnish. Thus, it is necessary to measure the degree of unsaturation in finish components and the usual indication of unsaturation is the iodine value.

Iodine value is a measure of general carbon to carbon unsaturation and is expressed as centigrams of iodine absorbed by one gram of sample. The sample is reacted with either Hanus iodine monobromide solution [68] or Wijis iodine monochloride solution [69][70] and the excess Hanus or Wijis solution is determined by titration with standard thiosulfate solution using starch as the indicator. The iodine monohalide (IX) is selective for unsaturated linkages.

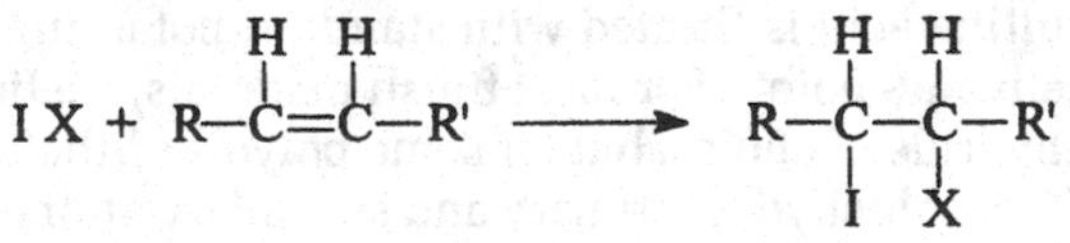

$$I\,X \; + \; R\!-\!\overset{\displaystyle H}{\underset{\displaystyle }{C}}\!=\!\overset{\displaystyle H}{\underset{\displaystyle }{C}}\!-\!R' \longrightarrow R\!-\!\overset{\displaystyle H}{\underset{\displaystyle I}{C}}\!-\!\overset{\displaystyle H}{\underset{\displaystyle X}{C}}\!-\!R'$$

As in the other analyses described above, a blank that does not contain the sample is run while the sample analysis is performed. The iodine value is calculated as:

$$Iodine\ Value = \frac{(B - V)\ x\ N\ x\ 12.69}{S} \tag{3-8}$$

where:

B	=	ml of $Na_2S_2O_3$ required for blank titration,
V	=	ml of $Na_2S_2O_3$ required for sample titration,
N	=	normality of the $Na_2S_2O_3$ solution,
12.69	=	milliequivalent weight of iodine in centigrams, and
S	=	sample weight in grams.

3.3.5 Other Specification Analyses

1. *Viscosity*. The specifications for viscosity are usually reported in centistokes (cSt) at a specific temperature; however, Saybolt Universal Seconds (SUS) are also used. The details for viscosity measurement are found in Section 3.1.2.

2. *Color*. Although the color of a material does not accurately describe its chemical composition, it can give a good indication of lot to lot variability. If the components are clear and pale in color, the specifications are normally expressed in American Public Health Association (APHA) units, while darker materials can be reported as Gardner Color. The APHA color system was developed for measuring the color of water and wastewater [71] and is a visual assessment of different concentrations of dilute solutions of either potassium chloroplatinate, K_2PtCl_6, [72] or a potassium chloro-platinate–cobalt chloride, $CoCl_2 \cdot 6H_2O$, mixture [73] in dilute hydrochloric acid. Either solution will yield exactly the same APHA color value. After a stock solution of the chloroplatinate is prepared, it is diluted to establish standards. The liquids are poured into 100 ml Nessler tubes and viewed vertically against a white background while compared to the sample being measured. The APHA color scale ranges from 1 to 500, with 1 being essentially water–white and 500 pale yellow.

The Gardner Color scale is based on comparing the color of transparent samples with arbitrarily numbered glass standards. These glass standards have chromaticity coordinates equivalent to solutions of potassium chloroplatinate (ranging from 1–8) and to ferric chloride–cobalt chloride solutions (ranging from 9–18)[72]. The sample to be measured is placed in a special cuvette in one side of a commercial, hand–held holder and the glass standards inserted in the other side. The glass standards are changed until a match is found [74]. In comparing these two color scales, a value of 500 APHA units is equivalent to 3 on the Gardner Scale.

3. *Water content*. The amount of water in this type of material is usually measured by titration with Karl Fischer reagent. The sample, with a weight selected to contain a maximum of 300 mg of water, is dissolved or dispersed in a suitable liquid and then titrated. Karl Fischer reagent is a mixture of iodine, sulfur dioxide, pyridine, and methanol. As long as any water is present, the iodine is reduced to colorless hydrogen iodide. The end point is the first appearance of free iodine, which can be determined either visually or electrometrically [75][76]. The equation for this reaction is:

$$\text{(pyridine)} \cdot I_2 + \text{(pyridine)} \cdot SO_2 + \text{(pyridine)} + H_2O \longrightarrow$$

$$2 \; \text{(pyridine)} \cdot HI + \text{(pyridine)} \cdot SO_3$$

and

$$\text{(pyridine)} \cdot SO_3 + ROH \longrightarrow \text{(pyridine)} \cdot HSO_4R$$

It must be emphasized that while titration techniques with the Karl Fischer reagent are some of the most widely used methods for measuring the water content of organic solids or liquids and most inorganic compounds, the reagent itself contains four toxic compounds. These are iodine, sulfur dioxide, pyridine, and methanol. The reagent should be dispensed in a well–ventilated area. Care must be exercised to avoid unnecessary inhalation of the reagent vapors or direct contact of the reagent with the skin. Following accidental spillage, wash with large quantities of water.

4. *Titer.* The solidification point of fatty acids is called the titer and is applicable to all fatty acids and many derivatives. Since saturated fatty acids solidify at a higher temperature than unsaturated fatty acids, this method can assist in the determination of the ratios of those two types of compounds. The sample is heated to about 130°C for a brief period on a hot plate to remove the water present. The melted sample is poured into a test tube, equipped with a thermometer and stirrer, and placed in a large beaker or titer apparatus. The beaker is immersed to about the level of the lip in a water bath and the sample cooled until solidification occurs [77]. Although this is a simple test, titer is frequently used in manufacturing specifications.

5. *Specific gravity.* The specific gravity, or weight per unit volume, can be measured by using either a hydrometer for approximations or a pycnometer for more accurate determinations. Hydrometers with various specific gravity ranges may be purchased from the American Society for

Testing and Materials [78]. Under most circumstances, however, the specific gravity is measured with a pycnometer. For materials with a kinematic viscosity of 40 stokes or less, a Leach type pycnometer is recommended, and for materials having a kinematic viscosity in excess of 40 stokes, or for solids or semi–solids at 25°C, a Hubbard type pycnometer should be used. If the pycnometers are calibrated with water at a specified temperature, such as 25°C, and the sample measured at the same temperature, the value is expressed as Specific Gravity, 25/25°C. If the temperatures for calibration and measurement are different, the specific gravity is reported as Specific Gravity, apparent, x/y°C where x is the temperature of the sample and y the temperature of the water [79].

6. *pH.* Since pH is an accurate measurement of the hydrogen ion concentration, within certain limits, it is widely used to characterize aqueous solutions. pH meters utilizing glass electrodes are readily available and the measurement itself is quite simple. A standard test method is available [80] that lists precautions, calibrations, and measurement of pH.

7. *Cloud point.* Many nonionic surfactants are less soluble at higher temperature than at lower temperature because of hydration and agglomeration of the micelles at the elevated temperature. The cloud point is measured by warming a 1% test solution of the pure surfactant until it becomes cloudy. Then, after removing from heat, record the temperature at which it becomes clear. Duplicate results in the same laboratory should not differ by more than 0.5°C [81].

8. *Turbidity at room temperature.* This is a simple visual observation of the sample at room temperature to see if it is clear or turbid.

3.4 ANALYSES FOR MATERIAL CHARACTERIZATION

Periodically, an investigator might wish to obtain a more thorough characterization of a chemical compound that would be revealed through the tests for their specifications. If one is purchasing a material from one supplier and wants to evaluate the same material from another source, it is necessary to study that material with more comprehensive techniques. Since fatty acids are derived from natural sources, they are almost never pure compounds, but mixtures. There are several grades of palmitic or stearic acids and each supplier might obtain their raw materials from different origins. For effective spin finish formulation, the fatty acid ratios must be maintained. Nonionic surfactants might contain different amounts of oxyalkylene groups (ethylene oxide or propylene oxide) that could significantly affect emulsification properties. Some of the oxyalkylene might not be reacted with the starting material, but could exist as free poly(oxyalkylene glycol). Although these properties are not normally included in the specifications, the results add important information for proper characterization.

1. *Fatty acid composition by gas-liquid chromatography.* The separation and identification of fatty acids and oils having 8 to 24 carbon atoms is accomplished by preparing the methyl esters of the acids [82], [83] and then injecting these esters into a gas chromatograph. Reagent grade fatty acids or fatty acid methyl esters may be purchased from specialty chemical supply houses for use as standards.

The gas chromatograph should have an oven capable of constant temperature operation between 190 and 210°C and equipped with a flame ionization or thermal conductivity detector. A 1.5 to 3.0 meter column should be packed with 20 weight percent diethylene glycol succinate polyester liquid phase on 80 to 100 mesh acid washed calcined diatomaceous earth. Helium carrier gas is quite satisfactory [84]. A comparison of the relative retention times of the sample with the standards will give information on the type of fatty acid present and the area under the peak is utilized to calculate the amount of acid in the sample. Summing and calculating these areas as a percent of fatty acid will give values quite close to true weight percent from which the mole percent can be determined.

2. *Oxyalkylene content of nonionic surfactants.* Oxyalkylene groups can be represented as:

$$\overset{R}{\underset{|}{}} $$
$$-O\,CH\,(CH_2)_nO-$$

where R can be a hydrogen atom and n can be one as in the case of the oxyethylene group. In the procedure proposed by Siggia [85], hydriodic acid is used to split the oxyalkene groups from poly(oxyalkylene glycol) ethers and esters with the liberation of an alkene and formation of free iodine. The reaction for poly(ethylene oxide), for example, is postulated to proceed as:

$$-(CH_2CH_2O)_x- + 2\,x\,HI \longrightarrow x\,I\,CH_2CH_2\,I + x\,H_2O$$

$$I\,CH_2CH_2\,I \xrightarrow{\text{decomp.}} H_2C{=}CH_2 + I_2$$

The excess iodine is titrated with 0.1 N potassium thiosulfate until the disappearance of the yellow color. At this time add a small amount of starch indicator and continue the titration slowly until the blue color has just disappeared [86]. It is essential that a blank without the sample be run in this analysis. The ethylene oxide content is calculated as:

$$\text{Ethylene Oxide, weight \%} = \frac{[(A - B) \times N \times 2.203]}{W} \qquad (3\text{--}9)$$

where:

A = ml of $Na_2S_2O_3$ solution required to titrate the sample,
B = ml of $Na_2S_2O_3$ solution required to titrate the blank,
N = normality of the $Na_2S_2O_3$ solution,
2.203 = molecular weight of ethylene oxide/20, and
W = grams of sample.

3. *Free poly(oxyethylene glycol)*. During the ethoxylation of fatty alcohols, acids or esters, it is possible for free poly(oxyethylene glycol) (PEG) to be formed. If the beginning raw materials have some moisture present or if water is introduced into the reaction vessel, the free glycol is formed rather than ethylene oxide reacting with the dry raw material. It can be very important for emulsifiers to have all of the ethylene oxide reacted with the base material or severe problems with emulsion quality might occur. The determination of free poly(oxyethylene glycol) is somewhat complex, but the method yields accurate results. The nonionic surfactant is dissolved in ethyl acetate and the free glycol removed by washing with a 30 weight percent solution of sodium chloride in water. Polyethylene glycols are recovered from the NaCl solution by chloroform extraction. Longman [87] also reports that the method can be used to separate poly(oxyethylene glycol) from poly(oxypropylene glycol) since under the test conditions, poly(oxypropylene glycol) is very soluble in ethyl acetate.

This procedure should be carried out in a large thermostatically controlled cabinet at 35 ± 1°C. After accurately weighing 5 ± 0.05 grams of sample, it is dissolved in ethyl acetate at 35 ± 1°C and transferred quantitatively into a 250 ml separatory funnel to a final volume of 75 ml. Several successive extractions are made with the NaCl solution until the PEG free condensate is isolated in the ethyl acetate and the PEG in the NaCl solution. The ethyl acetate is evaporated to obtain the amount of PEG free condensate. The NaCl solution is extracted with chloroform, and the chloroform evaporated to obtain the amount of free PEG. Calculations are:

$$PEG, \text{ weight \%} = \frac{100\,A}{W}$$

$$and \qquad (3\text{--}10)$$

$$PEG \text{ free condensate, weight \%} = \frac{100\,B}{W}$$

where:

A   =   weight in grams of PEG isolated,
B   =   weight in grams of condensate isolated, and
W   =   weight in grams of the sample.

REFERENCES

1. Anderson, H. C. (1966). Instrumentation, Techniques, and Applications of Thermogravimetry, *Techniques ans Methods of Polymer Evaluation, Volume 1, Thermal Analysis*, (P. E. Slade and L. T. Jenkins, eds.), Marcel Dekker, New York, pp. 87–111

2. Wunderlich, B. (1990). *Thermal Analysis*, Academic Press, Boston, pp. 371–410

3. Wendlandt, W. (1974). *Thermal Analysis*, John Wiley & Sons, New York, pp. 6–129

4. Doyle, C. D. (1966). Calculations in Thermogravimetric Analysis, *Techniques ans Methods of Polymer Evaluation, Volume 1, Thermal Analysis*, (P. E. Slade and L. T. Jenkins, eds.), Marcel Dekker, New York, pp. 113–216

5. Slade, P. E., unpublished data

6. Specialties from ICI for Fiber Producers (1987). *Laboratory Test Methods for Fiber Finishes: Heat Stability*, ICI Americas, Inc., Wilmington, DE

7. George, P. and Walsh, A. D. (1946). Decomposition of Tertiary–alkyl Peroxides, *Trans. Faraday Soc.*, **42**:94

8. Lloyd, W. G. (1956). The Low Temperature Autoxidation of Diethylene Glycol, *J. Am. Chem. Soc.*, **78**:72

9. Streitwieser, A. and Heathcock, C. H. (1985). *Introduction to Organic Chemistry*, Macmillian, New York, pp. 217–398

10. Lyne, D. G. (1955). The Dynamic Friction Between Cellulose Acetate Yarn and a Cylindrical Metal Surface, *J. Text. Inst.*, **46**: P112

11. Carpenter, D. K. and Westerman, L. (1975). Viscometric Methods of Studying Molecular Weight and Molecular Weight Distribution, *Polymer Molecular Weights, Part II*, (P. E. Slade, ed.), Marcel Dekker, New York, pp. 379–499

12. ASTM Method D 4878–88 (1991). Standard Test Method for Polyurethane Raw Materials: Determination of Viscosity of Polyols, *Annual Book of ASTM Standards, Volume 08.03*, ASTM, Philadelphia, pp 649–651

13. *Brookfield Synchro–lectric Viscometer* (1970), Brookfield Engineering Laboratories, Stoughton, MA, pp. 1–6

14. ASTM Method D 2196–86 (1991). Standard Test Method for the Rheological Properties of Non–Newtonian Materials by Rotational (Brookfield) Viscometer, *Annual Book of ASTM Standards, Volume 06.01*, ASTM, Philadelphia, pp. 277–279

15. ASTM Method D 2161–87 (1991). Standard Practice for Conversion of Kinematic Viscosity to Saybolt Universal Viscosity or to Saybolt Furol Viscosity, *Annual Book of ASTM Standards, Volume 05.02*, ASTM, Philadelphia, pp. 43–67

16. Howell, H. G. (1951). Inter–Fiber Friction, *J. Text. Inst.*, **42**: T521

17. Kamath, Y. K. (1990). Interfiber Friction in Relation to Spin Finish Distribution, Presented at the TRI 60th Annual Conference, Princeton, N.J., May 3

18. Röder, H. L. (1953). Measurement of the Influence of Finishing Agents on the Friction of Fibers, *J. Text. Inst..*, **44**: T247

19. Fort, T. and Olsen, J. S. (1961). Boundary Friction of Textile Yarns, *Textile Res. J.*, **31**: 1007

20. ASTM Method D 3412–89 (1991). Standard Test Method for Coefficient of Friction, Yarn to Yarn, *Annual Book of ASTM Standards, Volume 07.02*, ASTM, Philadelphia, pp. 11–15

21. Schick, M. J. (1973). Friction and Lubrication of Synthetic Fibers. Part IV. Effect of Fiber Material and Lubricant Viscosity and Concentration, *Textile Res. J.*, **43**: 342

22. Zaukelies, D. A. (1959). An Instrument for the Study of the Friction and the Static Electrification of Yarns, *Textile Res. J.*, **29**: 794

23. Rothschild Messinstrumente, Zürich (1995). *F–Meter R–1188, Electronic Instrument for Automatic Measurement of the Friction Coefficient*, Lawson–Hemphill Sales, Inc., Spartanburg, SC

24. Mayer, F. (1964). Gleiteigenshaften von Garmen, ihre Messung und Verbesserung, *SVF–Fachorgan*, Basel, Switzerland, **9**: 681

25. Specialties from ICI for Fiber Producers (1987). *Laboratory Test Methods for Fiber Finishes: Friction Test*, ICI Americas, Wilmington, DE

26. ASTM Method D 3103 (1991). Standard Test Method for Coefficient of Friction, Yarn to Solid Materials, *Annual Book of ASTM Standards, Volume 07.01*, ASTM, Philadelphia, pp. 848–854

27. Rothschild Messinstrumente, Zürich (1995). *Static–Voltmeter R–4021*, Lawson–Hemphill Sales, Inc., Spartanburg, S.C.

28. Kamath, Y. K., Hornby, S. B., Weigmann, H.–D., and Wilde, M. F. (1994). Wicking of Spin Finishes and Related Liquids into Continuous Filament Yarns, *Textile Res. J.*, **64**: 33

29. Washburn, E. W. (1921). The Dynamics of Capillary Flow, *Phys. Rev.*, **17**: 273

30. Hirt, D. E., Prud'homme, R. K., Miller, B., and Rebenfeld, L. (1990). Dynamic Surface Tension of Hydrocarbon and Fluorocarbon Surfactants Using the Maximum Bubble Pressure Method, *Colloids Surfaces*, **44**: 101

31. Keesom, W. H.(1921). Van der Waals Attractive Force, *Physik. Z.*, Leipzig, Germany, **22**: 129

32. Debye, P. J. (1920). Van der Waals Cohesive Forces, *Physik. Z.*, Leipzig, Germany, **21**: 178

33. London, F. (1937). The General Theory of Molecular Forces, *Trans. Faraday Soc.*, **33**: 8

34. Miller, B. and Young, R. A. (1975). Methodology for Studying the Wettability of Filaments, *Textile Res. J.*, **45**: 359

35. Kamath, Y. K., Dansizer, C. J., Hornby, S., and Weigmann, H.–D. (1987). Surface Wettability Scanning of Long Filaments by a Liquid Membrane Method, *Textile Res. J.*, **57**: 205

36. Wu, S. (1971). Calculation of Interfacial Tension in Polymer Systems, *J. Poly. Sci.: Part C*, **34**: 19

37.   Slade, P. E. and Hild, D. N. (1992). Surface Energies of Nonionic Surfactants Adsorbed onto Nylon Fiber and a Correlation with Their Solubility Parameters, *Textile Res. J.*, **62**: 535

38.   Owens, D. K. and Wendt, R. C. (1969). Estimation of the Surface Free Energy of Polymers, *J. App. Poly. Sci.*, **13**: 1741

39.   Zisman, W. A. (1964). Relation of Equilibrium Contact Angle to Liquid and Solid Constitution, *Contact Angle, Wettability, and Adhesion, ACS Advances in Chemistry Series, Vol. 43*, American Chemical Society, Washington, pp. 1–51

40.   Dalal, E. N. (1987). Calculation of Solid Surface Tension, *Langmuir*, **3**: 1009

41.   Kamath, Y. K., Dansizer, C. J., Hornby, S. B., and Weigmann, H.-D. (1991). Finish Distribution in Multifilament Yarn, *J. App. Poly. Sci.: Applied Polymer Symposium*, **47**: 281

42.   Gerold, E. A., Rebenfeld, L, Scott, M. G., and Weigmann, H.-D. (1979). PET Filaments with Radially–Differentiated Structure, *Textile Res. J.*, **49**: 652

43.   Weigmann, H.-D., Scott, M. G., Ribnick, A. S., and Matkowsky, R. D. (1977). Interactions of Nonaqueous Solvents with Textile Fibers. Part VIII: Mechanism of Dye Diffusion in Solvent–Treated Polyester Yarn, *Textile Res. J.*, **47**: 745

44.   AATCC, Pacific Section (1979). Effect of Surfactants and Heat on Dyeing and Physical Properties of Cotton, Nylon, and Polyester, *Textile Chem. Color.*, **11**: 20/35

45.   Needles, H. L., Berns, R. S., Lu, W.-C., Alger, K., and Varma, D. S. (1980). Effect of Nonionic Surfactants and Heat on Selected Properties of Nylon 6,6, *J. Appl. Poly. Sci.*, **45**: 1745

46.   Selivansky, D., Walsh, W. K., and Frushour, B. G. (1990). Interactions of Spin Finishes with Partially Oriented Poly(ethylene terephthalate) Fibers. Part I: The nature of Finish/Fiber Interactions, *Textile Res. J.*, **60**: 33

47.   AATCC Test Method 97–1989 (1995). Extractable Content of Greige and/or Prepared Textiles, *AATCC Technical Manual*, American Association of Textile Chemists and Colorists, Research Triangle Park, NC, p. 141

48.   Sax, N. I. and Lewis, R. J., Sr. (1989). *Dangerous Properties of Industrial Materials*, Seventh Edition, Volume III, Van Nostrand Reinhold, New York, pp. 3327, 1775

49.   ASTM Method D 2257–89 (1991). Standard Test Method for Extractable Matter in Textiles, *Annual Book of ASTM Standards, Volume 07.01*, ASTM, Philadelphia, pp. 622–625

50.   Drews, M. J., Ivey, K., and Lam, C. (1994). Supercritical Fluid Extraction as an Alternative to the Soxhlet Extraction of Textile Materials, *Textile Chem. Color.*, **26**: 29

51.   Pickering, W. F. (1971). *Modern Analytical Chemistry*, Marcel Dekker, New York, pp. 164–166

52. QP^{20+} *Pulse NMR for Quality Control* (1994). Oxford Instruments, Industrial Analysis Group, Concord, MA

53.   Oxford at Work (1994). *Spin Finish on Fiber by Pulsed NMR*, Oxford Instruments, Industrial Analysis Group, Concord, MA

54. Eckey, E. W. (1954). *Vegetable Fats and Oils*, Van Nostrand Reinhold, New York, pp. 217–240

55. American Oil Chemists' Society (1987). *Official Methods and Recommended Practices of the American Oil Chemists' Society*, (V. C. Mehlenbacher, ed.), American Oil Chemists' Society, Champaign, IL

56. International Union of Pure and Applied Chemistry, Applied Chemistry Division, Commission on Oils, Fats, and Derivatives (1987). *Standard Methods for the Analysis of Oils, Fats, and Derivatives, 7th Revision*, (C. Paquot and A. Hautfenne, eds.), Oxford, Oxfordshire, UK

57. Hamilton, R. J. and Rossell, J. B. (1986). *Analysis of Oils and Fats*, Elsevier Applied Science Publishers, New York

58. Mehlenbacher, V. C. (1960). *The Analysis of Fats and Oils*, Garrard Press, Champaign, IL

59. Noller, C. R. (1958). *Textbook of Organic Chemistry*, W. B. Saunders, Philadelphia, p. 132

60. ASTM Method D 1387–89 (1991). Standard Test Method for Saponification Number (Empirical) of Synthetic and Natural Waxes, *Annual Book of ASTM Standards, Volume 15.04*, ASTM, Philadelphia, p. 124

61. ASTM Method D 1962–85 (1991). Standard Test Method for Saponification Value of Drying Oils, Fatty Acids, and Polymerized Fatty Acids, *Annual Book of ASTM Standards, Volume 06.03*, ASTM, Philadelphia, pp. 265–266

62. ASTM Method D 1386–83 (1991). Standard Test Method for Acid Number (Empirical) of Synthetic and Natural Waxes, *Annual Book of ASTM Standards, Volume 15.04*, ASTM, Philadelphia, p. 123

63. ASTM Method D 1980–87 (1991). Standard Test Method for Acid Value of Fatty Acids and Polymerized Fatty Acids, *Annual Book of ASTM Standards, Volume 06.03*, ASTM, Philadelphia, pp. 227–228

64. ASTM Method D 4662–87 (1991). Standard Test Method for Polyurethane Raw Materials: Determination of Acid and Alkalinity Numbers for Polyols, *Annual Book of ASTM Standards, Volume 08.03*, ASTM, Philadelphia, pp. 571–572

65. ASTM Method E 222–88 (1991). Standard Test Method for Hydroxyl Groups by Acetic Anhydride Acetylation, *Annual Book of ASTM Standards, Volume 15.05*, ASTM, Philadelphia, pp. 264–269

66. ASTM Method D 1957–86 (1991). Standard Test Method for Hydroxyl Value of Fatty Oils and Acids, *Annual Book of ASTM Standards, Volume 06.03*, ASTM, Philadelphia, pp. 258–263

67. ASTM Method D 4274–88. (1991). Standard Test Method for Testing Polyurethane Raw Materials: Determination of Hydroxyl Numbers of Polyols, *Annual Book of ASTM Standards, Volume 08.03*, ASTM, Philadelphia, pp. 407–415

68. Warth, A. H. (1947). *The Chemistry and Technology of Waxes*, Van Nostrand Reinhold, New York, p. 326

69. ASTM Method D 1541–86 (1991). Standard Test Method for Total Iodine Value of Drying Oils and Their Derivatives, *Annual Book of ASTM Standards, Volume 06.03*, ASTM, Philadelphia, pp. 210–212

70. ASTM Method D 1959–85 (1991). Standard Test Method for Iodine Value of Drying Oils and Fatty Acids, *Annual Book of ASTM Standards, Volume 06.03*, ASTM, Philadelphia, pp. 261–263

71. Franson, M., ed. (1975). *Standard Methods for the Examination of Water and Waste Water, 14th Edition*, American Public Health Association, p. 65

72. ASTM Method D 4980–88 (1991). Standard Test Methods for Polyurethane Raw Materials: Determination of Gardner and APHA Color of Polyols, *Annual Book of ASTM Standards, Volume 08.03*, ASTM, Philadelphia, pp. 656–658

73. ASTM Method D 1209–84 (1991). Standard Test Method for Color of Clear Liquids (Platinum–Cobalt Scale), *Annual Book of ASTM Standards, Volume 06.01*, ASTM, Philadelphia, pp. 145–147

74. ASTM Method D 1544–80 (1991). Standard Test Method for Color of Transparent Liquids (Gardner Color Scale), *Annual Book of ASTM Standards, Volume 06.03*, ASTM, Philadelphia, pp. 213–214

75. ASTM Method E 203–75 (1991). Standard Test Method for Water Using Karl Fischer Reagent, *Annual Book of ASTM Standards, Volume 15.05*, ASTM, Philadelphia, pp. 255–263

76. ASTM Method D 4672–87 (1991). Standard Test Method for Polyurethane Raw Materials: Determination of Water Content of Polyols, *Annual Book of ASTM Standards, Volume 08.03*, ASTM, Philadelphia, pp. 591–593

77. ASTM Method D 1982–85 (1991). Standard Test Method for Titer of Fatty Acids, *Annual Book of ASTM Standards, Volume 06.03*, ASTM, Philadelphia, pp. 282–283

78. ASTM Method D 891–89 (1991). Standard Test Method for Specific Gravity, Apparent, of Liquid Industrial Chemicals, *Annual Book of ASTM Standards, Volume 15.05*, ASTM, Philadelphia, pp. 1–4

79. ASTM Method D 1963–85 (1991). Standard Test Method for Specific Gravity of Drying Oils, Varnishes, Resins, and Related Materials, *Annual Book of ASTM Standards, Volume 06.03*, ASTM, Philadelphia, pp. 267–269

80. ASTM Method E 70–90 (1991). Standard Test Method for pH of Aqueous Solutions With the Glass Electrode, *Annual book of ASTM Standards, Volume 15.05*, ASTM, Philadelphia, pp. 194198

81. ASTM Method D 2024–65 (1991). Standard Test Method for Cloud Point of Nonionic Surfactants, *Annual Book of ASTM Standards, Volume 15.04*, ASTM, Philadelphia, p. 204

82. ASTM Method D 2800–87 (1991). Standard Test Method for Preparation of Methyl Esters from Oils for Determination of Fatty Acid Composition by Gas–Liquid Chromatography, *Annual Book of ASTM Standards, Volume 06.03*, ASTM, Philadelphia, pp. 373–374

83. ASTM Method D 3457–87 (1991). Standard Test Method for Preparation of Methyl Esters from Fatty Acids for Determination of Fatty Acid Composition by Gas–Liquid Chromatography, *Annual Book of ASTM Standards, Volume 06.03*, ASTM, Philadelphia, pp. 416–417

84. ASTM Method D 1983–90 (1991). Standard Test Method for Fatty Acid Composition by Gas–Liquid Chromatography of Methyl Esters, *Annual Book of ASTM Standards, Volume 06.03*, ASTM, Philadelphia, pp. 284–287

85.   Siggia, S. (1963). *Quantitative Organic Analysis via Functional Groups, Third Edition*, John Wiley and Sons, New York, pp. 212–217

86.   ASTM Method D 2959–89 (1991). Standard Test Method for Ethylene Oxide Content of Polyethoxylated Nonionic Surfactants, *Annual Book of ASTM Standards, Volume 15.04*, ASTM, Philadelphia, pp. 313–316

87.   Longman, G. F. (1975). *The analysis of Detergents and Detergent Products*, John Wiley and Sons, New York, pp. 312–316

CHAPTER 4
YARN TO HARD SURFACE LUBRICANTS FOR FIBER FINISHES

Perhaps the most important reason for applying a finish to synthetic fibers is to control the tension during the spinning process, and also the friction in subsequent processing. These yarn manufacturing procedures could either be at the locations where the fibers were initially spun or at their textile customers' plants. It is very important to remember that the goal is to *control* tension, and not necessarily to reduce the tension to the lowest possible value. The proper choice of a lubricant for the spin finish will allow one to achieve this control of tension needed for each process that the yarn will undergo. We must consider, however, that the finish normally will be applied as an oil–in–water emulsion to many yarn types and that the emulsifier chosen will also affect friction and thus the tension. In many processes, a compromise must be made between low friction at one step and a higher friction at another. Examples of this conundrum could be in yarn texturing. When the yarn traverses through guides and the texturizer heater, low friction is required. As the yarn then passes through the actual texturing device, either a pin or a friction disk assembly, it is necessary for the yarn to have a higher friction for proper bulk development. Sometimes lubricants are selected to volatilize at one manufacturing step, leaving another component to control the tension at a later stage in that process.

In choosing lubricants, the formulator must also consider that about twenty to thirty percent of the finish is absorbed into some fibers within a few weeks. If a substantial amount of the lubricant migrates into the yarn, the frictional properties will change from those measured immediately after spinning. Since many different types of synthetic fibers are stored in warehouses for times that may vary from weeks to years, we must allow for the absorption mechanism to proceed, *for it will*.

4.1 WATER–INSOLUBLE LUBRICANTS

4.1.1 Petroleum Based Lubricants

Mineral Oil, CAS# 804247–5

Mineral oil was one of the initial components used in spin finishes for fiber lubrication [1]. Legends in the finish community relate that mineral oil was the lubricant of choice for finishes used in the production of artificial silk, now known as viscose rayon, in the early 1900s. This fiber was tested in experimental parachutes during World War I, replacing the more generally used real silk. In 1941, a patent was issued to the National Oil Products Company [2] for a spin finish containing 70% mineral oil by weight for softening and lubricating synthetic yarn. Since then, mineral oil has been used extensively for lubricating synthetic fibers that were fully drawn before shipment to a textile fabrication plant. After the development of partially oriented polyester (POY) yarn and partially oriented nylon (PON) yarn, mineral oil became less important because it tended to absorb into their more open crystalline structure.

Table 4–1

White Mineral Oils

U.S.P. Mineral Oil		Light Mineral Oil, NF	
Product	Viscosity, cSt @ 40°C	Product	Viscosity, cSt @ 40°C
Kaydol®	64–70	Rudol®	38–30
Orzol®	61–64	Ervol®	24–26
Britol®	57–60	Benol®	18–20
Gloria®	39–42	Blandol®	14–17
Protol®	35–37	Carnation®	11–14
		Klearol®	7–10

White oils are refined mineral oils, consisting of saturated aliphatic and non–polar alicyclic hydrocarbons, and are available in several grades and viscosities. Table 4–1 lists the white mineral oils available from the Petroleum Specialties Group of the Witco Corporation [3] and gives their viscosities. Other technical grades are also available.

Currently, mineral oil is used as a lubricant for fiberglass, and to a lesser degree as a wool processing assistant, as a finish for acetate yarn, and a small amount for fully drawn synthetic fibers. The major use for mineral oil, however, is in coning oils. After synthetic fibers are textured to develop bulk and loft for converting into softer fabric, the yarn is lubricated with a coning oil for assisting in the next steps in fabric preparation. These coning oils are added to the yarn without solubilization or emulsification, an approach known as neat application. Mineral oil is blended with emulsifiers to assist in coning oil removal.

4.1.2 Natural Product Glycerides and Esters

Most of the components that are common in spin finishes are made from the fatty acids isolated from natural products. These fatty acids are normally obtained by the saponification of natural fats or oils, which exist as glyceride esters. The shortest carbon chain length normally found in a natural fat is butyric, or butanoic acid, with four carbons. Table 4–2 [4] lists butyric and some longer chain fatty acids giving their common name, the IUPAC (International Union of Pure and Applied Chemistry) designation, the number of carbons and the degree of unsaturation. The natural plant or animal fats or oils from which the acids are extracted are recorded in Table 4–3 [4], with the amount of each acid present. It is interesting that all natural fatty acids contain an even number of carbon atoms since the biosynthesis of these acids are built up two carbons at a time from acetic acid units. These units are products of acetyl CoA, the thiol ester derived from acetic acid and coenzyme A [4].

The selection of the lubricant for inclusion in a finish composition is completely dependent upon the processing conditions for the yarn and the end use of the product. If the yarn is to be prepared under conditions in which heated surfaces are present, lubricants made from saturated fatty acids are needed to prevent varnish formation. As an example of the range of evaporation rates, Figure 4–1 displays a chart of the weight loss versus time for a 1 gram sample of selected esters held at 200°C [5]. Sometimes, however, a lower viscosity oil is needed and unsaturated materials are chosen since double bonds reduce viscosity. Generally, one must compromise between thermal stability and viscosity to achieve optimum results. It must be remembered that the fatty acids used to prepare the lubricants are never pure, but a mixture of various chain lengths, with the exact composition dependent upon the acid.

Table 4–2
Common Fatty Acids From Natural Sources

Fatty Acid Common Name	Fatty Acid IUPAC Name	Number of Carbon Atoms	Number of Double Bonds
Butyric	Butanoic	4	0
Caproic	Hexanoic	6	0
Caprylic	Octanoic	8	0
Capric	Decanoic	10	0
Lauric	Dodecanoic	12	0
Myristic	Tetradecanoic	14	0
Palmitic	Hexadecanoic	16	0
Stearic	Octadecanoic	18	0
Oleic	9–Octadecenoic	18	1
Linoleic	9,12–Octadeca-dienoic	18	2
Linolenic	9,12,15–Octadecatrienoic	18	3
Ricinoleic	12-Hydroxy-9-octadecenoic	18	1

(Ref. 4. Reprinted by permission of Prentice–Hall, Englewood Cliffs, NJ)

4.1.2.1 Esters of Myristic Acid

Isopropyl Myristate, CAS# 110–27–0

Myristic, or tetradecanoic, acid is probably the shortest chain fatty acid to be esterified with an alcohol for use as an ingredient in a fiber finish composition. It is found in butter fat, coconut oil, and palm kernel oil at levels below twenty percent. One commercial ester is isopropyl myristate sold by Inolex with the designation Lexolube® IPM [6] and Stepan using the trade name of Kessco® IPM [7]. Some common properties are given in Table 4–4.

Table 4-3

Fatty Acid Composition of Fats and Oils

Fat or Oil	Saturated Acids, %							Unsaturated Acids, %			Di–	Tri–
								Mono–				
	C_8	C_{10}	C_{12}	C_{14}	C_{16}	C_{18}	$>C_{18}$	C_{16}	C_{18}	$>C_{18}$	C_{18}	C_{18}
Beef Tallow			0.2	2–3	25–30	21–26	0.4–1	2–3	39–42	0.3	2	
Butter Fat	1–2	2–3	1–4	8–13	25–32	8–13	0.4–2	1–2	2–5	22–29	3	
Castor Oil(a)					0–1				0–9		3–7	
Coconut Oil	5–9	4–10	44–51	13–18	7–10	1–4			5–8	0–1	1–3	
Corn Oil				0–2	8–10	1–4		1–2	30–50	0–2	34–56	
Cottonseed Oil				0–3	17–23	1–3			23–44	0–1	34–55	
Lard				1	25–30	12–16		2–5	41–51	2–3	3–8	
Olive Oil			0–1	0–2	7–20	1–3	0–1	1–3	53–86	0–3	4–22	
Palm Oil				1–6	32–47	1–6			40–52		2–11	
Palm Kernel Oil	2–4	3–7	45–52	14–19	6–9	1–3	1–2	0–1	10–18		1–2	
Peanut Oil				0.5	6–11	3–6	5–10	1–2	39–66		17–38	
Soybean Oil				0.3	7–11	2–5	1–3	0–1	22–34		50–60	2–10

(a) 80–92% ricinoleic, C_{18} monounsaturated acid with hydroxyl group.

(Reference [4]. Reprinted by permission of Prentice–Hall, Inc., Englewood Cliffs, NJ)

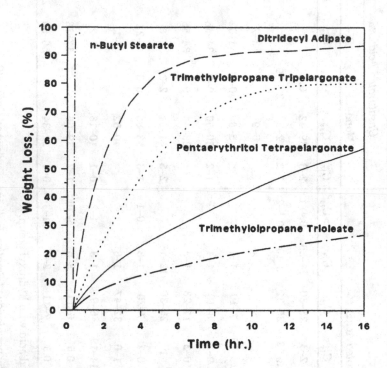

Figure 4–1. Isothermal (200°C) Volatility of One Gram of Selected Esters.
(Reference [5]. Reprinted by permission of Inolex Chemical Company)

Table 4–4

Typical Properties of Isopropyl Myristate

Property	Value
Iodine Value	<1
Acid Value	<1
Color, APHA	20
Viscosity, cSt @ 25°C	5–9
10 g. Evaporation Loss, (4 Hrs @ 250°C), Wt.%	99.9

The structural formula of isopropyl myristate is:

$$CH_3-\underset{\underset{H}{|}}{\overset{\overset{CH_3}{|}}{C}}-O-\overset{\overset{O}{||}}{C}-CH_2(CH_2)_{11}CH_3$$

Cetyl Myristate, CAS# 259–01–1

Another ester of myristic acid that is available for use in fiber lubrication is cetyl myristate. The primary uses are as a base in stick cosmetics and as a thickener in creams and lotions, but it can be an effective fiber lubricant. It is supplied as a flaked solid by Stepan under the name of Kessco® 654 [8][12]. The typical properties of cetyl myristate are given in Table 4–5 and the structural formula is:

$$CH_3(CH_2)_{14}CH_2-O-\overset{\overset{O}{||}}{C}-CH_2(CH_2)_{11}CH_3$$

Table 4–5

Typical Properties Cetyl Myristate

Property	Value
Acid Value	2 max
Saponification Value	116–124
Melting point, °C	47–53
Volatility, 1 g. sample, % loss, 1 Hr. @ 160°C	1.1

4.1.2.2 Esters of Palmitic Acid

Palmitic acid, the fatty acid lubricating moiety in palmitate esters, is sold in four grades [9], with the actual palmitic acid content ranging from

66% to 98%. These palmitic acids are most probably made by distillation under vacuum, either in a batch or continuous process, of saponified animal fats or palm oil. This removes the residual glycerides and improves the color. The major impurity is stearic acid with some minor amounts of myristic, pentadecanoic, and margaric also present. Table 4–6 presents the composition and additional data for these grades of palmitic acids. Obviously, one would like for a vendor to use the best grade possible, but the cost–quality factor must be considered.

Table 4–6

Properties of Palmitic Acids

Commercial Acid	% Palmitic	% Stearic	Titer	Iodine Number	Acid Number
cp palmitic	92	5	59–61	0.5	216–220
cp palmitic, 97%	98	2	61.6	0.5	216–220
cp palmitic, 80%	81	15	56–58	1.0	214–218
palmitic–eutectic	66	30	53–55	0.5	211–213

(Ref. [9]. Copyright 1978. Reprinted by permission of John Wiley & Sons, Inc.)

Isopropyl Palmitate, CAS# 142–91–6

After commercial production of nylon began in about 1938, it was soon discovered that this new synthetic fiber could be made into a fine denier yarn that was quite suitable for use in the construction of hosiery for women. Some lubricants in use at the time had viscosities that would result in a higher processing friction than desired. While ethyl palmitate and propyl palmitate were investigated, the lubricant most frequently used was isopropyl palmitate. It had a low viscosity so that friction was not a problem, but it was very volatile. On the 15 denier monofilament fully drawn flat yarns used at the time, finish absorption was not an obstacle, but isopropyl palmitate is currently not employed as a finish component in the quantity that it was in the past.

Isopropyl palmitate is sold by Henkel as Emerest® 2316 [10], by Inolex as Loxolube® IPP [11], and by Stepan as Kessco® IPP [12]. Selected properties for this material are listed in Table 4–7.

Table 4–7

Typical Properties of Isopropyl Palmitate

Property	Value
Acid Value	1.0 max
Saponification Value	182–191
Iodine Value	0.5 max
Specific Gravity, 25/25°C	0.850–0.855
Color, APHA	20 max
Viscosity, cSt @ 25°C	5
Isopropyl Palmitate, Wt. %	78 min

Isopropanol is usually prepared by the hydrogenation of acetone [13] [17]. This alcohol is then reacted with palmitic acid to form the ester. The structure is:

$$CH_3-\underset{\underset{H}{|}}{\overset{\overset{CH_3}{|}}{C}}-O-\overset{\overset{O}{||}}{C}-CH_2(CH_2)_{13}CH_3$$

It must be noted that this structure is not entirely correct, since it shows only palmitic acid, and not the mixture containing the impurities of stearic acid and the other minor acid ingredients. All of the other structures presented in this book will assume pure components and not their naturally occurring mixtures.

Octyl Palmitate, CAS# 29806–73–3

Octyl alcohol, with a longer alkyl chain than isopropyl alcohol, is also commercially available. Although commonly known as octyl palmitate, the ester is actually made from 2–ethylhexanol instead of n–octnol because of the much lower cost of the branched alcohol. Octyl palmitate is a low viscosity ester that is found in coning oils and carding lubricants as well as in spin finishes. It is manufactured by Inolex as Lexolube® EHP [14] and Stepan and sold under the name Kessco® OP [15][16]. The chemical structure of octyl palmitate is:

$$CH_3(CH_2)_3\overset{\overset{\displaystyle C_2H_5}{|}}{\underset{\underset{\displaystyle H}{|}}{C}}CH_2-O-\overset{\overset{\displaystyle O}{\|}}{C}-CH_2(CH_2)_{13}CH_3$$

Typical properties of octyl palmitate are shown in Table 4–8.

Table 4–8

Typical Properties of Octyl Palmitate

Property	Value
Acid Value	3 max
Saponification Value	50
Color, APHA	20 max
Viscosity, cSt @ 25°C	15

Cetyl Palmitate, CAS# 540–10–3

Another commercially available ester of palmitic acid is cetyl palmitate, an ingredient of many cosmetic preparations. Although this ester is the major component of spermaceti, the wax obtained from the head of the sperm whale, the current ester is synthetic. Cetyl alcohol is manufactured by the reduction of palmitic acid with a nickel catalyst [17].

$$C_{15}H_{31}\text{-}\overset{\overset{\displaystyle O}{\|}}{C}-OH \xrightarrow[\substack{3,000\text{-}5,000 \\ \text{lb./sq. in.}}]{H_2\ (Ni)} C_{16}H_{33}OH$$

The structure of cetyl palmitate is:

$$CH_3(CH_2)_{14}CH_2-O-\overset{\overset{\displaystyle O}{\|}}{C}-CH_2(CH_2)_{13}CH_3$$

Cetyl palmitate is a high temperature, high smoke point solid lubricant melting at about 53°C. It is sold by Henkel as Cutina® CP [18], Stepan as Kessco® 653 [12][19], and Witco as Kemester® CP [20][21]. Some properties are listed in Table 4–9.

Table 4–9

Typical Properties of Cetyl Palmitate

Property	Value
Acid Value	2 max
Saponification Value	109–117
Iodine Value	1 max

4.1.2.3 Esters of Stearic Acid

The primary source of stearic acid is the saponification of beef tallow. The separation of tallow fatty acids into solid (saturated) and liquid (unsaturated) fractions cannot be accomplished through distillation as with lower molecular weight liquid oils because of the similarity of the boiling points. The main commercial separation technique is known as pressing.

In this procedure, the mixed solids are cooled to about 40°F, and partial solidification is accomplished. A pressing of the cakes thus formed removes the liquid fraction. By two or three subsequent melting–cooling–pressing cycles, followed by a bleaching step, a stearic acid free of unsaturation is obtained [22]. Several grades are listed in Table 4–10 [9].

Table 4–10

Grades of Stearic Acid

Grade	% Palmitic	% Stearic	% Oleic	Titer, °C	Iodine #
Commercially pure	7	90		65–68	1 max
Single–pressed	52	38	5	53–54	5–10
Double–pressed	52	39	4	54–55	4.5–7
Triple–pressed	52	43		55–56	0.5 max
Hydrogenated tallow	29	66		57–61	1 max
Stearic–palmitic	21	72		60–64	1 max

(Ref. [9]. Copyright 1978. Reprinted by permission of John Wiley & Sons, Inc.)

Usually, triple–pressed stearic is the fatty acid grade sold for use in finish components, so actually these components contain more palmitic acid than stearic; however, the name of stearate is still used.

Butyl Stearate, CAS# 123–95–5

Butyl stearate is considered by many finish formulators as the standard product for effective fiber–hard surface lubrication. It is readily available, inexpensive, and is an excellent lubricant. The main disadvantages of butyl stearate are the high volatility and ease of absorption into the fiber. This component is manufactured by esterifying one of the grades of stearic acids, Table 4–10, with *n*–butyl alcohol, which is prepared by the reductive carbonylation of propylene. The viscosity of this product is about 12 cSt at 25°C with a pour point of 20°C, so in cold weather it can be a semi–solid. Butyl stearate has a structure of:

$$CH_3-(CH_2)_3-O-\overset{\displaystyle O}{\overset{\displaystyle \|}{C}}-(CH_2)_{16}-CH_3$$

Typical properties of butyl stearate are presented in Table 4–11.

Table 4–11

Typical Properties of Butyl Stearate

Property	Value
Saponification Number	165–180
Color, APHA	150 max
Viscosity, 25°C, cSt	12
Iodine Value	1 max
Acid Value	1.0 max
Hydroxyl Value	2.0 max
Moisture, % by Weight	0.1 max
10 g. Evaporation Loss (4 Hrs. @ 250°C), Wt. %	99

Butyl stearate is sold by Henkel under the name Emerest® 2326 [23][24], Inolex as Lexolube® NBS [25][26], Stepan as Kessco® BS [12][16] and by Witco as Kemester® 5510 [20].

Isobutyl Stearate, CAS# 646–13–9

Isobutyl stearate (or octadecanoic acid, 2–methylpropyl ester) is similar to butyl stearate, but it has a lower pour point, 10°C, and a melting point of 15°C. These properties follow the normal pattern in which branched chain molecules have lower melting points than straight chain molecules. The viscosity of isobutyl stearate (10 cSt at 25°C) is slightly lower than of butyl stearate (12 cSt at 25°C), again showing the effect of chain branching [27]–[29]. The structure is:

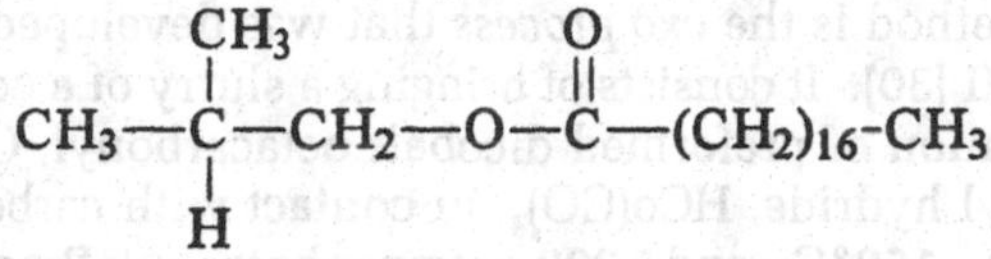

$$CH_3-\underset{\underset{H}{|}}{\overset{\overset{CH_3}{|}}{C}}-CH_2-O-\overset{\overset{O}{||}}{C}-(CH_2)_{16}-CH_3$$

Isobutyl alcohol can be prepared by the hydroboration–oxidation of isobutylene in the following reaction corresponding to anti–Markovnikov addition of water to the carbon–carbon double bond [4].

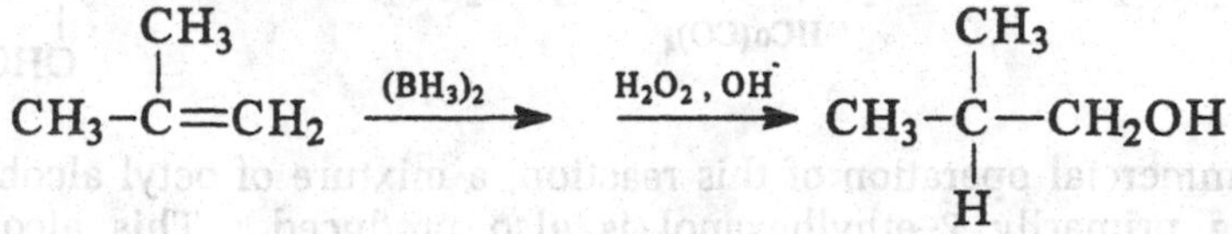

$$CH_3-\overset{\overset{CH_3}{|}}{C}=CH_2 \xrightarrow{(BH_3)_2} \xrightarrow{H_2O_2, OH^-} CH_3-\underset{\underset{H}{|}}{\overset{\overset{CH_3}{|}}{C}}-CH_2OH$$

Properties of isobutyl stearate are listed in Table 4–12, and it may be purchased from Henkel as Emerest® 2324 [23][29], Inolex as Lexolube® BS–Tech [26][28], Stepan as Kessco® IBS [12][16], or Witco as Kemester® 5415 [20].

Table 4–12

Typical Properties of Isobutyl Stearate

Property	Value
Saponification Value	165–175
Acid Value	3 max
Hydroxyl Value	4 max
Iodine Number	3 max
Viscosity, cSt at 25°C	10
Color, Gardner	1 max
10 g. Evaporation Loss (4 Hrs. @ 250°C), Wt. %	99

2–Ethylhexyl Stearate, CAS# 22047–49–0

The ester, 2–ethylhexyl stearate, is produced in a normal esterification reaction between 2–ethylhexanol and one of the stearic acid grades given by Table 4–10, with the standard grade being triple pressed. 2–Ethylhexanol is manufactured by a process that is very valuable in commercial organic synthesis. This method is the *oxo process* that was developed in Germany during World War II [30]. It consists of bringing a slurry of a cobalt salt and an olefin, or a solution of preformed dicobalt octacarbonyl, $Co_2(CO)_8$, and cobalt tetracarbonyl hydride, $HCo(CO)_4$, in contact with carbon monoxide and hydrogen at 150°C and 200 atmospheres. Propylene gives *n*–butyraldehyde and *i*–butyraldehyde in a ratio of 3:2.

$$CH_3\overset{\overset{\displaystyle H}{|}}{C}=\!=\!CH_2 + CO + H_2 \xrightarrow[HCo(CO)_4]{Co_2(CO)_8} CH_3CH_2CH_2CHO \text{ and } CH_3\overset{\overset{\displaystyle H}{|}}{\underset{\underset{\displaystyle CHO}{|}}{C}}CH_3$$

In the commercial operation of this reaction, a mixture of octyl alcohols containing primarily 2–ethylhexanol is also produced. This alcohol probably arises by aldol addition of two moles of *n*–butyraldehyde, dehydration, and hydrogenation.

$$2\ CH_3CH_2CH_2CHO \longrightarrow CH_3CH_2CH_2-\overset{\overset{\displaystyle OH}{|}}{\underset{\underset{\displaystyle H}{|}}{C}}-\overset{\overset{\displaystyle H}{|}}{\underset{\underset{\displaystyle C_2H_5}{|}}{C}}-CHO$$

$$\Big\downarrow -H_2O$$

$$CH_3CH_2CH_2-\overset{}{\underset{\underset{\displaystyle H}{|}}{C}}=\overset{}{\underset{\underset{\displaystyle C_2H_5}{|}}{C}}-CHO \xrightarrow{H_2} CH_3(CH_2)_3-\overset{\overset{\displaystyle H}{|}}{\underset{\underset{\displaystyle C_2H_5}{|}}{C}}-CH_2OH$$

2–Ethylhexyl stearate is somewhat more heat stable than butyl stearate and has about the same viscosity (about 15 cSt at 25°C *versus* 12 cSt at 25°C). It is sold by Henkel as Stantex® EHS [23][31], Hoechst–Celanese as Afilan® EHS [32], and Inolex as Lexolube® T–110 [26][33]. A composite list of properties is presented by Table 4–13.

Table 4–13
Typical Properties of 2–Ethylhexyl Stearate

Property	Property
Acid Value	0–4
Hydroxyl Number	5 max
Viscosity @ 25°C, cSt	15
Gardner Color	3 max
Saponification Number	140–155
Iodine Number	5 max
10 g. Evaporation Loss (4 hrs. @ 250°C), Wt. %	98

Tridecyl Stearate, CAS# 31556–45–3

Tridecyl alcohol is a branched 13 carbon material made by the condensation of propylene. When propylene is passed over a granular, solid, phosphoric acid catalyst (a calcined mixture of phosphoric acid and diatomaceous earth) at approximately 200°C and 150 p.s.i., a mixture of

$$\text{x } CH_3\overset{\overset{\displaystyle H}{|}}{C}{=}CH_2 \xrightarrow{H_3PO_4} CH_3\overset{\overset{\displaystyle H}{|}}{\underset{\underset{\displaystyle CH_3}{|}}{C}}{-}\left[CH_2\overset{\overset{\displaystyle H}{|}}{\underset{\underset{\displaystyle CH_3}{|}}{C}}{-}\right]_{x\text{-}2}CH{=}CHCH_3$$

and

$$CH_3\overset{\overset{\displaystyle H}{|}}{\underset{\underset{\displaystyle CH_3}{|}}{C}}{-}\left[CH_2\overset{\overset{\displaystyle H}{|}}{\underset{\underset{\displaystyle CH_3}{|}}{C}}{-}\right]_{x\text{-}2}CH_2CH{=}CH_2$$

higher alkenes is produced [30]. When x = 4, a tetramer with a twelve–carbon alkene is produced in large volumes for use in the detergent industry. It is the alky– moiety in alkylbenzene sulfonates. The third carbon is added in a typical oxo process in which one carbon atom is added to the chain to form the aldehyde that is then reduced to the alcohol through hydrogenation. A thirteen–carbon chain alcohol is thus produced in large volumes for use in nonionic detergents. Tridecyl alcohol is esterified with one of the stearic acid grades, probably triple pressed, to yield tridecyl stearate. Tridecyl stearate is another in the series of low viscosity esters that are excellent lubricants. Since the starting materials are manufactured in large volumes, the price of tridecyl stearate is lower than many other esters. The structure of tridecyl stearate is:

$$CH_3-\underset{\underset{CH_3}{|}}{\overset{\overset{H}{|}}{C}}-CH_2-\underset{\underset{CH_3}{|}}{\overset{\overset{H}{|}}{C}}-CH_2-\underset{\underset{CH_3}{|}}{\overset{\overset{H}{|}}{C}}-CH_2-\underset{\underset{CH_3}{|}}{\overset{\overset{H}{|}}{C}}-CH_2-O-\overset{\overset{O}{||}}{C}-CH_2(CH_2)_{15}CH_3$$

Tridecyl stearate may be purchased from Henkel as Emerest® 2308 [23][34], from Hoechst–Celanese as Afilan® TDS [35], from Inolex as Lexolube® B–109 [26][36], and from Witco as Kemester® 5721 [20]. Typical properties of tridecyl stearate are listed in Table 4–14.

Table 4–14

Typical Properties of Tridecyl Stearate

Property	Value
Acid Value	1 max
Hydroxyl Value	4.0 max
Iodine Number	1 max
Saponification Number	118–128
Water, %	0.1 max
Viscosity @ 25°C, cSt	25
Color, APHA	70 max
10 g. Evaporation Loss (4 Hrs. @ 250°C), Wt. %	20

Isoceytyl Stearate, CAS# 25339–09–7

Although isocetyl stearate has long saturated carbon chains in both the alcohol and acid sections of the molecule, it is a low viscosity ester because of branching of the alcohol. The structure of isocetyl alcohol is not completely understood, but it has methyl branching on the cetyl alcohol chain. Isocetyl stearate has a melting point of about –10°C and is quite heat stable. Properties of isocetyl stearate are given in Table 4–15.

Table 4–15

Typical Properties of Isocetyl Stearate

Property	Value
Color, APHA	100 max
Water Content, %	0.5 max
Saponification Value	110–120
Hydroxyl Value	4 max
Acid Value	1 max
Iodine Value	1 max
Viscosity, cSt at 25°C	31–42

It is sold by Hoechst–Celanese as Afilan® ICS [37] and by Witco as Kemester® 5822 [20].

Ethylene Glycol Monostearate, CAS# 111–60–4

Although the most common use of ethylene glycol is as an antifreeze additive for water–cooled internal combustion engines, this polyhydric alcohol is an important chemical intermediate and solvent. One of the commercial methods for preparing ethylene glycol is the air oxidation of ethylene in the presence of a silver catalyst to form ethylene oxide and then hydrolysis of the oxide to the alcohol [17]. The reaction may be represented as:

$$\begin{array}{c} CH_2 \\ \| \\ CH_2 \end{array} + 1/2\ O_2 \xrightarrow[250°C]{Ag\ cat.} \begin{array}{c} CH_2 \\ | \\ CH_2 \end{array}\!\!>\!\!O \xrightarrow[H^+]{H_2O} \begin{array}{c} CH_2OH \\ | \\ CH_2OH \end{array}$$

Ethylene glycol monostearate (stearic acid, 2–hydroxyethyl ester) is a solid lubricant sold by Inolex as Lexemul® EGMS [38] and by Stepan as Kessco® EGMS [39] and by Witco as Kemester® 6000 SE [20], with a structure of:

$$
\begin{array}{c}
\quad\ \ \overset{\displaystyle H}{|}\ \ \overset{\displaystyle H}{|}\ \qquad \overset{\displaystyle O}{\|} \\
HO-C-C-O-C-CH_2(CH_2)_{15}CH_3 \\
\quad\ \ \underset{\displaystyle H}{|}\ \ \underset{\displaystyle H}{|}
\end{array}
$$

Some of the properties of ethylene glycol monostearate are listed in Table 4–16.

Table 4–16

Typical Properties of Ethylene Glycol Monostearate

Property	Value
Water Content, %	0.5 max
Saponification Value	179–195
Acid Value	5 max
Iodine Value	1 max
Melting Point, °C	57±5

Ethylene Glycol Distearate, CAS# 627–83–8

This ester is similar to monostearate described above, except two molecules of stearic acid are reacted with the ethylene glycol. The diester is sold by Inolex and called Lexemul® EGDS [40], Stepan with the name Kessco® EGDS [39], and Witco as Kemester® EGDS [27]. It has a structure of:

$$
\begin{array}{c}
\qquad\quad \overset{\displaystyle O}{\|} \qquad\qquad\qquad\qquad \overset{\displaystyle O}{\|} \\
CH_3(CH_2)_{15}CH_2-C-O-CH_2CH_2-O-C-CH_2(CH_2)_{15}CH_3
\end{array}
$$

Both the monostearate and distearate esters can be used as softeners in many textile applications, as well as lubricants. The typical properties of the distearate ester are listed in Table 4–17.

Table 4–17

Typical Properties of Ethylene Glycol Distearate

Property	Value
Water Content, %	1 max
Saponification Value	190–200
Acid Value	6 max
Iodine Value	1 max
Melting Point, °C	62

Propylene Glycol Monostearate, CAS# 1323–39–3

Propylene glycol, or 1,2–propanediol, is prepared by a process called hydroxylation, or the conversion of alkenes into 1,2–diols, or dihydroxy alcohols containing two –OH groups on adjacent carbons. These materials are also known as glycols. In the hydroxylation reaction, two hydroxyl groups are added to a double bond in the presence of oxidizing agents. Two of the most commonly used are cold alkaline potassium permanganate ($KMnO_4$) or peroxy acids, such as peroxyformic acid [4]. These reagents are stereoselective and stereospecific. When potassium permanganate is the oxidizing agent, the hydroxyl groups are added to form the *syn–diol*, while peroxy acid oxidation results in diols with the hydroxyl groups in the *anti–* position. The reaction, for example using $KMnO_4$, to prepare propylene glycol follows this route:

$$CH_3-\overset{\overset{\displaystyle H}{|}}{C}=CH_2 \quad \xrightarrow[H_2O]{KMnO_4} \quad CH_3-\overset{\overset{\displaystyle H}{|}}{\underset{\underset{\displaystyle OH}{|}}{C}}-\overset{}{\underset{\underset{\displaystyle OH}{|}}{C}H_2}$$

One mole of propylene glycol reacts with one mole of stearic acid resulting in the product, propylene glycol monostearate. This product is a solid at 25°C with a pour point of 36°C. If the stearic acid reacts with the primary hydroxyl group, the structure is:

$$CH_3-\overset{\overset{\displaystyle H}{|}}{\underset{\underset{\displaystyle OH}{|}}{C}}-\overset{\overset{\displaystyle H}{|}}{\underset{\underset{\displaystyle H}{|}}{C}}-O-\overset{\overset{\displaystyle O}{\|}}{C}-CH_2(CH_2)_{15}CH_3$$

In actual practice, however, the stearic acid could react with either hydroxyl group and the resultant material is probably a mixture of the primary and secondary monostearate. It is sold by Henkel as Emerest® 2380 [23][41], by Inolex with the name Lexemul® P [42], and by Stepan using the designation Kessco® PGMS [39]. Typical properties are given in Table 4–18.

Table 4–18

Typical Properties of Propylene Glycol Monostearate

Property	Value
Color, Gardner	2 max
Saponification Value	171–183
Acid Value	3 max
Iodine Value	0.5 max
Melting Range, °C	33.5–38.5

Glycerol Monostearate, CAS# 123-94-4

Although the glycerol esters can be used as softeners or emulsifiers and also lubricants, they are included in this section discussion of stearate esters. Glycerol is obtained by saponification of fats and reacted with stearic acid to obtain this solid lubricant. The structure of glycerol monostearate is:

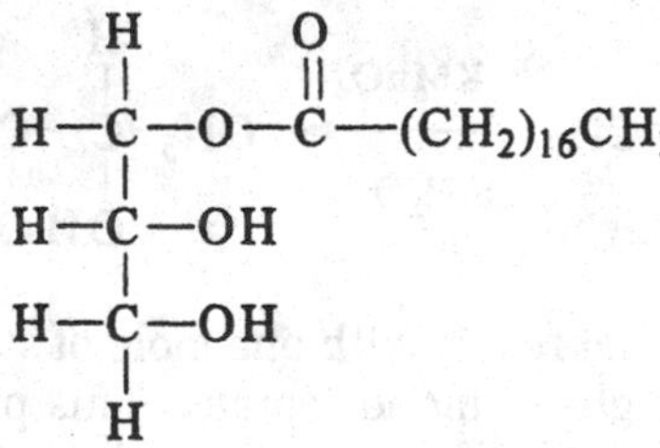

$$
\begin{array}{c}
\quad\quad H \quad\quad O \\
\quad\quad | \quad\quad\; || \\
H-C-O-C-(CH_2)_{16}CH_3 \\
\quad\quad | \\
H-C-OH \\
\quad\quad | \\
H-C-OH \\
\quad\quad | \\
\quad\quad H
\end{array}
$$

Glycerol monostearate is sold by Abitec as Capmul® GMS [43], by Henkel as Emerest® 2401 [23][44], by Inolex as Lexemul® 515 [26], by Stepan as Kessco® GMS [45], and by Witco as Kemester® 6000 SE [20]. Typical properties are shown in Table 4–19.

Table 4–19

Typical Properties of Glycerol Monostearate

Property	Value
Color, Gardner	3 max
Saponification Value	162–176
Acid Value	3 max
Iodine Value	1 max
Melting Range, °C	57–60
Water, %	1 max

Glycerol Monoisostearate, CAS# 123–94–4

While glycerol monostearate is a solid at room temperature, the isostearate version is a liquid with a pour point of 5°C and a viscosity of 260 cSt at 100°F. Isostearic acid is a C_{18} fatty acid with a methyl group randomly distributed on the alkyl chain. This chain branching reduces friction and imparts unique lubricating properties during fiber processing. The ester is sold by Henkel as Emerest®2410 [23][46]. Some properties are listed in Table 4–20.

Table 4–20

Typical Properties of Glycerol Monoisostearate

Property	Value
Moisture, %	0.5 max
Saponification Value	162–172
Acid Value	5 max
Iodine Value	3 max

4.1.2.4 Esters of Oleic Acid

The oleate esters are excellent lubricants with low viscosities resulting from the unsaturation in the long carbon chain. Unfortunately, this unsaturation can lead to varnish formation at high temperatures. For this reason oleate esters are not the preferred choice for applications involving prolonged exposure to elevated temperatures. Although the structures drawn in this and other sections show unsaturated fatty acids in a straight line, actually the actual structure is quite different. Almost invariably, the double bonds have a cis or Z configuration [17]. The most probable structure of oleic acid is:

Methyl Oleate, CAS# 112–62–9, Propyl Oleate, CAS# 111–59–1, and Butyl Oleate, CAS# 142–77–8

Oleic acid is a major constituent in several fats and oils, including beef tallow, corn oil, cottonseed oil, lard, olive oil, palm oil, and peanut oil. It is easily prepared by saponification of one of these glyceride fatty materials. Normal esterification yields one of the above lubricants.

These short chain alcohol oleate esters have similar properties with low viscosities and can be formulated into finishes with very low friction. Methyl oleate is sold by Stepan as Stepan® C–68 [47] and Witco as Kemester®104 [20], propyl oleate by Henkel as Emerest®2302 [23][48], and butyl oleate by Witco as Kemester® 4000 [20][27]. As an example of these three, propyl oleate has a structure of:

The typical properties of propyl oleate are given in Table 4–21.

Table 4–21

Typical Properties of Propyl Oleate

Property	Value
Color, Gardner	6 max
Saponification Value	170–180
Acid Value	6.0 max
Iodine Value	77–83
Viscosity, cSt @ 100°F	5

Oleyl Oleate, CAS# 91697–49–3

Oleyl alcohol may be prepared by the reduction of oleic acid, either through hydrogen reduction over a nickel catalyst or by a reaction using lithium aluminum hydride. The ester of oleyl alcohol and oleic acid may be formed through normal esterification reactions. It is sold by Henkel as Rilanit OLO [49], by Inolex as Oleyl Oleate [26], and by Witco as Starfol® OO [20]. Since it has a viscosity of 25 cSt at 25°C, it can be used as a high smoke point replacement for mineral oil or butyl stearate. Some of the normal properties of oleyl oleate are listed in Table 4–22.

Table 4–22

Typical Properties of Oleyl Oleate

Property	Value
Hydroxyl Value	< 12
Saponification Value	100–110
Acid Value	< 1
Iodine Value	87–97
Viscosity, cSt @ 25°C	25

Glycerol Monooleate, CAS# 111–03–5 and Glycerol Trioleate, CAS# 122–32–7

Although the glycerol oleates are widely used in cosmetic formulations, they are effective fiber lubricants and softeners. The monooleate is used in mold release agents, a vehicle for agricultural insecticides, and as a rust preventive additive for compounded oils. It may be described chemically as:

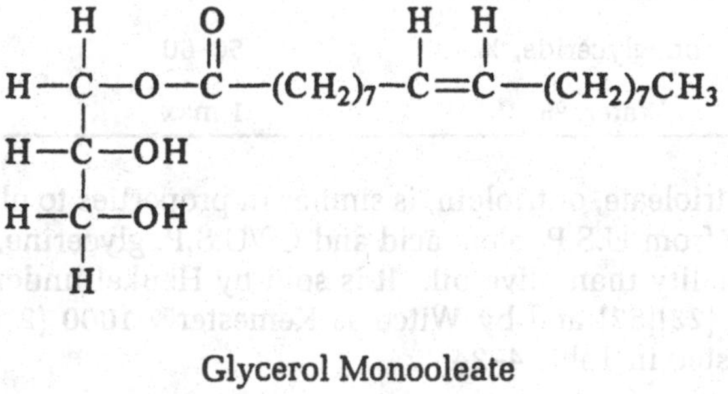

$$H-\overset{\overset{\displaystyle H}{|}}{\underset{\underset{\displaystyle H-\overset{\displaystyle H}{\underset{|}{C}}-OH}{|}}{C}}-O-\overset{\overset{\displaystyle O}{||}}{C}-(CH_2)_7-\overset{\overset{\displaystyle H}{|}}{C}=\overset{\overset{\displaystyle H}{|}}{C}-(CH_2)_7CH_3$$

Glycerol Monooleate

$$
\begin{array}{l}
\overset{\displaystyle H}{\underset{\displaystyle |}{}} \quad \overset{\displaystyle O}{\underset{\displaystyle ||}{}} \qquad\quad \overset{\displaystyle H}{\underset{\displaystyle |}{}} \;\; \overset{\displaystyle H}{\underset{\displaystyle |}{}} \\
H-C-O-C-(CH_2)_7-C=C-(CH_2)_7CH_3 \\
\overset{\displaystyle |}{\underset{\displaystyle }{}} \qquad\; \overset{\displaystyle O}{\underset{\displaystyle ||}{}} \qquad\quad \overset{\displaystyle H}{\underset{\displaystyle |}{}} \;\; \overset{\displaystyle H}{\underset{\displaystyle |}{}} \\
H-C-O-C-(CH_2)_7-C=C-(CH_2)_7CH_3 \\
\overset{\displaystyle |}{\underset{\displaystyle }{}} \qquad\quad\; \overset{\displaystyle O}{\underset{\displaystyle ||}{}} \qquad\quad \overset{\displaystyle H}{\underset{\displaystyle |}{}} \;\; \overset{\displaystyle H}{\underset{\displaystyle |}{}} \\
H-C-O-C-(CH_2)_7-C=C-(CH_2)_7CH_3 \\
\overset{\displaystyle |}{\underset{\displaystyle H}{}}
\end{array}
$$

Glycerol Trioleate

Glycerol monooleate is supplied by Abitec as Capmul® GMO [43], by Henkel as Emerest® 2421 [23][50], by Stepan as Kessco® GMO [51], and by Witco as Kemester®2000 [20][27]. Typical properties of this ester are given in Table 4–23.

Table 4–23

Typical Properties of Glycerol Monooleate

Property	Value
Color, Gardner	11
Saponification Value	165–175
Acid Value	6
Iodine Value	72–78
Viscosity, cSt @ 100°F	91
Monoglyceride, %	50–60
Water, %	1 max

Glycerol trioleate, or triolein, is similar in properties to olive oil, but since it is made from U.S.P. oleic acid and CP/U.S.P. glycerine, it is more consistent in quality than olive oil. It is sold by Henkel under the name Emerest® 2423 [23][52] and by Witco as Kemester® 1000 [27]. Typical properties are listed in Table 4–24.

Table 4–24

Typical Properties of Glycerol Trioleate

Property	Value
Color, Gardner	5
Hydroxyl Value	10 max
Acid Value	5
Iodine Value	85–91
Viscosity, cSt @ 100°F	43

4.1.2.5 Coconut Oil and Esters of Coconut Fatty Acids

Coconut Oil, CAS# 8001–31–8

This component is food grade coconut oil with the composition described in Table 4–3. One principal advantage for using coconut oil is the low degree of unsaturation in the fatty acids of the glyceride, as reflected in the low iodine number. This makes coconut oil an excellent choice for a lubricant if the yarn is to be in contact with hot surfaces. The approximate structure, assuming all lauric acid, of coconut oil is:

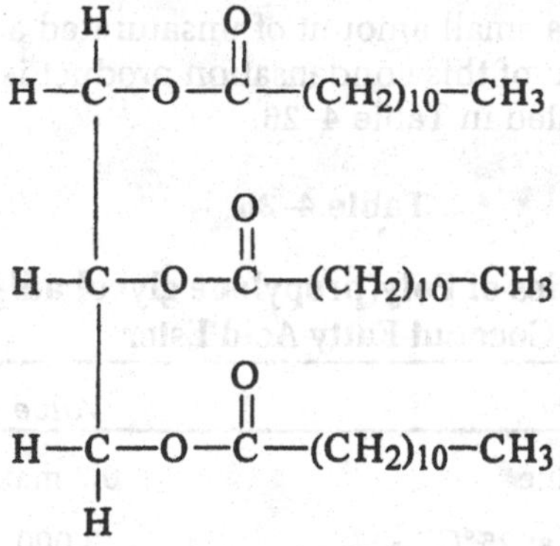

$$
\begin{array}{c}
H \\
| \\
H-C-O-\overset{\displaystyle O}{\overset{\|}{C}}-(CH_2)_{10}-CH_3 \\
| \\
H-C-O-\overset{\displaystyle O}{\overset{\|}{C}}-(CH_2)_{10}-CH_3 \\
| \\
H-C-O-\overset{\displaystyle O}{\overset{\|}{C}}-(CH_2)_{10}-CH_3 \\
| \\
H
\end{array}
$$

After coconuts ripen, the nuts are husked and the white meat, containing about 50% water and 30 to 40% oil, is removed and dried to produce a product called copra. Dry copra contains 65–70% oil and the coconut oil is obtained primarily by processing in mechanical screw presses. After pressing, some of the remaining oil can be extracted by using aliphatic hydrocarbon solvents. Food grade oil is prepared by removing some free fatty acids and flavoring chemicals. Coconut oil is sold for fiber uses by Abitec Corporation as Pureco® 76 [53]. Typical properties of food grade coconut oil are given in Table 4–25.

Table 4–25

Typical Properties of Coconut Oil

Property	Value
Saponification Number	248–264
Melt Point, °F	76–80
Acid Value	0.1 max
Viscosity at 100°F, cSt	28–33
Iodine Value	0–12
Color, Gardner	3 max

Poly(propylene glycol adipate)diol, Coconut Fatty Acid Ester,
CAS# 68583–87–9

This material is a complex ester derived form the co–condensation of propylene glycol and adipic acid esterified with the coconut fatty acids obtained from the saponification of coconut oils. These include C_8 through C_{18} saturated acids and a small amount of unsaturated acids. The number average molecular weight of this condensation product is about 2000. The typical properties are listed in Table 4–26.

Table 4–26

Typical Properties of Poly(propylene glycol adipate)diol, Coconut Fatty Acid Ester

Property	Value
Acid Value	0.7 max
Viscosity, cSt at 25°C	3,000
Specific Gravity, 25/25°C	1.1
Water Content, Wt. %	0.1 max
10 g. Evaporation Loss, 4 Hrs. @ 250°C, Wt. %	5

Since this is such a complex product, a structural formula to represent it is not appropriate. Poly(propylene glycol adipate)diol, coconut fatty acid ester is sold by Inolex as Lexolube® Z–100 [26][54].

Trimethylolethane Tricaprylate/Caprate (C$_8$–C$_{10}$), CAS# 69226–98–8

     Trimethylolethane, 2–hydroxymethyl–2–methyl–1,3–propanediol, is a white, odorless, crystalline trihydric alcohol [55]. It is produced by the reaction of formaldehyde with propionaldehyde. 2,2–Dimethylolpropion-aldehyde is formed in a base catalyzed aldol reaction and reacts with another molecule of formaldehyde in a crossed Cannizzaro reaction to complete the process.

$$CH_3CH_2CHO \ + \ 2\,HCHO \ \xrightarrow{\ OH^-\ } \ CH_3-\overset{\displaystyle CH_2OH}{\underset{\displaystyle CH_2OH}{C}}-CHO$$

$$CH_3-\overset{\displaystyle CH_2OH}{\underset{\displaystyle CH_2OH}{C}}-CHO \ \xrightarrow[\text{HCHO}]{\text{NaOH}} \ CH_3-\overset{\displaystyle CH_2OH}{\underset{\displaystyle CH_2OH}{C}}-CH_2OH$$

     Both caprylic (octanoic), C$_8$, and capric (decanoic), C$_{10}$, acids are the two lowest molecular weight fractions of coconut fatty acids (Table 4–3) and can easily be separated from the whole acid mixture. These react with trimethylolethane to form the final ester. The structure is:

$$
\begin{array}{l}
\quad\quad\overset{\displaystyle O}{\overset{\displaystyle \|}{CH_2-O-C}}-(CH_2)_{6\text{-}8}-CH_3 \\
\quad| \\
CH_3-C-CH_2-O-\overset{\displaystyle O}{\overset{\displaystyle \|}{C}}-(CH_2)_{6\text{-}8}-CH_3 \\
\quad| \\
\quad\quad CH_2-O-\overset{\displaystyle O}{\overset{\displaystyle \|}{C}}-(CH_2)_{6\text{-}8}-CH_3
\end{array}
$$

TME C_8–C_{10} ester is a liquid product with excellent lubricating properties. It is manufactured and sold by ICI Surfactants as Emkarate™ 2030 (or G–3892) [56] and by Inolex as Lexolube® 3N–309 [26][57]. The typical properties of trimethylolethane tricaprylate/caprate are presented in Table 4–27.

Table 4–27
Typical Properties of TME C_8–C_{10} Ester

Property	Value
Acid Value	1.0 max
Iodine Number	1.0 max
Saponification Number	310–330
Water, %	0.1 max
Viscosity @ 25°C, cSt	30–40
Color, APHA	100 max

Trimethylolpropane Tricaprylate/Caprate (C_8–C_{10}), *CAS# 11138–60–6*

An ester that is quite similar to TME C_8–C_{10}, trimethylolpropane tricaprylate/caprate (TMP C_8–C_{10}), the TMP version is also a good lubricant with approximately the same viscosity and volatility. Trimethylolpropane is synthesized by a process similar to TME, but the starting material is butyraldehyde instead of propionaldehyde. The structure is:

$$
\begin{array}{c}
\text{CH}_2\text{-O} - \overset{\overset{\displaystyle O}{\|}}{\text{C}} - (\text{CH}_2)_{6\text{-}8} - \text{CH}_3 \\
| \qquad\qquad\qquad O \\
\text{CH}_3-\text{CH}_2\text{-C}-\text{CH}_2-\text{O}-\overset{\|}{\text{C}}-(\text{CH}_2)_{6\text{-}8}\text{-CH}_3 \\
| \qquad\qquad O \\
\text{CH}_2-\text{O}-\overset{\|}{\text{C}}-(\text{CH}_2)_{6\text{-}8} - \text{CH}_3
\end{array}
$$

The TMP C_8–C_{10} ester is a product of Inolex or Stepan and sold by Inolex as Lexolube® 3N–310 [26][58] and by Stepan as Kessco® 887 [59]. Representative properties of TMP C_8–C_{10} ester are presented in Table 4–28.

Table 4–28

Typical Properties of TMP C_8–C_{10} Ester

Property	Value
Acid Value	0.5 max
Saponification Number	310–330
Water, %	0.1 max
Viscosity @ 25°C, cSt	35
Iodine Value	1 max
Color, APHA	100 max
10 g. Evaporation Loss, (4 Hrs. @ 250°C), Wt.%	5.6

Pentaerythritol Tetracaprylate/Caprate (C_8–C_{10}), CAS# 3008–50–2

Pentaerythritol is a very important tetrahydric alcohol prepared by the reaction of an aqueous solution of acetaldehyde with an excess of formalin in the presence of lime. The reaction consists of an aldol addition followed by a crossed Cannizzaro reaction [30].

$$3\ HCHO\ +\ CH_3CHO\ \xrightarrow{\ Ca\,(OH)_2\ }\ HOCH_2-\underset{\underset{CH_2OH}{|}}{\overset{\overset{CH_2OH}{|}}{C}}-CHO$$

$$HOCH_2-\underset{\underset{CH_2OH}{|}}{\overset{\overset{CH_2OH}{|}}{C}}-CHO\ \xrightarrow[\ Ca\,(OH)_2\ ]{\ HCHO\ }\ HOCH_2-\underset{\underset{CH_2OH}{|}}{\overset{\overset{CH_2OH}{|}}{C}}-CH_2OH$$

The pure tetrahydric alcohol is an odorless, white, crystalline compound. Pentaerythritol usually crystallizes in a tetragonal form and this highly symmetrical compound melts at 261–262°C. Pentaerythritol tetraesters of the short chain fatty acids form readily, but stearic hindrance prevents complete esterification of the longer chain acids.

The structural formula of pentaerythritol tetracaprylate/caprate may be represented as:

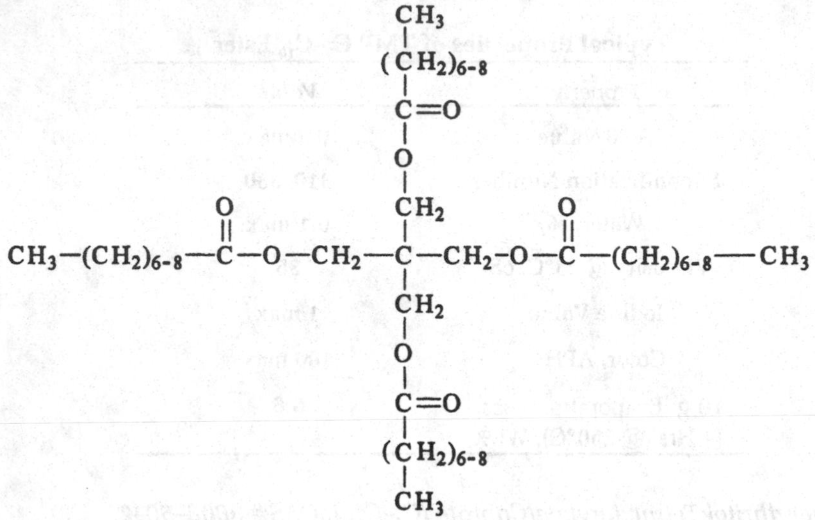

$$
\begin{array}{c}
CH_3 \\
| \\
(CH_2)_{6-8} \\
| \\
C=O \\
| \\
O \\
| \\
\end{array}
$$

$$
\begin{array}{ccc}
& O & CH_2 & O \\
& \parallel & | & \parallel \\
CH_3-(CH_2)_{6-8}-C-O-CH_2-C-CH_2-O-C-(CH_2)_{6-8}-CH_3 \\
& & | \\
& & CH_2 \\
& & | \\
& & O \\
& & | \\
& & C=O \\
& & | \\
& & (CH_2)_{6-8} \\
& & | \\
& & CH_3
\end{array}
$$

This ester is sold by Inolex as Lexolube ® 4N–415 [26] and by Stepan as Kessco® 874 [59]. The normal properties are listed in Table 4–29.

Table 4–29

Typical Properties of Pentaerythritol C_8–C_{10} Ester

Property	Value
Acid Value	1 max
Viscosity @ 25°C, cSt	56–62
Iodine Value	1 max
Color, APHA	100 max

4.1.3 Esters of Synthetic Monocarboxylic Acids

Perhaps the principal synthetic monocarboxylic acid used as a component in fiber finishes is pelargonic. The name pelargonic, or more properly nonanoic, acid is derived from *Pelargonium*, plants of a family that includes the geranium. These plants are some of the few natural substances in which this C_9 acid is found. It is manufactured commercially by two routes. The first is oxidation of oleic acid with potassium permanganate or ozone, with azelaic acid also formed in that reaction. The reaction is [30]:

$$CH_3(CH_2)_7\overset{\overset{H}{|}}{C}=\overset{\overset{H}{|}}{C}(CH_2)_7\overset{\overset{O}{||}}{C}OH \xrightarrow{\text{Ox.}} CH_3(CH_2)_7\overset{\overset{O}{||}}{C}OH + HO\overset{\overset{O}{||}}{C}(CH_2)_7\overset{\overset{O}{||}}{C}OH$$

Oleic Acid   Pelargonic Acid   Azelaic Acid

The second approach for preparing pelargonic acid is by adding a carbon atom to 1–octene by the oxo process [60]. Details of this process are presented in the section on 2–ethylhexyl stearate on page 84. The starting alkene can itself be made through the oligomerization of ethylene. Either of these processes yields a carboxylic acid of high purity and one that forms esters that are very valuable in fiber processing.

Propylene Glycol Dipelargonate, CAS# 41395–83–9

Propylene glycol dipelargonate is another ester that was designed especially for cosmetic use, but it can be useful as a low viscosity fiber lubricant. The glycol itself is manufactured by hydroxylation of propylene (see section on propylene glycol monostearate) and then esterified with two moles of pelargonic acid. The structure of this ester is:

$$CH_3(CH_2)_7-\overset{\overset{O}{||}}{C}-O-\overset{\overset{\overset{CH_3}{|}}{}}{\underset{\underset{H}{|}}{C}}-CH_2-O-\overset{\overset{O}{||}}{C}-(CH_2)_7CH_3$$

This ester is sold by Henkel with the trade name Emerest® 2388 [23][61]. Some properties are found in Table 4–30.

Table 4–30

Typical Properties of Propylene Glycol Dipelargonate

Property	Value
Acid Value	0.5 max
Saponification Number	305–320
Viscosity @ 25°C, cSt	11
Color, Gardner	1
Iodine Value	1 max

Triethylene Glycol Dipelargonate, CAS# 106–06–9

Ethylene glycol reacts with successive molecules of ethylene oxide to give diethylene glycol and triethylene glycol, some higher condensation products, and finally high molecular weight polyethylene glycol. When the trimer, triethylene glycol, reacts with two molecules of pelargonic acid, the ester is produced. It can be used as a component in defoamers, as a flotation agent, and as a base for fiber lubricants. This product is sold by Henkel as Emery® 6783 [23][62], by Inolex with the chemical name Triethylene glycol di (C$_9$) [26], and by Stepan as Plasticizer SC–B® [63]. Normal properties are listed in Table 4–31 and the structure is:

$$CH_3(CH_2)_7-\overset{\overset{\displaystyle O}{\|}}{C}-O-(CH_2)_2O(CH_2)_2O(CH_2)_2-O-\overset{\overset{\displaystyle O}{\|}}{C}-(CH_2)_7CH_3$$

Table 4–31

Typical Properties of Triethylene Glycol Dipelargonate

Property	Value
Acid Value	1 max
Saponification Number	257–263
Viscosity @ 25°C, cSt	11
Color, Gardner	1
Hydroxyl Value	5 max

Trimethylolethane Tripelargonate, CAS# 10535–50–9 and Trimethylolpropane Tripelargonate, CAS# 126–57–8

These two esters of pelargonic, or nonanoic, acid are very similar to each other and to the tromethylolethane (TME) and to the trimethylol propane (TMP) esters of C$_8$–C$_{10}$ fatty acids. They all have low viscosities resulting in low fiber to hard surface friction yet possess low volatility and excellent heat stability. The pelargonic acids utilized to prepare these esters are quite pure so that unsaturation is almost nonexistent and thus varnish formation on hot surfaces is very low. TME tripelargonate is produced by Inolex and sold as Lexolube® 3P–309 [26]. The TMP ester is marketed by

Henkel as Emery® 6701 [23][64], by Hoechst–Celanese as Afilan® TMPP [65], by Inolex as Lexolube® 3P–310 [26], and by Stepan as Kessco® 887 C–9 [66]. The structure of TMP tripelargonate, as an example of this group, is:

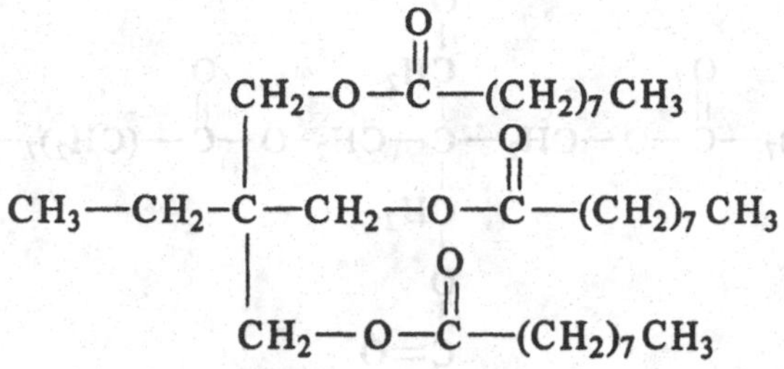

$$CH_2-O-\overset{\overset{\displaystyle O}{\|}}{C}-(CH_2)_7\,CH_3$$

$$CH_3-CH_2-\overset{|}{\underset{|}{C}}-CH_2-O-\overset{\overset{\displaystyle O}{\|}}{C}-(CH_2)_7\,CH_3$$

$$CH_2-O-\overset{\overset{\displaystyle O}{\|}}{C}-(CH_2)_7\,CH_3$$

Some properties of TMP tripelargonate are shown in Table 4–32.

Table 4–32

Typical Properties of Trimethylolpropane Tripelargonate

Property	Value
Acid Value	0.5 max
Saponification Number	295–310
Viscosity @ 25°C, cSt	35
Color, APHA	100 max
Hydroxyl Value	3 max
10 g. Evaporation Loss (24 hrs. @ 200°C), Wt. %	25

Pentaerythritol Tetrapelargonate, CAS# 14450–05–6

One of the leading low friction and low volatility esters in fiber lubrication is pentaerythritol tetrapelargonate. Other than in fiber lubrication, this type of compound is also effective as metal lubricants and in synthetic motor oils. The structural formula of pentaerythritol tetrapelargonate is shown below and the properties are listed in Table 4–33.

$$
\begin{array}{c}
CH_3 \\
| \\
(CH_2)_7 \\
| \\
C=O \\
| \\
O \\
| \\
CH_2 \\
\end{array}
$$

$$CH_3-(CH_2)_7-\overset{\overset{\textstyle O}{\|}}{C}-O-CH_2-\overset{\overset{\textstyle |}{CH_2}}{\underset{\overset{\textstyle |}{CH_2}}{C}}-CH_2-O-\overset{\overset{\textstyle O}{\|}}{C}-(CH_2)_7-CH_3$$

$$
\begin{array}{c}
O \\
| \\
C=O \\
| \\
(CH_2)_7 \\
| \\
CH_3 \\
\end{array}
$$

Table 4–33

Typical Properties of Pentaerythritol Tetrapelargonate

Property	Value
Acid Value	0.5 max
Saponification Number	310–330
Viscosity @ 25°C, cSt	62
Color, APHA	50 max
Hydroxyl Value	4 max
Iodine Number	1 max
10 g. Evaporation Loss (24 hrs. @ 200°C), Wt. %	14

The ester is made and sold by Henkel as Emerest® 2486 [23][67], Hoechst–Celanese as Afilan® PP [68], by Inolex as Lexolube® 4P–415 [26], and by Stepan as Kessco® 874 C–9 [69].

Octyl Isononanoate, CAS# 71566–49–9

Octyl isononanoate is an excellent example of the effects of chain branching on viscosity. This is one of the lowest viscosity esters of any that are available, especially since it contains seventeen carbons in the chain. The acid, isononanoic, is prepared by an interesting series of reactions.

When isobutylene is passed into cold 60% sulfuric acid and the solution heated to 100°C, a mixture of dimers and trimers (about 4:1), along with smaller amounts of higher polymers, is formed [30]. The mixture of dimers is called diisobutylene and it consists of four parts of 2,4,4–trimethyl–1–pentene and one part of 2,4,4–trimethyl–2–pentene.

$$2 \ (CH_3)_2C{=}CH_2 \ \underset{}{\overset{H_2SO_4}{\rightleftharpoons}} \ \underset{20\,\%}{(CH_3)_3C\overset{\overset{\textstyle H}{|}}{C}{=}C(CH_3)_2} \ \text{and} \ \underset{80\,\%}{(CH_3)_3CCH_2\overset{\overset{\textstyle CH_3}{|}}{C}{=}CH_2}$$

The mixture of diisobutylenes can then be used in the oxo process to synthesize 3,5,5–trimethylhexanal, the so–called nonyl aldehyde. If the double bond is not in the terminal position, isomerization to terminal takes place before reaction [30].

$$(CH_3)_3CCH_2\overset{\overset{\textstyle CH_3}{|}}{C}{=}CH_2$$

$$\text{and} \quad + \ CO + H_2O \ \longrightarrow \ (CH_3)_3CCH_2\overset{\overset{\textstyle CH_3}{|}}{\underset{\underset{\textstyle H}{|}}{C}}CH_2CHO$$

$$(CH_3)_3C\overset{\overset{\textstyle H}{|}}{C}{=}C(CH_3)_2$$

The aldehyde is then easily oxidized to isononanoic acid. The acid is esterified with 2–ethylhexanol to produce the final ester. The structure is:

$$CH_3(CH_2)_3\overset{\overset{\textstyle C_2H_5}{|}}{\underset{\underset{\textstyle H}{|}}{C}}CH_2{-}O{-}\overset{\overset{\textstyle O}{||}}{C}{-}CH_2\overset{\overset{\textstyle CH_3}{|}}{\underset{\underset{\textstyle H}{|}}{C}}CH_2C(CH_3)_3$$

This ester is a product of Stepan under the name Kessco® Octyl Nanoate [16] with some of the properties listed in Table 4–34.

Table 4–34

Typical Properties of Octyl Isononanoate

Property	Value
Acid Value	1 max
Saponification Number	202–210
Viscosity @ 25°C, cSt	5.1
Iodine Number	1 max
1 g. Volatility (1 hr. @ 160°C), Wt. %	100

4.1.4 Esters of Synthetic Dicarboxylic Acids

The primary synthetic dicarboxylic acid used in fiber lubrication is adipic acid. This acid is readily available at low cost and very high purity since it is synthesized as one of the monomers in the polymerization of nylon 6,6. In addition to being a component in nylon polymer and in fiber lubricants, adipic acid has many other uses, including an ingredient in food to add tartness. There are two commercial routes to prepare adipic acid: by the oxidation of cyclohexane [70] or by the carbonylation of tetrahydrofuran [30]. These reactions may be presented as:

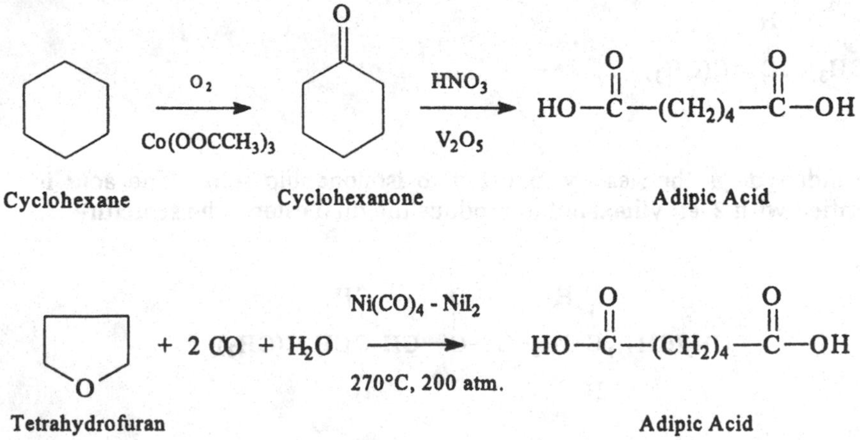

Cyclohexane   Cyclohexanone   Adipic Acid

Tetrahydrofuran   Adipic Acid

Although many esters could be synthesized from adipic acid, only a few are used in fiber finish formulations. A very low viscosity (4 cSt @ 25°C), highly volatile (100% weight loss from a 1 gram sample @ 200°C for 16 hours) ester, diisopropyl adipate, is produced by Inolex as Lexol® DIA [26] as well as an adipate ester of oleyl alcohol (43 cSt @ 25°C, 30% weight loss from a 1 gram sample @ 200°C for 16 hours) as Lexolube ® 2X–114 [26]. The two most widely used esters, however, are the dioctyl/decyl adipate and ditridecyl adipate.

Dioctyl/Decyl Adipate, CAS# 132752–18–2

Dioctyl/decyl adipate is sold by Hoechst–Celanese as Afilan® ODA [71] and by Inolex under the designation di ($C_8 C_{10}$) adipate [26]. Table 4–35 gives the typical properties of dioctyl/decyl adipate and the structure is:

$$CH_3(CH_2)_{6-8}-O-\overset{\overset{\displaystyle O}{\|}}{C}-CH_2)_4-\overset{\overset{\displaystyle O}{\|}}{C}-O-(CH_2)_{6-8}CH_3$$

Table 4–35

Typical Properties of Dioctyl/Decyl Adipate

Property	Value
Acid Value	0.25 max
Saponification Number	280–300
Viscosity @ 25°C, cSt	16
Color, APHA	50 max
Hydroxyl Value	4 max
Iodine Number	1 max
Moisture, %	0.05 max

Ditridecyl Adipate, CAS# 16958–92–2

Tridecyl alcohol is a readily available compound (see section on tridecyl stearate, page 85) that easily reacts with adipic acid to form the ester. The methyl branching on the alcohol carbon chain results in a lower than expected viscosity for such a long aliphatic carbon arrangement. This material possesses excellent lubrication and heat stability requirements. The typical properties of ditridecyl adipate are listed in Table 4–36.

Table 4–36

Typical Properties of Ditridecyl Adipate

Property	Value
Acid Value	0.05 max
Saponification Number	217–227
Viscosity @ 25°C, cSt	27–32
Color, APHA	50 max
Hydroxyl Value	1 max
Moisture, %	0.05 max
10 g. Evaporation Loss (4 Hrs. @ 250°C), Wt.%	13

It is manufactured by Hoechst–Celanese as Afilan® TDA [72] and by Inolex under the trade name Lexolube® 2X–109 [26][73]. The structural formula of this component is:

$$CH_3(\overset{\displaystyle H}{\underset{\displaystyle CH_3}{C}}-CH_2)_4-O-\overset{\displaystyle O}{C}-(CH_2)_4-\overset{\displaystyle O}{C}-O-(CH_2-\overset{\displaystyle H}{\underset{\displaystyle CH_3}{C}})_4CH_3$$

4.1.5 Polymers

Several polymeric materials have been used as finish components, with varying degrees of success. Generally, polymers cause very high friction, and commonly act as an adhesive to bind filaments together. The polymers that have been used are listed below.

Microcrystalline Wax, CAS# 63231–60–7

Petroleum–based waxes have been used as finish components for many years. Paraffin wax and microcrystalline wax were among the very first components tested for spin finishes. Microcrystalline wax is an oxidized/chemically reacted hydrocarbon wax that has its principal end use as an ingredient in floor polish. These products are manufactured by air oxidation, catalytic oxidation, or other chemical modifications to form esters, fatty acids, and salts that impart saponifiability and polarity to the basic hydrocarbon. It was initially added to finishes to improve molecular orientation prior to drawing [74] and for making rope [75]. The chemical structure of microcrystalline wax has not been completely defined; however, the typical properties are shown in Table 4–37.

Table 4–37
Typical Properties of Microcrystalline Wax

Property	Value
Acid Value	14
Saponification Number	25–35
Melting Point, °F	190–215
Viscosity, cSt @ 99°C	14–18
Color, Gardner	2 max

It is sold by Witco as Multiwax® ML–445 [3] and Petrolite as C–7500 Polymer® [76].

Polyvinyl Alcohol, CAS# 25213–24–5

Polyvinyl alcohols (PVA) are widely used as a textile warp sizing agent, in which it imparts high abrasion resistance. The PVA films on these textiles have good elongation, tensile strength, and flexibility. Partially hydrolyzed grades, which possess increased polyester adhesion, can lead to superior weaving characteristics [70]. In addition to use in the paper industry in coating and sizing formulations, polyvinyl alcohol is valuable in adhesive formulations where it can act as the primary binding agent or as a compounding material. PVA adhesives bond especially well to cellulosic surfaces.

Polyvinyl alcohol is a water soluble material prepared by the direct hydrolysis of polyvinyl acetate or through an ester interchange reaction [77]. The monomeric vinyl acetate can be made through several procedures, but one commercial route is by the catalyzed addition of acetic acid to acetylene [30]. The sequence of reactions to make polyvinyl alcohol are:

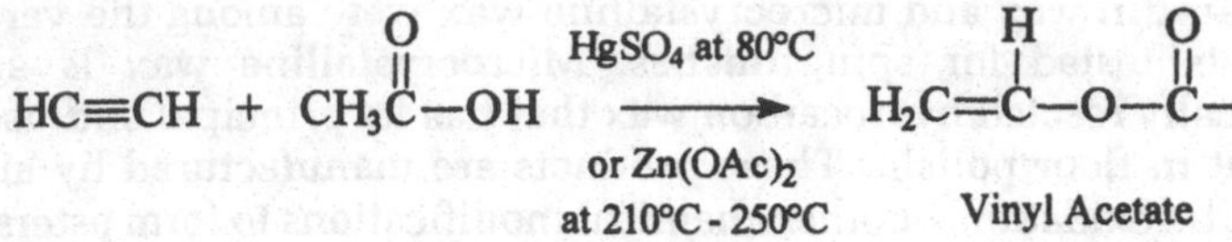

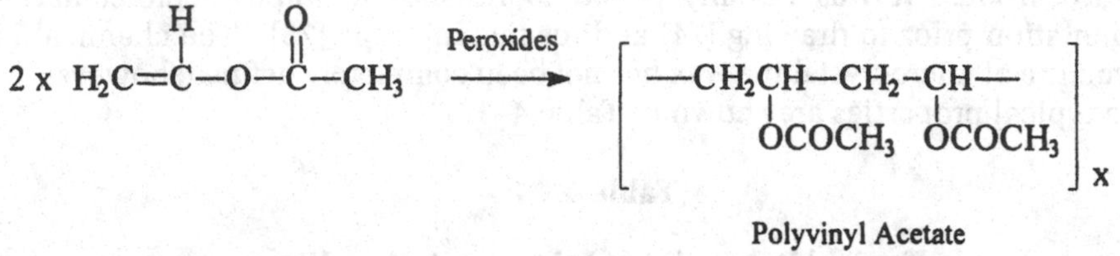

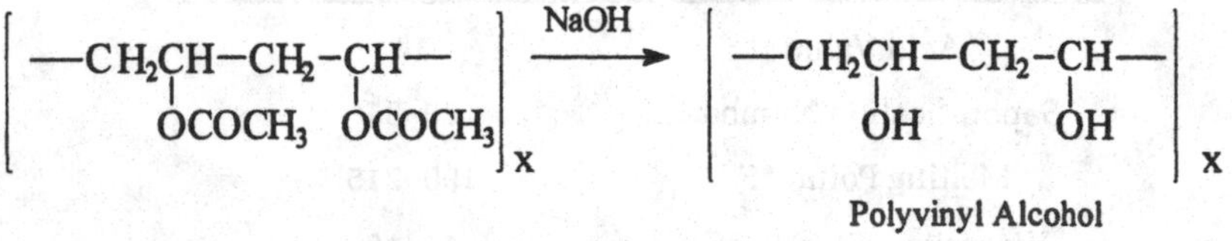

$$HC{\equiv}CH + CH_3\overset{\overset{\displaystyle O}{\|}}{C}{-}OH \xrightarrow[\substack{\text{or } Zn(OAc)_2 \\ \text{at } 210°C - 250°C}]{\text{HgSO}_4 \text{ at } 80°C} H_2C{=}\overset{\overset{\displaystyle H}{|}}{C}{-}O{-}\overset{\overset{\displaystyle O}{\|}}{C}{-}CH_3$$

Vinyl Acetate

$$2 \text{ x } H_2C{=}\overset{\overset{\displaystyle H}{|}}{C}{-}O{-}\overset{\overset{\displaystyle O}{\|}}{C}{-}CH_3 \xrightarrow{\text{Peroxides}} \left[\begin{array}{c} {-}CH_2CH{-}CH_2{-}CH{-} \\ \quad | \qquad\qquad | \\ \quad OCOCH_3 \quad OCOCH_3 \end{array} \right]_x$$

Polyvinyl Acetate

$$\left[\begin{array}{c} {-}CH_2CH{-}CH_2{-}CH{-} \\ \quad | \qquad\qquad | \\ \quad OCOCH_3 \quad OCOCH_3 \end{array} \right]_x \xrightarrow{\text{NaOH}} \left[\begin{array}{c} {-}CH_2CH{-}CH_2{-}CH{-} \\ \quad | \qquad\qquad | \\ \quad OH \qquad\quad OH \end{array} \right]_x$$

Polyvinyl Alcohol

The typical properties are given in Table 4–38.

Table 4–38

Typical Properties of Polyvinyl Alcohol

Property	Value
Hydrolysis, %	73–78
Residual Polyvinyl Acetate, %	37–42
pH (4% in Water)	6
Viscosity (4% in Water) cSt, 20°C	2

Polyvinyl alcohol for fiber finishes is sold by DuPont as Elvanol 51–05 [78] and by Air Products as Airvol® 203 [79]. The main uses for this polymer was on nylon tire yarn to increase the tensile strength, abrasion resistance, and durability [1] and as lubricants on glass fiber [80][81].

Emulsifiable Polyethylene, CAS# 68441–17–8

Normal polyethylene is not acceptable as a finish component because it cannot be emulsified. The producer of emulsifiable polyethylene has made a low molecular weight polymer that has then been air oxidized to add functional end groups that can be used in oil–in–water emulsions. The typical properties of emulsifiable polyethylene are listed in Table 4–39.

Table 4–39

Typical Properties of Emulsifiable Polyethylene

Property	Value
Color, Gardner	2 max
Acid Number	14.0–18.5
Density, 25/25°C	0.942
Weight Average Molecular Weight	6,100
Number Average Molecular Weight	3,200
Viscosity, cSt @ 125°C	690–1275

The main use for this polymer is as a component in water emulsion floor polishes where it imparts excellent slip resistance and outstanding toughness. Its use in spin finishes was limited mainly to tire yarn to improve durability by reducing fiber to fiber friction [82], but it was also a component to reduce the tackiness of elastic fibers [83]. The polymer selected for fiber finish use has a low molecular weight and is sold by Eastman as Epolene® E–10 [84].

4.1.6 Silicones

Silicone polymers are not known as widely as many other lubricants because they are used sparingly on yarn that must be processed through a dyeing step. Most silicone oils block dye absorption and can lead to non–uniformly dyed products. These oils are quite valuable in several areas, including sewing thread lubrication, knitting lubricants, polypropylene texturizing, rope making, and on spandex fibers. In one primary application, industrial sewing thread, silicones provide the lubrication necessary to prevent the needles from excessive heat being generated during the sewing operation.

A wide range of silicone polymers are available, including

poly(dimethyl siloxane) of several viscosities, poly(dimethyl–methyl:phenyl siloxane), hydrocarbon modified silicones, and silicone polyglycol copolymers [85].

Poly(dimethyl siloxane), CAS# 63148–62–9

This is the simplest of the silicone polymers, with a structure of:

$$
(CH_3)_3SiO \longrightarrow \left[\begin{array}{c} CH_3 \\ | \\ SiO \\ | \\ CH_3 \end{array} \right]_x \longrightarrow Si(CH_3)_3
$$

It is sold by Dow Corning as Dow Corning® 200 [85] in several viscosities ranging from 20 cSt at 25°C to 12,500 cSt at 25°C. All of the products are colorless liquids.

Poly(methyl/phenyl siloxane)

Improved heat stability but higher fiber to steel friction may be obtained by replacing one of the methyl groups with a phenyl radical. This material has a structure of:

$$
(CH_3)_3SiO \longrightarrow \left[\begin{array}{c} CH_3 \\ | \\ SiO \\ | \\ CH_3 \end{array} \right]_x \longrightarrow \left[\begin{array}{c} CH_3 \\ | \\ SiO \\ | \\ \bigcirc \end{array} \right]_y \longrightarrow Si(CH_3)_3
$$

These silicone polymers are light amber in color and vary in molecular weight and thus viscosity. Dow Corning® 510 , CAS# 33204–76–1, has a viscosity of 50 cSt at 25°C, Dow Corning® 550, CAS# 33204–76–1, a viscosity of 125 cSt at 25°C, and Dow Corning® 710, CAS# 63148–58–3, a viscosity of 500 cSt at 25°C [85].

Oleophillic Silicone/Organic Copolymers, CAS# 68440-90-4

Unlike the previously discussed silicone products, the polymers made with an eighteen-carbon chain alkyl group replacing one of the methyl groups are compatible with esters and mineral oils as well as being somewhat compatible with isopropanol. One such product, Dow Corning® T-4-0096 [85], has a viscosity of 50 cSt at 25°C that results in low fiber to metal friction. The structure of this polymer is:

$$
(CH_3)_3SiO \underbrace{\begin{bmatrix} CH_3 \\ | \\ SiO \\ | \\ CH_3 \end{bmatrix}}_{x} \underbrace{\begin{bmatrix} CH_3 \\ | \\ SiO \\ | \\ C_{18}H_{37} \end{bmatrix}}_{y} Si(CH_3)_3
$$

Poly(glycol/silicone copolymers)

Two special fiber finish components are sold by Dow Corning, they are Dow Corning® FF-400, CAS# 70914-12-4, and FF-414, CAS# 67762-97-4 [85]. These polymers have a glycol group replacing one of the methyl units on the silicone chain resulting in a water dispersible or soluble chemical component. The glycol groups are either ethylene or propylene glycol, or a combination of the two. FF-400 is light amber in color, has a viscosity of 300 cSt at 25°C (and relatively low friction) and is dispersible in water only with high shear. This product probably contains more propylene oxide than the other product. FF-414 is completely compatible in water and has a viscosity of 465 cSt at 25°C that results in a high friction component similar to polyethylene glycol. The general chemical structure is:

$$
(CH_3)_3SiO \underbrace{\begin{bmatrix} CH_3 \\ | \\ SiO \\ | \\ CH_3 \end{bmatrix}}_{x} \underbrace{\begin{bmatrix} CH_3 \\ | \\ SiO \\ | \\ R \end{bmatrix}}_{y} Si(CH_3)_3
$$

where R is either ethylene and/or propylene glycol.

4.2 WATER–SOLUBLE LUBRICANTS

The types of water–soluble lubricants that act as neither good emulsifiers nor cohesive agents are very limited. The only ones that have been used as finish components are the random and the block ethylene oxide–propylene oxide copolymers. Although these two types of polymers may contain the same basic chemical substances, they are prepared in different ways. When a random copolymer is synthesized, ethylene oxide and propylene oxide are added to the reaction vessel and allowed to polymerize. Block copolymers are made by first preparing low molecular weight fractions of each monomer and then reacting these low MW units. The copolymer types will be grouped together as random and block, since the principal differences within each group are in the molecular weights.

During the 1960s, polyester double–knit fabric was produced in every knitting mill in the country and this fabric was used in all types of apparel. The volume of polyester filament yarn consumed by these mills was so large that fiber producers were challenged to devise a new type of yarn that could be textured on high–speed equipment. This new polyester filament was partially oriented yarn (POY) that was drawn on texturing machines in addition to the bulk development. In POY friction twist texturing the yarn was taken off the spun package, through a tension control device and then across a long heater plate. On some equipment this heater was 5 to 6 feet long and the temperatures above 200°C. Many finishes contained a large amount of unsaturated material that developed varnish deposits on these heaters. The only way to remove this deposit was to hydrolyze it with a strong caustic solution, such as oven cleaner. The use of these types of products created serious safety problems, and it became the responsibility of the finish formulators to improve these compositions.

Ethylene oxide–propylene oxide lubricants exhibit a wide range of fiber to hard surface friction, which varies with running speed and molecular weight. The low friction at high yarn velocities, combined with high smoke point and resistance to absorption, make this type of lubricant especially suitable for high–speed, high–temperature fiber processes, such as false–twist texturing. Frequently, a combination of lubricants is selected to give the desired level of friction control.

It had been known for several years that chemicals containing ethylene oxide decompose to leave very little residues except for the initiating molecule. The degradation of these molecules goes through a free radical mechanism, called autoxidation [70][86], as outlined below.

$$-O-CH_2CH_2-O-CH_2CH_2-O-$$

$$\downarrow R\cdot\ or\ O_2$$

$$-O-CH_2CH_2-O-CHCH_2-O-$$

$$\downarrow O_2\ and/or\ RH$$

$$-O-CH_2CH_2-O-\underset{|}{\overset{|}{C}}HCH_2-O-$$
$$OOH$$

The hydroperoxide moiety then decomposes and the adjacent C–C bond is broken as:

$$-O-CH_2CH_2-O-\underset{|}{C}HCH_2-O-$$
$$OOH$$

$$\downarrow Heat$$

$$-O-CH_2CH_2-O-\underset{|}{C}HCH_2-O-\ +\ \cdot OH$$
$$\underset{\bullet}{O}$$

$$\downarrow$$

$$-O-CH_2CH_2-O-\underset{|}{C}=O\ +\ \cdot CH_2-O-\ \longrightarrow\ CH_2O$$
$$H$$

The unzipping of the poly(ethylene oxide) chain continues with the evolution of formaldehyde and the formation of other free radicals. If the chain is poly(propylene oxide) the material evolved is acetaldehyde. Thus, minimal residues remain to cause varnishes.

Random EO/PO Copolymers:
50–HB Series, CAS# 9038–95–3 and 75–H Series, CAS# 9003–11–6

Although random EO–PO copolymers can be manufactured by many companies that perform reactions of ethylene oxide or propylene oxide with some starting moiety, the major supplier is Union Carbide. These are sold as the Ucon® family of lubricants. Typical of this family of water–soluble materials are the 50–HB (50% EO and 50% PO) and 75–H (75% EO and 25% PO) series. These are self scouring, polyoxyalkylene ether fluids that are effective on a wide range of synthetic and natural fibers. Their complete water solubility allows thorough removal in conventional scouring processes, thus reducing dye spotting and barré that are common problems with emulsions and knitting lubricants.

The major differences in the compounds in each series are in their molecular weights, thus viscosity and frictional variability. Typical properties of the Ucon® 50–HB lubricants are listed in Table 4–40 [87][88].

Table 4–40

Typical Properties of Ucon® 50–HB Lubricants

Ucon Lubricant	Viscosity @ 40°C, cSt	Average Molecular Weight	Pour Point, °F (ASTM)	Flash Point, °F, Cleveland Open Cup
50–HB–55	8.33	270	−85	200
50–HB–100	19.1	520	−60	385
50–HB–170	33.8	750	−45	400
50–HB–260	52.0	970	−40	460
50–HB–400	80.0	1,230	−36	480
50–HB–660	132	1,590	−30	445
50–HB–2000	398	2,660	−25	440
50–HB–3520	700	3,380	−20	445
50–HB–5100	1,015	3,930	−20	450

Although 50–HB–5100 has been used in several finish compositions, its use should be curtailed. Union Carbide has reported a serious toxicity hazard when rats inhaled a mist of this product. Additional work was performed under the direction of The American Fiber Manufacturers Association Toxicology Subcommittee. A synopsis of the committee report follows [89].

This study evaluated five different test materials (Ucon 50HB2000, Ucon 75H1400, Pluronic L–64, Pluronic L–31, and

Pluronic 17 R–1) in comparison to an air control and a positive control material (Ucon 50HB5100) Whole body liquid droplet aerosol exposures were conducted six hours per day, five days per week for two consecutive weeks. The exposure level for the positive control was 55 mg/m^3, while the various test materials were evaluated at 100 mg/m^3 (97–103 mg/m^3 was the range of mean exposure concentrations for the five materials). Aerosol particle size (equivalent aerodynamic diameter) was approximately 2.3 microns for the Ucon test materials, and 1.8 microns for the Pluronic test materials. The geometric standard deviation was approximately 1.9 for all test materials. Each group consisted of ten male albino rats.

After three exposures, 9 of 10 rats exposed to Ucon 50HB5100, and 6 of 10 rats exposed to Ucon 50HB2000 had died. At necropsy, congestion, consolidation and red discoloration of the lungs was noted. A moderate to severe alveolitis characterized by intra–alveolar edema, hemorrhage, and fibrin deposition was observed microscopically. Surviving animals from these two groups were sacrificed after five days of exposure. At necropsy these animals exhibited elevated lung weights, and similar macroscopic and microscopic lesions.

Animals exposed to the other test materials survived with essentially no signs of toxicity through the ten exposure days. Body weights throughout the exposure period were similar between the control group and the experimental groups. Hematological evaluations did not reveal any test–material–related abnormalities. Organ weights were similar between the control group and experimental groups. There were no exposure–related macroscopic abnormalities noted at necropsy. A minimum degree of alveolitis was noted microscopically in most animals from the exposure group.

No exposure–related differences were noted in pharmacotoxic signs, body weights, organ weights, macroscopic observations at necropsy, or microscopic evaluations for those animals permitted a two–week post–exposure recovery period.

In summary, based on the results of this study the test materials Ucon 75H1400, Pluronic L–64, Pluronic L–31, and Pluronic 17 R–1 at 100 mg/m^3 were similar to each other in that only a slight alveolitis was noted after two weeks of exposure, and the response subsided by two weeks post–exposure. In contrast, Ucon 50HB2000 at 100 mg/m^3 caused death in approximately half the animals after five exposures, and Ucon 50HB5100 at 55 mg/m^3 caused death in 9 of 10 animals after five exposures.

The toxicity of Ucon® 50HB5100 is apparently molecular weight related and is limited to the random copolymers. The effect of Ucon® 50HB660 at 500 mg/m³ is about the same as the effect of Ucon® 50HB5100 at 5 mg/m³ with the same protocol. Ucon® 50HB260 produces no effect at all.

The typical properties of the 75–H series of polymers are shown in Table 4–41.

Table 4–41

Typical Properties of Ucon® 75–H Lubricants

Ucon Lubricant	Viscosity @ 40°C, cSt	Average Molecular Weight.	Pour Point, °F (ASTM)	Flash Point, °F, Cleveland Open Cup
75–H–450	90.9	980	5	465
75–H–1400	282	2,470	40	520
75–H–9500	1,800	6,950	40	510
75–H–90,000	17,850	15,000	40	538

Random EO–PO copolymers may be represented by the structure:

$$R\,O-(CH_2CH_2O)_x-(CH_2\overset{\overset{\displaystyle CH_3}{|}}{C}HO)_y-H$$

For the 50–HB series x and y are equivalent and for the 75–H series x is three times y.

Block EO/PO Copolymers, CAS# 9003–11–6

Block EO–PO copolymers, like the random version, are manufactured by many suppliers; however, the principal source is BASF. They are sold under the Pluronic® and Tetronic® names for a variety of end uses. These include defoaming/antifoaming, dispersion, foaming, emulsification, lubrication, and wetting end uses. The chemical structure of these polymers is:

$$HO-(CH_2CH_2O)_x-(CH_2\overset{\overset{\displaystyle CH_3}{|}}{C}HO)_y-(CH_2CH_2O)_x-H$$

This structure shows that Pluronic® lubricants have the hydrophobe, polyoxypropylene, between two hydrophilic groups, polyoxyethylene.

In fiber lubrication these polymers are most frequently used as hydrodynamic lubricants. High speed lubricants should possess low internal cohesion, or resistance to shear (i.e., low viscosity), since viscous resistance may be one source of hydrodynamic friction. In practice, however, frictional phenomena are complicated, and high molecular weight polyether waxes may be more advantageous in friction reduction than low viscosity liquids. Commonly, Pluronic® type lubricants are more widely used than some of the other block copolymers.

In the tables that follow, the names of the Pluronic® materials follow this convention; a letter, followed by two or three numbers. In them L indicates a liquid, P a paste, and F a solid. The first one or two numbers show the molecular weight of the hydrophobe, polyoxypropylene. The last number is the percent polyoxyethylene. As an example, L31 means a liquid, 1 is 10% polyoxyethylene and 3 the lowest molecular weight polyoxypropylene unit, a molecular weight of 950 [90]. L-61, L-62, L-121, and L-122 are also produced by Ethox [91] and sold with those designations. Another block EO/PO copolymer is Afilan® 4PF [92] made by Hoechst–Celanese. Various properties of Pluronic® and Ethox® lubricants are presented in Tables 4–41, 4–42, and 4–43, while Afilan® 4PF properties are given in Table 4–44.

Table 4–42

Typical Properties of Liquid Pluronic® and Ethox® Lubricants

Product	Viscosity, 25°C, cPs	Average Mol. Wt.	Pour Point, °C
L31	175	1,100	−32
L35	375	1,900	7
L42	280	1,630	−26
L43	310	1,850	−1
L44	440	2,200	16
L61	325	2,000	−29
L62	450	2,500	−4
L63	490	2,650	10
L64	850	2,900	16
L81	475	2,750	−37

Table 4–42, Continued

Product	Viscosity, 25°C, cPs	Average Mol. Wt.	Pour Point, °C
L92	700	3,650	7
L101	800	3,800	−23
L121	1,200	4,400	5
L122	1,750	5,000	20

Table 4–43

Typical Properties of Solid Pluronic® Lubricants

Product	Viscosity, 77°C, cPs	Average Mol. Wt.	Melt. Point, °C
F38	260	4,700	48
F68	1,000	8,400	52
F77	480	6,600	48
F98	2,700	13,000	58
F108	2,800	14,600	57
F127	3,100	12,600	56

Table 4–44

Typical Properties of Paste Pluronic® Lubricants

Product	Viscosity, 60°C, cPs	Average Mol. Wt.	Melt. Point, °C
P65	180	3,400	27
P75	250	4,150	27
P84	280	4,200	34
P85	310	4,600	34
P103	285	4,950	30
P104	390	5,900	32
P105	750	6,500	35
P123	350	5,750	31

Table 4–45

Typical Properties of Afilan® 4PF

Property	Value
Color, Gardner	2 max
Pour Point, °C	20
Specific Gravity, 25/25°C	1.05
Water, %	1.0 max
Viscosity, cSt @ 25°C	950

Variations of the Pluronic® block EO–PO lubricants are the reverse Pluronic® R components [78]. These lubricants are prepared by adding ethylene oxide to ethylene glycol to provide a hydrophilic moiety of a specified molecular weight. Propylene oxide is then added to obtain hydrophobic blocks on the outside of the molecule. The molecular structure of these reverse blocks is:

$$\underset{\displaystyle |}{CH_3} \qquad\qquad\qquad \underset{\displaystyle |}{CH_3}$$
$$HO-(CHCH_2O)_x-(CH_2CH_2O)_y-(CH_2CHO)_x-H$$

Reversing the hydrophobic and hydrophilic blocks creates lubricants similar to the Pluronic® surfactants, but with some important differences. The Pluronic® R polymers have lower foaming, greater defoaming, and reduced gelling tendencies. Representative properties of Pluronic® R and Ethox® polymers are shown by Table 4–46. Since these polymers are terminated by

Table 4–46

Typical Properties of Reverse Pluronic® and Ethox® R Lubricants

Product	Viscosity	Physical Form, 20°C	Average Molecular Weight	Melt/Pour Point, °C
10R5	440*	Liquid	1,950	15
10R8	400***	Solid	4,550	46
12R3	340*	Liquid	1,800	−20

Table 4–46, Continued

Product	Viscosity	Physical Form, 20°C	Average Molecular Weight	Melt/Pour Point, °C
17R1	300*	Liquid	1,900	−27
17R2	450*	Liquid	2,150	−25
17R4	600*	Liquid	2,650	18
17R8	1,600***	Solid	7,000	53
22R4	950*	Liquid	3,350	24
25R1	460*	Liquid	2,700	−27
25R2	680*	Liquid	3,100	−5
25R4	1,110*	Liquid	3,600	25
25R5	370**	Paste	4,250	30
25R8	2,600***	Solid	8,550	54
31R1	660*	Liquid	3,250	−25
31R2	850*	Liquid	3,300	9
31R4	300**	Paste	4,150	24

*25°C, cPs
**60°C, cPs
***77°C, cPs

secondary hydroxyl groups, they have lower reactivity than the primary hydroxyl terminating Pluronic® lubricants. Another producer of reverse EO/PO copolymers is Ethox Chemical [91]. In the preceding table, the numerical digits before the R in the product identification show the relative molecular weight of the polyoxypropylene (10 = 1000 and 31 = 3100) and the digit after the R indicates the percent polyoxyethylene (1 = 10% and 8 = 80%).

REFERENCES

1. Redston, J. P., Bernholz, W. F., and Schlatter, C. (1973). Chemicals Used as Spin Finishes for Man-Made Fibers, *Textile Res. J.*, **43**: 325

2. Kapp, R. and Steik, K. T., assignors to National Oil Products Co., Harrison, NJ (1941). Composition and Process for Treating Fibrous Materials, *U.S. Patent 2,268,141*

3. Petroleum Specialties Group, Witco Refined Products. (1994). *White Oils, Petrolatums, Microcrystalline Waxes, Petroleum Distillates*, Witco Corporation, Greenwich, CT

4. Morrison, R. T. and Boyd, R. N. (1983). *Organic Chemistry*, Fourth Edition, Allyn and Bacon, Boston, pp. 1041, 1159, 470, 382–383

5. Inolex Product Literature (1994). *Isothermal Volatility of Lexolube® Esters*, Inolex Chemical Company, Philadelphia

6. Inolex Product Bulletin (1995). *Loxolube® IPM*, Inolex Chemical Company, Philadelphia

7. Stepan Company (1995). Products for the Textile and Fiber Industries, Catalog on Computer Disk, *Kessco® IPM*, Stepan Company, Northfield, IL

8. Stepan Product Bulletin (1995). *Kessco® Cetyl Myristate 654*, Stepan Company, Northfield, IL

9. Perkins, G. P. (1978). Analysis and Standards of Fatty Acids, *Kirk–Othmer Encyclopedia of Chemical Technology*, Third Edition, Vol. 4, John Wiley & Sons, New York, pp. 845–853

10. Henkel Corporation, Textile Chemicals, Data Sheet (1989). *Emerest® 2316. Isopropyl Palmitate*, Henkel Corporation, Charlotte, NC

11. Inolex Product Bulletin (1995). *Lexolube® IPP*, Inolex Chemical Company, Philadelphia

12. Stepan Industrial Specialties (1994). Products for the Textile Industry, Product Directory, *Kessco® Alcohol Esters*, Stepan Company, Northfield, IL

13. Burgo, R. and Tuszynski, W. (1995). Inolex Chemical Company, personal communication

14. Inolex Product Bulletin (1994). *Lexolube® EHP*, Inolex Chemical Company, Philadelphia

15. Stepan Company (1995). Products for the Textile and Fiber Industries, Catalog on Computer Disk, *Kessco® Octyl Palmitate*, Stepan Company, Northfield, IL

16. Stepan Textile Products, Technical Information (1995). *Alcohol Esters*, Stepan Company, Northfield, IL

17. Brewster, R. Q. (1953). *Organic Chemistry*, Second Edition, Prentice–Hall, Englewood Cliffs, NJ, pp. 192, 353, 309, 154

18. Henkel KGaA Product Data Sheet (1995). *Ester einwertiger Alkohole, Cutina® CP*, Henkel KGaA, Düsseldorf, Federal Republic of Germany

19. Stepan Product Bulletin (1995). *Kessco® Cetyl Palmitate 653*, Stepan Company, Northfield, IL

20. Fiber Production Auxiliaries (1994). *Fiber and Yarn Production: Lubrication*, Witco Corporation, Greenwich, CT

21. Humco Chemical Technical Information (1983). *Kemester® CP*, Humco Chemical Division, Witco Corporation, Memphis, TN

22. Satkowski, W. B., Huang, S. K., and Liss, R. L. (1966). Polyoxyethylene Esters of Fatty Acids, in *Nonionic Surfactants* (M. J. Schick, ed.), Marcel Dekker, Inc., New York, pp. 142–174

23. Textile Chemicals, Technical Bulletin 103A (1993). *Alkyl and Glycerol Esters*, Henkel Corporation, Textile Chemicals, Charlotte, NC, p. 28–29

24. Henkel Corporation Data Sheet TC–0230 (1989). *Emerest® 2326, Butyl Stearate*, Henkel Corporation, Charlotte, NC

25. Inolex Product Bulletin (1994). *Lexolube® NBS*, Inolex Chemical Company, Philadelphia

26. Inolex Chemical Company (1994). *Directory of Chemicals Available or Possible from Inolex Chemical Company*, Inolex Chemical Company, Philadelphia

27. Humko Chemical Technical Information (1994). *Humko Chemical Oleo Esters for Textile and Fiber Industries*, Humko Chemical Division, Witco Corporation, Memphis, TN

28. Inolex Product Bulletin (1994). *Lexolube® BS–Tech*, Inolex Chemical Company, Philadelphia

29. Henkel Corporation, Textile Chemicals, Data Sheet TC–0228 (1989). *Emerest® 2324, Isobutyl Stearate*, Henkel Corporation, Charlotte, NC

30. Noller, C. W. (1958). *Textbook of Organic Chemistry*, W. B. Saunders, Philadelphia, pp. 172–173, 45, 538, 555, 527

31. Henkel Corporation, Textile Chemicals (1992). *Stantex® EHS, 2–Ethylhexyl Stearate*, Henkel Corporation, Charlotte, NC

32. Hoechst–Celanese Specialty Chemicals Group, Product Data Sheet (1993). *Afilan® EHS*, Hoechst–Celanese Corporation, Charlotte, NC

33. Inolex Product Bulletin (1994). *Lexolube® T–110*, Inolex Chemical Company, Philadelphia

34. Henkel Corporation, Textile Chemicals Data Sheet TC–0227 (1989). *Emerest® 2308, Tridecyl Stearate*, Henkel Corporation, Charlotte, NC

35. Hoechst–Celanese Specialty Chemicals Group, Product Data Sheet (1993). *Afilan® TDS*, Hoechst–Celanese Corporation, Charlotte, NC

36. Inolex Product Bulletin (1994). *Lexolube® B–109*, Inolex Chemical Company, Philadelphia

37. Hoechst–Celanese Specialty Chemicals Group, Product Data Sheet (1993). *Afilan® ICS*, Hoechst–Celanese Corporation, Charlotte, NC

38. Inolex Product Bulletin (1994). *Lexemul® EGMS*, Inolex Chemical Company, Philadelphia

39. Stepan Textile Products, Technical Information (1995). *Glycol Esters*, Stepan Company, Northfield, IL

40. Inolex Product Bulletin (1994). *Lexemul® EGDS*, Inolex Chemical Company, Philadelphia

41. Henkel Corporation, Textiles Chemicals (1983). *Emerest® 2380 and Emerest® 2318, Propylene Glycol Stearates*, Henkel Corporation, Charlotte, NC

42. Inolex Product Bulletin (1994). *Lexemul® P*, Inolex Chemical Company, Philadelphia

43. Abitec Performance Products (1994). *Mono– and Digylcerides*, Abitec Corporation, Columbus OH

44. Henkel Corporation, Textile Chemicals (1989). *Emerest ® 2401, Glycerol Monostearate*, Henkel Corporation, Charlotte, NC

45. Stepan Industrial Specialties (1994). Products for the Textile Industry, Product Directory, *Kessco® Glycerol Monostearate*, Stepan Company, Northfield, IL

46. Henkel Corporation, Textile Chemicals Data Sheet TC–0130 (1989). *Emerest® Glycerol Monoisostearate*, Henkel Corporation, Charlotte, NC

47. Stepan Textile Products, Technical Information (1995). *Methyl Esters*, Stepan Company, Northfield, IL

48. Henkel Corporation, Textile Chemicals (1989). *Emerest® 2302, Propyl Oleate*, Henkel Corporation, Charlotte, NC

49. Henkel KGaA Data Sheet (1995). *Estereinwertiger Alkohole, Rilanit® OLO*, Henkel KGaA, Düsseldorf, Federal Republic of Germany

50. Henkel Corporation, Textile Chemicals Data Sheet TC–0232 (1989). *Emerest® 2421, Glycerol Monooleate*, Henkel Corporation, Charlotte, NC

51. Stepan Industrial Specialties (1994). Products for the Textile Industry, Product Directory, *Kessco® Glycerol Monooleate*, Stepan Company, Northfield, IL

52. Henkel Corporation, Textile Chemicals Data Sheet TC–0253 (1989). *Emerest® 2423, Glycerol Trioleate*, Henkel Corporation, Charlotte, NC

53. Abitec Performance Products (1994).*Vegetable Oils*, Abitec Corporation, Columbus, OH

54. Inolex Product Bulletin (1994). *Lexolube® Z–100*, Inolex Chemical Company, Philadelphia

55. Weber, J. and Daley, J. (1978). Other Polyhdric Alcohols, *Kirk–Othmer Encyclopedia of Chemical Technology, Volume 1*, Third Edition, John Wiley & Sons, New York, pp. 778–789

56. ICI Product Information Bulletin (1992). *Emkarate™ 2030 (G–3892)*, ICI Surfactants, Charlotte, NC

57. Inolex Product Bulletin (1994). *Lexolube® 3N–309*, Inolex Chemical Company, Philadelphia

58. Inolex Product Bulletin (1994). *Lexolube® 3N–310*, Inolex Chemical Company, Philadelphia

59. Stepan Industrial Specialties (1994). Products for the Textile Industry, Product Directory, *Polyol Esters*, Stepan Company, Northfield, IL

60. Hill, R. H., assignor to The Standard Oil Company of Indiana, Chicago, IL (1957), Method for Preparing Carboxylic Acids, *U. S. Patent 2,815,355*

61. Henkel Corporation Textile Chemicals (1983). *Emerest® 2388, Propylene Glycol Dipelargonate*, Henkel Corporation, Charlotte, NC

62. Henkel Corporation, Textile Chemicals (1989). *Emery® 6783, Triethylene Glycol Dipelargonate*, Henkel Corporation, Charlotte, NC

63. Stepan Textile Products, Technical Information (1995). *Miscellaneous Esters*, Stepan Company, Northfield, IL

64. Henkel Corporation, Textile Chemicals Data Sheet TC–0172 (1989). *Emery® 6701, Trimethylol Propane Tripelargonate*, Henkel Corporation, Charlotte, NC

65. Hoechst–Celanese Specialty Chemicals Group, Product Data Sheet (1993), *Afilan® TMPP*, Hoechst–Celanese Corporation, Charlotte, NC

66. Stepan Textile Products, Technical Information (1995). *Trimethylolpropane Esters*, Stepan Company, Northfield, IL

67. Henkel Corporation, Textile Chemicals Data Sheet TC–0140 (1989). *Emerest® 2486, Pentaerythritol Tetrapelargonate*, Henkel Corporation, Charlotte, NC

68. Hoechst–Celanese Chemical Specialty Group, Product Data Sheet (1993). *Afilan® PP*, Hoechst–Celanese Corporation, Charlotte, NC

69. Stepan Textile Products, Technical Information (1995). *Pentaerythritol Esters*, Stepan Company, Northfield, IL

70. Streitwieser, A. and Heathcock, C. H. (1985). *Introduction to Organic Chemistry*, Macmillan, New York, pp. 863, 400, 217, 510

71. Hoechst–Celanese Specialty Chemicals Group, Product Data Sheet (1993). *Afilan® ODA*, Hoechst–Celanese Corporation, Charlotte, NC

72. Hoechst–Celanese Specialty Chemicals Group, Product Data Sheet (1993). *Afilan® TDA*, Hoechst–Celanese Corporation, Charlotte, NC

73. Inolex Product Bulletin (1994). *Lexolube® 2X–109*, Inolex Chemical Company, Philadelphia

74. Barrett, H. D., Estes, R. T., and Stowe, G. C., assignors to Monsanto Chemical Company, St. Louis, MO (1963). Yarn Manufacture and Products Obtained Thereby, *U.S. Patent 3,113,369*

75. Patterson, H. T. and Strohmaier, A. J., assignors to E. I. du Pont de Nemours and Co., Wilmington, DE. (1964). Rope Finish, *U.S. Patent 3,155,537*

76. Petrolite MSDS (1990). *C–7500 Polymer®*, Petrolite Corporation, St. Louis, MO

77. Denny, R. G. (1970). Polyvinyl Acetate and Polyvinyl Alcohol, in *Process Economics Program, Report No. 57*, Stanford Research Institute, Menlo Park, CA, pp. 90–92

78. Teot, A. S. (1978). Resins, Water Soluble, *Kirk–Othmer Encyclopedia of Chemical Technology*, Third Edition, Volume 20, John Wiley & Sons, New York, pp. 210–213

79. Air Products (1993). *Airvol® Polyvinyl Alcohol*, Air Products and Chemicals, Inc., Allentown, PA

80. Golosova, L. V. and Zatsepin, K. S. (1969). Lubricating and Sizing Agent for Glass Fiber, *U.S.S.R.Patent 235,907*

81. Marzocchi, A. and Rammel, G. E., assignors to Owens–Corning Fiberglass, Toledo, OH (1969). Glass Fiber Lubricant Coatings, *U.S. Patent 3,462,254*

82. Ross, S. E., assignor to AlliedSignal, Inc., Morristown, NJ. (1963). Polyethylene Wax Coating of Continuous Synthetic Filament Yarns, *U.S. Patent 3,103,448*

83. Boe, N. and Duke, B. H., assignor to Monsanto Company, St. Louis, MO. (1972). Spinning and Separation of Segmented Elastomer Fibers, *German Patent 2,205,157*

84. Eastman (1994). *Epolene® Waxes: Low–Molecular–Weight Polyolefin Resins for Industrial Applications*, Eastman Chemical Company, Kingsport, TN

85. Dow Corning (1974). *Information about Fiber, Yarn, and Thread Finishes*, Dow Corning Corporation, Midland, MI

86. Slade, P. E., unpublished data

87. Union Carbide, Textiles and Paper Chemicals, Product Information (1985). *Ucon® 50–HB Lubricants*, Union Carbide Corporation, Danbury, CT

88. Union Carbide Chemicals and Plastics (1992), *Ucon® Fluids and Lubricants*, Union Carbide Chemicals and Plastics, Inc., Specialty Chemicals Division, Danbury, CT

89. Meeting Summary, American Fiber Manufacturers Association Toxicology Subcommittee, November 13, 1990, Washington, DC

90. BASF (1989). *Pluronic® and Tetronic® Surfactants*, BASF Corporation, Mount Olive, NJ

91. Ethox Product Brochure (1994). *Block Copolymers*, Ethox Chemicals, Inc., Greenville, SC

92. Hoechst–Celanese Specialty Chemicals Group, Product Data Sheet (1993). *Afilan® 4PF*, Hoechst–Celanese Corporation, Charlotte, NC

CHAPTER 5

COHESIVE AGENTS AND FIBER TO FIBER LUBRICANTS

Fiber to fiber lubricants fall into two categories, cohesive agents that increase fiber friction and softeners that reduce that friction. The chemical structures of these components differ and are discussed in the following segments of this chapter.

5.1 ATTRACTIVE FORCES AT INTERFACES

Forces existing between different objects in our natural world vary greatly in type and strength. From gravity to intra–atomic, both attractive and repulsive forces maintain cohesion and order in all materials. There is only one major *repulsive* interaction, that is, the highly unlikely situation of two molecules attempting to occupy the same space at the same time. The energy term for this interaction is very large as the molecules attempt to approach each other and are thus repelled, otherwise there is little effect at relatively long distances. In contrast, there are many different types of *attractive* forces; in chemistry these can be divided into two main classes, primary and secondary forces. Primary chemical forces are ionic, covalent, and coordinate molecular bonding, but when materials are attracted to polymer surfaces we must consider other types of forces.

In very few instances can the attraction between a polymer surface and some other substances be attributed to primary bonding. In most situations the forces are quasi–primary chemical, often called hydrogen bonding, or else one of the secondary forces. These secondary, or van der Waals', forces are divided into three additional groups; Keesom, Debye, and London dispersion forces. Keesom forces [1] are interactions between permanent dipoles where one dipole tends to align the other into an energetically favored arrangement. When a permanent dipole causes the polarization of another molecule (either polar or non–polar) these interactions are called Debye forces [2]. If a situation exists in which there is an instantaneous dissymmetry of electrons in one molecule polarizing the electron cloud in another nearby molecule, thus inducing dipoles of opposite polarity, the forces are called London dispersion forces [3]. Table 5–1 shows the relative bond energies of the various types of molecular forces [4].

Table 5–1

Comparison of Molecular Bond Energies

Type of Force	Energy, Kcal/mole
Primary Chemical Bonds	
Ionic	140–250
Covalent	15–170
Secondary Intermolecular Bonds	
Hydrogen Bonds	up to 12
Dipole–Dipole	up to 10
London Dispersion	up to 0.5

(Reference [4])

The circumstances necessary for the formation of different types of secondary bonding and some typical examples are presented below.

1. *Hydrogen Bonding.* Several requirements for hydrogen bonding are: (1) A highly electronegative atom, such as O, Cl, F, or N, or a strongly electronegative group such as $-CCl_3$ or $-CN$ with a hydrogen atom attached must be present. The electronegative atom or group polarizes the electron cloud of the molecule in the direction away from the proton. (2) Another electronegative atom, which may or may not be in a molecule of the same species as the

first atom or group, is also necessary. Hydrogen bonds are most likely to form when the latter substance has a lone pair of exposed electrons. A classical example would be the bond formed between the nitrogen of an amide group on a nylon surface and a molecule of water [4].

2. *Dipole–Dipole Attraction*. If a molecule is unsymmetrical, that is, the molecular structure is such that some atoms have a different electronegativity than the other atoms, that molecule is said to possess a dipole moment. Nitrobenzene, for instance, has a large dipole moment because the $-NO_2$ group has a vastly different electronegativity than the phenyl ring to which it is attached. When two molecules with permanent dipoles approach each other, the dipoles align with opposite charges in the approaching molecules and dipole–dipole bonds are formed. A disperse dye may be attracted to a fiber surface by this mechanism.

3. *Dipole–Induced Dipole Attraction*. For bonding of this type to happen, one moiety must have a permanent dipole and the other an electron rich cloud that can be deformed by the permanent dipole into a new arrangement of dissymmetry. An example of these attractive forces would be the adsorption of acetone onto a polystyrene surface.

4. *London Dispersion Forces*. Forces of this type depend on an electron rich area instantaneously deforming into a temporary dipole and this dipole inducing a dipole in another molecule. Although these forces are among the weakest found in nature, many examples exist. The strength of polyethylene largely depends of London dispersion forces preventing the tangled polyethylene chains from sliding by each other.

When a spin finish, in an oil–in–water emulsion, is applied to yarn, the oil components and the water are all attracted to the fiber surface by one or more of the mechanisms listed above. Many finish oil blends contain components that, as they are adsorbed, change the fiber to fiber friction more than they affect fiber to metal friction. These components are called cohesive agents if they increase the fiber to fiber friction and softeners if they reduce the friction. They are generally classified as nonionic surfactants.

A surfactant, or surface–active agent, can be defined as a substance that, when present in low concentrations in a system, has the property to adsorb onto the surfaces or interfaces of the system and to alter the surface or interfacial energy of those surfaces [5]. The term *interface* indicates a boundary between any two immiscible phases; the term *surface* denotes an interface where one phase is a gas, usually air.

Interfacial free energy is defined as the minimum amount of work required to create that interface. The interfacial free energy per unit area is the property measured when one determines the interfacial tension between the two phases. If the surface tension of a liquid is measured, we are assessing the interfacial free energy per unit area of the boundary between the liquid and the air above it. When an interface is expanded, the minimum amount of work required to create an additional quantity of that interface is the product of the interfacial tension, γ, times the increase in the area of the interface; $W = \gamma\,(A)$. A surface–active agent is then defined as a substance that, at low concentrations, adsorbs at some or all of the interfaces in the system and significantly changes the amount of work required to expand those interfaces. Surfactants commonly act to reduce the interfacial free energy rather than to increase it.

Surface–active agents have a characteristic molecular configuration consisting of a structural group that has very little attraction for a polar solvent, known as a lyophobic group, together with a group that has a strong attraction to that solvent, called the lyophilic group. This is known as an *amphipathic* structure. When a surfactant is dissolved in a solvent, the presence of the lyophobic group in the interior of the solvent causes a disruption of the solvent liquid structure, increasing the free energy of the system. In an aqueous solution of a surfactant, the water is distorted by the lyophobic (hydrophobic) groups of the surfactant, and results in an increase in the free energy of the system, meaning that less work is needed to bring a surfactant molecule to the surface than is required for a water molecule. The surfactant therefore concentrates at the surface. Since less work is now needed to bring the molecule to the surface, the presence of the surfactant decreases the work needed to create a unit area of surface, defined as either the surface tension or the surface free energy. On the other hand, the presence of the lyophilic (hydrophilic) group prevents the surfactant from being completely expelled from the solvent and to form a separate phase, or layer, since that expulsion would require the dehydration of the hydrophilic group. The amphipathic structure of the surfactant creates not only the concentration at the surface and reduction of the surface tension of the solvent, but also the orientation of the molecule at the surface with its hydrophilic group in the aqueous phase and the hydrophobic group away from it [5].

5.2 FINISH COMPONENTS AFFECTING FIBER COHESION

Many surfactants are sold by finish component suppliers that are water dispersible or soluble and nonionic, but are neither primary emulsifiers nor superior fiber to metal lubricants. Some examples are ethoxylated (or ethoxylated–propoxylated) alcohols and poly(ethylene glycol), or PEG, esters of fatty acids. They do add some emulsification properties to a finish oil blend and are frequently mixed with an emulsifier to improve the quality of the emulsion. Some of these materials are fair to good fiber to metal (F–M) lubricants, but they probably cannot be used alone to reduce this type of friction. These compounds are primarily included in the finish oil mixture to control the fiber to fiber (F–F) friction, which is commonly called cohesion.

During the process of spinning a synthetic fiber, a finish is applied to the yarn. In many circumstances this finish is applied as an oil–in–water emulsion. The oil phase of the emulsion consists of lubricants, emulsifiers, cohesive agents, antistats and lesser amounts of other components. An interface between the polymer surface in the yarn bundle and the bulk phase of the emulsion is created and the emulsion (both oil and water phases) is attracted to the polymer surface. Since the melt–spun yarn is very dry at this point in the process, some water is absorbed into the fiber and the rest of the emulsion is dispersed throughout the yarn bundle by capillary action. Between the point of finish application and the winding of the yarn onto the package most of the water evaporates or is removed by another mechanism because the bobbin, when taken from the spinning machine, contains only the equilibrium amount of water. Equilibrium moisture, or moisture regain for several types of fibers are listed in Table 5–2 [6].

Table 5–2

Equilibrium Moisture for Different Fiber Types

Fiber Type	Moisture Regain	Fiber Type	Moisture Regain
Wool	16.0	Triacetate	3.5
Silk	14.0	Qiana	2.5
Rayon	13.0	Acrylic	1.5
Mercerized Cotton	11.0	Spandex	1.3
Raw Cotton	7.0	Polyester	0.5
Acetate	6.5	Olefin	0.01
Nylon	4.5	Glass	0

(Ref. [6]. Reprinted by permission of Macmillian Publishing Co.)

At some time, either while the water is present or after it has evaporated and only the finish oil remains, the components are adsorbed onto the polymer surface. The various components must compete for sites on that surface and will orient themselves to yield the optimum interfacial free energy for that system. Movement of these molecules from the emulsion occurs because there is a decrease in the total free energy caused by either an increase in bonding energy at the fiber interface or because the surfactant finds a lower free energy environment at that interface than in the bulk of the emulsion. The mechanism of this free energy of bonding in nonionic surfactants adsorbing onto fiber surfaces can be explained by either: (1) polar attraction of the poly(alkylene oxide) or hydrophilic portion of the molecule to the highly polar groups on the fiber surface, (2) an induced dipole attraction of the hydrophobic (or non–poly(alkylene oxide)) portion of the molecule to the aliphatic chains on the fiber surface, or (3) perhaps some hydrogen bonding of the ether oxygen of the poly(alkylene oxide) unit to the nitrogen, oxygen, or other group in the polymer. In most of the adsorption processes of surfactants being attracted to many different fibers, all three mechanisms may occur simultaneously [7]. This adsorption is in direct competition for space on the surface with the attraction of the lubricant molecule to the aliphatic chain portion on the yarn surface through a London dispersion mechanism. As the finishes adsorb onto the fiber surface, the arrangement of molecules will alter the frictional properties.

Although many different chemicals may be classified as cohesive agents or softeners, Table 5–3 lists some typical surfactants that are widely used for those end uses. Several general observations about the effects of chemical structure can be noted from this table.

1. Most ethoxylated alcohols or acids that result in high F–F friction have saturated, linear alkyl chains of nine to fourteen carbons. Some exceptions to this generalization include ethoxylated oleyl alcohol ethers, where higher stick–slip is probably due to the *cis*–configuration of the unsaturated chain. There are a few other exceptions, but not many.

2. Short poly(ethylene oxide) chain lengths, ten ethylene oxide units or less, result in higher F–F friction for nylon, polyester, and polypropylene than longer chain lengths. Some components with longer ethylene oxide chains are effective for rayon and acrylic fibers.

3. Alcohols seem to produce higher F–F friction than their acid counterparts.

4. Branching in the hydrophobic carbon chain reduces F–F and F–M friction.

5. There is no apparent correlation between F–M friction and F–F friction.

Table 5-3

Finish Component Property Summary

Base Chemical	Moles EO	Acid Ester	Alcohol Ether	Fiber–Fiber Friction		Fiber–Metal Friction
				Level	Stick–Slip	
Castor Oil	200	✓		Low	Low	High
Sorbitan Monostearate	20	✓		Low	Medium	High
Sorbitan Monostearate	0	✓		Medium	Low	Medium
Sorbitan Tristearate	20	✓		Medium	Medium	Medium
Tridecyl Alcohol	6		✓	Medium	Medium	Medium
Lauric Acid	9, Methyl Capped	✓		Medium	Medium	Low
Glycerol Monolaurate	20	✓		Medium	High	Medium
Oleyl Alcohol	20		✓	Medium	High	Low
Lauryl Alcohol	4		✓	High	High	Low
C_{12}–C_{13} Alcohol	6, Methyl Capped		✓	High	High	Low

(Reference [25]. Reprinted by permission of ICI Americas)

In selecting finish components for optimum cohesion, it appears that one should choose a hydrophobe based on fatty acids or alcohols isolated from coconut oil, a C_9 (pelargonic) or a C_{12} (lauric) acid or alcohol. Hydrophobic chains with a number of carbon atoms less than nine are not effective, probably because of the orientation of the molecule on the fiber surface. Longer carbon chains, such as stearates, generally reduce the level of fiber to fiber friction and significantly reduce the stick–slip. The poly(ethylene oxide) repeat unit also affects the cohesion. A chain length of 4–6 repeat units has the highest F–F friction and the highest stick–slip. Long

POE chains reduce the level of friction and the stick–slip. It is intriguing that a nonionic surfactant prepared by the polyoxyethylenation of a C_{12-15} alcohol mixture with 9–11 moles of ethylene oxide yield the best detergents for soil removal from nylon or polyester [5], and also provide good fiber cohesion. This detergency does not hold for cotton, for which anionic surfactants are the most effective.

One of the problems that has been associated with this type of nonionic surfactant is the tendency for them to attack polyurethane. In the processing of yarn, many kinds of textile equipment use polyurethane drive belts, texturing discs, and rolls. If polyurethane contacts yarn, especially in a warm environment, and if these cohesive agents are components of the spin finish they will absorb into the polyurethane and reduce the hardness. The so–called soft forms of this polymer, the poly(ether–isocyanates), are more susceptible to this attack than harder types made from poly(ester-isocyanates) [8]. The problem occurs when the hydroxyl end group of the poly(ethylene oxide) chains is attracted to the polyurethane and partially dissolves the urethane polymer [9]. A test for this absorption has been proposed by ICI Specialty Chemicals [10]. Absorption can be reduced by the addition of propylene oxide to the ethylene oxide feed stream or by end–capping the hydroxyl group with an ethyl or larger moiety.

5.3 NONIONIC SURFACTANTS AS COHESIVE AGENTS

Although almost all of the cohesive agents included in the subsequent list can be effective emulsifiers, the primary objective in their inclusion in a finish oil mixture is to increase fiber to fiber friction and stick–slip. This section will include fatty alcohol and acid that have saturated, linear chains of 14 or fewer carbon atoms and reacted with up to 15 moles of ethylene oxide. In addition to this grouping, oleyl alcohol ethoxylates and other derivatives are incorporated as cohesive agents.

5.3.1 Alcohol Ether Ethoxylates

(POE) Decyl Alcohols, CAS# 26183–52–8

The ethoxylated decyl alcohols have primary application in the textile industry as wetting agents in fabric preparation, but they are also employed as cohesive agents. They are not, however, as effective as some other alcohol ethers.

Decyl alcohol, or 1–decanol, may be prepared by the hydrogen reduction of decanoic, or capric, which is one of the low molecular weight fatty acid fractions in coconut oil. Another route in the synthesis of 1–decanol is by the Zeigler process [11]. The first step in the Zeigler process is the reaction of ethylene with triethyl aluminum under conditions of about 100°C and 1000 p.s.i. to obtain a high molecular weight product.

$$Al \Big\langle \begin{array}{l} CH_2CH_3 \\ CH_2CH_3 \\ CH_2CH_3 \end{array} + n \ CH_2CH_2 \longrightarrow Al \Big\langle \begin{array}{l} (CH_2CH_2)_x CH_2CH_3 \\ (CH_2CH_2)_y CH_2CH_3 \\ (CH_2CH_2)_z CH_2CH_3 \end{array}$$

This alkyl aluminum is then oxidized with air to form an alcoholate:

$$Al \Big\langle \begin{array}{l} (CH_2CH_2)_x \ CH_2CH_3 \\ (CH_2CH_2)_y \ CH_2CH_3 \\ (CH_2CH_2)_z \ CH_2CH_3 \end{array} + 3/2 \ O_2 \longrightarrow Al \Big\langle \begin{array}{l} O(CH_2CH_2)_x \ CH_2CH_3 \\ O(CH_2CH_2)_y \ CH_2CH_3 \\ O(CH_2CH_2)_z \ CH_2CH_3 \end{array}$$

The aluminum alcoholate is hydrolyzed with a dilute acid forming straight–chain alcohols.

$$Al \Big\langle \begin{array}{l} O(CH_2CH_2)_x \ CH_2CH_3 \\ O(CH_2CH_2)_y \ CH_2CH_3 \\ O(CH_2CH_2)_z \ CH_2CH_3 \end{array} + 3 \ H_2O \xrightarrow{\ H^+ \ } \begin{array}{l} HO(CH_2CH_2)_x CH_2CH_3 \\ and \\ HO(CH_2CH_2)_y CH_2CH_3 \\ and \\ HO(CH_2CH_2)_z CH_2CH_3 \end{array}$$

$$+ \ Al(OH)_3$$

Proper adjustment of reaction conditions can shift the distribution of alcohol chain lengths to higher or lower molecular weights to yield products for different end uses. Lower molecular weight alcohols, C_6 through C_{10} chain lengths, are principally used to make esters for plasticizers. The higher alcohols, C_{12} through C_{20}, are produced for the detergent industry [11].

As with all other alcohols, decyl alcohol is reacted with ethylene oxide in the presence of an alkaline, probably potassium hydroxide, catalyst to make the poly(oxyethylene) alcohol ether [12]. The structural formula for these compounds is:

$$CH_3(CH_2)_8CH_2O(CH_2CH_2O)_xH$$

Commercially available decyl alcohol ethers include ones where x equals four, six, or nine. All these are effective wetting agents as determined by the Draves test [13]. In this test, a 5–gram skein of cotton yarn is loaded with a 3–gram weight and immersed in a tall cylinder of surfactant solution or some other liquid. As the solution displaces the air in the skein, when sufficient air has been replaced by the liquid, the skein will sink. The wetting agents that are more effective have shorter times for sinking to occur.

1. POE (4) Decyl Alcohol. This short EO chain component is dispersible in water and mineral oil [14], and is a good wetting agent at concentrations of 0.1% or less. It is also an intermediate in the manufacture of anionic surfactants. The ether sold by Ethox under the designation Ethal® DA–4 [14] [15] and by Henkel with a trade name Trycol® 5950 [16][17]. Typical properties of POE (4) decyl alcohol are listed in Table 5–4.

Table 5–4

Typical Properties of POE (4) Decyl Alcohol

Property	Value
Acid Value	1 max
Hydroxyl Value	160–175
Color, APHA	100 max
pH, 5% in Distilled Water	6–8
Moisture, %	1 max
Draves Wetting, Seconds (0.1% @ 25°C)	3

2. POE (6) Decyl Alcohol. This version of poly(oxyethylene) decyl alcohol is widely used as a wetting agent in textile fabric dyeing systems. It is completely water soluble and is dispersible in mineral oil [16]. Although POE (6) decyl alcohol acts to provide some cohesion in fiber to fiber friction, it probably would not be the compound of choice for that use. It is sold by Ethox as Ethal® DA–6 [14][18], by Henkel as Trycol® 5952 [16][19], and Stepan as Stepantex® DA–6 [20]. Some properties of this compound are presented in Table 5–5.

Table 5–5

Typical Properties of POE (6) Decyl Alcohol

Property	Value
Acid Value	1 max
Hydroxyl Value	120–135
Cloud Point, 1%, °C	39–45
Color, APHA	100 max
pH, 5% in Distilled Water	5–8
Moisture, %	3 max
Draves Wetting, Seconds (0.1% @ 25°C)	6

3. POE (9) Decyl Alcohol. The increased level of ethylene oxide in this material causes it to become insoluble in mineral oil while maintaining the water solubility [16]. The cohesive ability is marginal. Various attributes of POE (9) decyl alcohol are given in Table 5–6.

Table 5–6

Typical Properties of POE (9) Decyl Alcohol

Property	Value
Hydroxyl Value	92–111
Cloud Point, 1%, °C	76–85
Color, APHA	50 max
pH, 5% in Distilled Water	5–8
Moisture, %	1 max
Draves Wetting, Seconds (0.1% @ 25°C)	15

It is manufactured by Ethox and sold as Ethal® DA–9 [14][21] and by Henkel with a trade name of Trycol® 5956 [16][22].

(POE) Lauryl Alcohols, CAS# 9002–92–0

Dodecanoic acid, commonly called lauric acid, is one of the major fatty acids in coconut oil and palm kernel oil. After the hydrolysis of the raw glycerides, a mixture of free fatty acids is formed that can be separated by fractional distillation, yielding individual carboxylic acids of over 90% purity [23].

Fats and oils may also be converted by transesterification into the methyl esters of the carboxylic acids by allowing methanol to react with them in the presence of a basic catalyst. The methyl esters of the acids are fractionally distilled and hydrolyzed to prepare acids of very high purity. The reaction may be represented as:

$$
\begin{array}{c}
\quad\quad\quad\underset{\|}{\overset{O}{}} \\
H_2C-O-C-R \\
| \quad\quad\;\underset{\|}{\overset{O}{}} \\
H\,C-O-C-R' + CH_3OH \quad \xrightarrow{\text{base}} \quad
\begin{array}{c}
CH_2OH \quad RCOOCH_3 \\
| \\
CHOH \quad + R'COOCH_3 \\
| \\
CH_2OH \quad R''COOCH_3
\end{array} \\
| \quad\quad\;\underset{\|}{\overset{O}{}} \\
H_2C-O-C-R''
\end{array}
$$

An alternate procedure to the hydrolysis of the methyl fatty ester and reduction of the acid to the linear alcohol is the catalytic reduction of the ester directly to the alcohol. This synthesis technique requires more severe conditions than most reduction using a catalyst of mixed oxides known as copper chromite, $CuO.CuCr_2O_4$ [23]. For example, lauryl alcohol may be prepared following the equation:

$$
CH_3(CH_2)_{10}COOCH_3 \quad \xrightarrow[\text{150°C, 5000 psi}]{H_2\,,\;CuO.CuCr_2O_4} \quad CH_3(CH_2)_{10}CH_2OH + CH_3OH
$$

Other possible routes in preparing lauryl alcohol include chemical reduction by use of lithium aluminum hydride, or through the Zeigler process. Poly(oxyethylene) lauryl alcohol ethers are manufactured by adding ethylene oxide to the alcohol catalyzed by alkaline compounds. The general structure of the POE lauryl alcohols is:

$$CH_3(CH_2)_{10}CH_2O(CH_2CH_2O)_xH$$

Ethylene oxide content varying between x equaling 3 and 23 is available from component producers.

1. POE (3) Lauryl Alcohol. This compound is sold as an oil–soluble emulsifier with the primary application for compounding into coning oil formulations. Since the addition of another ethylene oxide unit results in a product with significant cohesive properties, it is reasonable to assume that POE (3) lauryl alcohol would also result in high fiber to fiber friction. It is sold by Ethox as Ethal® 326 [14][24]. The aspects of this compound are listed in Table 5–7.

Table 5–7

Typical Properties of POE (3) Lauryl Alcohol

Property	Value
Acid Value	1 max
Hydroxyl Value	165–175
Cloud Point, 1%, °C	<25
Color, APHA	50 max
pH, 5% in Distilled Water	5.0–6.5
Moisture, %	1 max

2. POE (4) Lauryl Alcohol. According to data developed by ICI [25] and displayed in Table 5–3, the four EO mole alcohol ether in a finish results in high fiber to fiber friction and high stick–slip on nylon, polyester, and acetate yarn and high stick–slip on polypropylene and rayon fiber. This material is one of the most effective cohesive agents in the component library. It is water and mineral oil dispersible and a very good mineral oil emulsifier, so it is also effective in coning oil preparations to assist in removal of the oils in scouring procedures. When blended with other emulsifiers the POE (4) ether is a valuable emulsifier for fatty acid esters. This product is manufactured by Ethox and supplied under the name Ethal® LA–4 [14][26], by Henkel with the designation Trycol® 5882 [16][27], by ICI with the name Brij® 30 [28], and by Witco with an identification of Witconol® 5966 [29]. A composite of a list of the properties of this component is displayed in Table 5–8.

Table 5–8

Typical Properties of POE (4) Lauryl Alcohol

Property	Value
Acid Value	1 max
Hydroxyl Value	150–165
Cloud Point, 1%, °C	<25
Color, APHA	50 max
pH, 5% in Distilled Water	5.0–6.5
Refractive Index	1.4450–1.4550
Moisture, %	1 max

3. POE (7) Lauryl Alcohol. The seven EO mole modification of lauryl alcohol is an excellent detergent and oil dispersant, so it is probably also effective in fiber cohesion. It is furnished by Ethox as Ethal® LA–7 [14][30] and has manufacturing specifications similar to some of the other lauryl alcohol ethoxylates. Common properties are itemized in Table 5–9.

Table 5–9

Typical Properties of POE (7) Lauryl Alcohol

Property	Value
Acid Value	1 max
Hydroxyl Value	113–122
Cloud Point, 1%, °C	48–52
Color, APHA	100 max
pH, 5% in Distilled Water	4.5–6.5
Moisture, %	1 max

4. POE (9) Lauryl Alcohol. The two additional moles of EO transform the alcohol ethoxylates from liquids to pastes. Other than that change, this item is quite similar to the preceding compound. It is made and sold by Ethox as Ethal® 926 [14][31]. The specifications are registered in Table 5–10.

Table 5-10

Typical Properties of POE (9) Lauryl Alcohol

Property	Value
Acid Value	1 max
Hydroxyl Value	96–108
Cloud Point, 1%, °C	75–80
Color, APHA	50 max
pH, 5% in Distilled Water	5.0–6.5
Moisture, %	1 max

5. POE (23) Lauryl Alcohol. Although the significantly larger level of ethylene oxide in this ether does not induce cohesive properties in nylon, polyester, polypropylene, or acrylic yarns, it is more effective for rayon and acetate. Data in Table 5-2 show [25] that for these more hydrophilic fibers the stick–slip values are high. The longer ethylene oxide chains are probably attracted to the fiber surface by a quasi–chemical or dipole mechanism because of the water in the fiber, resulting in higher cohesion. This product is also an excellent detergent and co–emulsifier and finds application as a rewetting agent with damp fabrics. The normal state at room temperature is a solid with a melting point of about 40°C. A component furnished by Henkel as Trycol® 5964 [16][32] and by ICI as Brij® 35 [28]. Representative properties of POE (23) lauryl ether are given in Table 5-11.

Table 5-11

Typical Properties of POE (23) Lauryl Alcohol

Property	Value
Acid Value	1 max
Hydroxyl Value	—
Cloud Point, 5% Saline, °C	93
Color, Gardner	1
pH, 5% in Distilled Water	5.0–6.5
Moisture, %	1 max

POE Oleyl Alcohols, CAS# 9004–98–2

Exceptions to the general rule that alkyl chain lengths of nine to fourteen carbon atoms result in higher fiber to fiber friction and more pronounced stick–slip are the ethoxylated oleyl alcohols. This is probably the result of the *cis*–configuration of the unsaturated chain, represented structurally as [33]:

$$CH_2OH$$
$$CH_3$$

Oleyl alcohol can be synthesized using the techniques described for decyl and lauryl alcohol and is ethoxylated by the standard procedures. There are three commercial compounds available with 10, 20, and 23 moles of ethylene oxide reacted with the alcohol.

Unfortunately, the unsaturated alkyl chain will cross–link in oxidative conditions of heat and the presence of air and form undesirable varnishes. Use of these excellent cohesive agents should be limited to processes in which heat is not a factor.

1. POE (10) Oleyl Alcohol. This product can be quite valuable if a medium level of fiber to fiber friction (F–F) is wanted and the formulator desired high stick–slip. If POE (10) oleyl alcohol is applied to nylon, polyester, polypropylene, or rayon yarn those results can be expected. High F–F and high stick–slip is found on acetate, but only medium stick–slip on acrylic fiber [25]. The specifications for this ether are recorded in Table 5–12.

Table 5–12

Typical Properties of POE (10) Oleyl Alcohol

Property	Value
Acid Value	0.5 max
Hydroxyl Value	83–95
Cloud Point, 1%, °C	56–62
Color, Gardner	4 max
pH, 5% in Distilled Water	5.0–6.5
Moisture, %	3 max

Manufactured by Ethox and ICI, it is sold with the trade names of Ethal® OA–10 [14][34] and Brij® 96 [28].

2. POE (20) Oleyl Alcohol. The 20–mole EO homolog in the oleyl alcohol series is made by Henkel using the name Trycol® 5971 [16] and ICI as Brij® 98 [28]. It has slightly different frictional properties than the 10–mole version. Fibers with this component in a finish are reported [25] to have medium fiber to fiber friction but high stick–slip on nylon, rayon, and polyester. The stick–slip on polypropylene, acrylic, and acetate is only at a medium level, so it is slightly less effective as the ethylene oxide content is increased. Water solubility is also changed since the 10–mole material is only soluble in hot water, but the additional ten moles of EO produce a compound with water solubility at room temperature. Characteristic properties are given in Table 5–13.

Table 5–13

Typical Properties of POE (20) Oleyl Alcohol

Property	Value
Cloud Point, 5% Saline, °C	87
Color, Gardner	1 max
Melting Point, °C	43
pH, 5% in Distilled Water	5.0–6.5
Moisture, %	3 max

3. POE (23) Oleyl Alcohol. The addition of three additional moles of ethylene oxide results in few changes to the preceding material, except that it is somewhat more hydrophilic. This detail is revealed in solubility data since the 20–mole EO alcohol is dispersible in butyl stearate although the 23–mole EO alcohol is insoluble in the same solvent, while water solubility is maintained. The frictional properties are about the same for both components. The compound is manufactured by Ethox and sold as Ethal® OA–23 [14][35] and by Henkel using the designation Trycol® 5972 [16]. Various properties are shown in Table 5–14.

Table 5–14

Typical Properties of POE (23) Oleyl Alcohol

Property	Value
Acid Value	1.0 max
Hydroxyl Number	56–66
Cloud Point, 5% Saline, °C	89
Color, Gardner	2 max
Melting Point, °C	47
pH, 5% in Distilled Water	5.0–6.5
Moisture, %	1.5 max

5.3.2 Derivatives of Alcohol Ether Ethoxylates

There are many reasons for placing some organic functional groups on the end of the ethylene oxide chain reacted with either alcohols or acids. These include the reduction of the attack on polyurethane and an increase in the fiber to fiber friction and a high level of stick–slip [36]. The addition of these groups to the chain is commonly called *capping*. Normally these capping groups are alkyl or propylene oxide.

Although alkyl groups such as ethyl, isopropyl or t–butyl have been utilized, one of the most common alkyl capping moieties is methyl. A reaction that may be used to prepare an alkyl capped ethoxylated alcohol is called the Williamson synthesis [23]. A generalized reaction would be:

$$RO(CH_2CH_2)_x H + Na \longrightarrow RO(CH_2CH_2O)_x^- Na^+ + 1/2\, H_2$$

The first step in this reaction is the addition of sodium metal to the ethoxylated alcohol to form a sodium alkoxide. This alkoxide is then treated with an alkyl halide to form the final methyl capped alkoxylate.

$$RO(CH_2CH_2O)_x^- Na^+ + R'Br \longrightarrow RO(CH_2CH_2O)_x R' + Na^+ Br^-$$

In some instances where the presence of trace amounts of halide ion might interfere with fiber processing or if there are other concerns about methyl as the capping group, an alternate procedure has been suggested [37]. Reactions to synthesize these compounds are analogous to the preparation of acetals or formals [33]. The reaction proceeds as:

$$H_2C{=}O \;+\; 2\,H\;OCH_2CH_2OCH_3 \longrightarrow H_2C(OCH_2CH_2OCH_3)_2 \;+\; H_2O$$

The di–(methylglycol)–formal produced in this reaction is then reacted with an ethoxylated fatty alcohol to form the final end–capped product.

$$H_2C(OCH_2CH_2OCH_3)_2 \;+\; RO(CH_2CH_2O)_xH \xrightarrow{\;H^+\;}$$

$$RO(CH_2CH_2O)_xOCH_2OCH_2CH_2OCH_3 \;+\; HOCH_2CH_2OCH_3$$

Propylene oxide end caps are prepared my adding this alkoxy compound to the reaction vessel after the ethylene oxide reaction is complete.

POE (5) Lauryl Alcohol, Formal Methoxyethylene Glycol Capped,
CAS# 73507–40–1

This compound, prepared by following the procedure of Kleber [37], is an excellent cohesive agent for most fiber types, but especially on yarn for which a low viscosity material is required. A probable structure of the alcohol derivative is:

$$CH_3(CH_2)_{11}O{-}(CH_2CH_2O)_5{-}CH_2{-}O{-}CH_2CH_2OCH_3$$

Although the base alcohol is written as lauryl, it is probably the C_{12}–C_{14} mixture normally designated as lauryl alcohol [37]. Classical properties are presented by Table 5–15.

Table 5–15

Typical Properties of POE (5) Lauryl Alcohol, Formal Methoxyethylene Glycol Capped

Property	Value
Density, g/cm³	0.97
Melting Point, °C	5
pH, 5% in Distilled Water	7–9
Flash Point, Open Cup, °C	215

This alcohol is a product of Hoechst–Celanese and it is sold as Afilan® GEV [38].

POE (6) Lauryl Alcohol, Methyl Capped, CAS# 68908–51–0

This alcohol derivative is somewhat similar to the preceding product, and, as expected, is also a very good cohesive agent. It is probably made by the Williamson synthesis. The structure is:

$$CH_3(CH_2)_{11}O(CH_2CH_2O)_6CH_3$$

As with other lauryl alcohols, the chain length is probably twelve to fourteen carbon atoms. The material is sold by ICI as Cirrasol® G–3886 [39][40]. Sundry properties are listed in Table 5–16.

Table 5–16

Typical Properties of POE (6) Lauryl Alcohol, Methyl Capped

Property	Value
Acid Number	3 max
Hydroxyl Number	35 max
Color, Gardner	2 max
Viscosity, cSt @ 25°C	22
Water, %	2 max

Formulators should include this compound in a finish with caution, since aerosols of the material are reported to have an anaesthetic action on mucous membranes [41]. In most situations aerosols are not formed, but if the finish is applied to the yarn immediately prior to passing through a device for air interlacing, the formation of droplets is possible. Vapors created by exposure to heated surfaces are not considered hazardous.

(POE)$_x$(POP)$_y$ Pentaerythritol, Formal Methoxyethylene Glycol Capped, CAS# Proprietary

One of the most innovative cohesive agents/boundary lubricants synthesized in recent years is this polyol derivative. It is manufactured by reacting ethylene oxide and propylene oxide with pentaerythritol and then capping with the formal methoxyethylene glycol [37]. The most probable structure is:

$$C[CH_2O(CH_2CH_2O)_x(CH_2\underset{\underset{H}{|}}{\overset{\overset{CH_3}{|}}{C}}O)_yCH_2OCH_2CH_2OCH_3]_4$$

This is a product from Hoechst–Celanese called Afilan® ET [42] and some properties are displayed in Table 5–17

Table 5–17

Typical Properties of (POE)$_x$(POP)$_y$ Pentaerythritol, Formal Methoxyethylene Glycol Capped

Property	Value
Pour Point, °C	–6
Flash Point, °C	250
Density, 20°C (g/cm³)	1.09
pH, 1% in Distilled Water	7
Solubility in Water	Soluble @ 20°C

5.3.3 Ethoxylated Fatty Acids and Polyethylene Glycol Acid Esters

A preceding section (5.3.1) describes one type of nonionic surfactant, those based on an alcohol hydrophobes. Another important hydrophobe are fatty acids. The surfactant is synthesized by reacting the fatty acid with ethylene oxide or by esterifying with a previously prepared polyethylene glycol The reaction of ethylene oxide with the fatty acid to form POE esters is [43]:

$$RCOOH + n \; H_2C \underset{O}{\overset{}{\diagdown \diagup}} CH_2 \xrightarrow{\;\;OH^-\;\;} RCOO(CH_2CH_2O)_nH$$

This alkali catalyzed esterification of the fatty acids by ethylene oxide is accompanied by a transesterification reaction so that the product is a mixture of the polyethylene glycol monoester, the POE diester, and polyethylene glycol.

$$2 \, R\,COO(CH_2CH_2O)_nH \rightleftharpoons RCOO(CH_2CH_2O)_nOCR + H\,O(CH_2CH_2O)_nH$$

If the diester and free polyethylene glycol are not minimized by the manufacturer, the effectiveness as a cohesive agent could be jeopardized. An analytical procedure for measuring the amount of free polyethylene glycol in poly(ethylene glycol) esters was described in Chapter 3.

An alternate technique to prepare polyoxyethylene esters of fatty acids is by esterification with polyethylene glycol, which is available in a number of molecular weight ranges.   The reaction of these acids with poly–ethyleneglycol to form the ester and water is reversible, and since polyethylene glycol has two terminal hydroxy groups available for ester formation, the diester is also formed [43].

$$RCOOH + HO(CH_2CH_2O)_nH \rightleftharpoons RCOO(CH_2CH_2O)_nH + H_2O$$

$$2\,RCOOH + HO(CH_2CH_2O)_nH \rightleftharpoons RCOO(CH_2CH_2O)_nOCR + 2\,H_2O$$

Relative contents of monoester and diester in the equilibrium mixture depends upon the ratio of the reactants. An excess of polyethylene glycol is normally used to insure a high yield of the monoester, while a higher molar ratio of the fatty acid results in a predominance of the diester.

To insure a shift of the equilibrium reaction to maximize the product yield, the water generated is removed by either forming an azeotrope with a water–immiscible solvent, a water miscible solvent followed by drying, or by reacting under vacuum. The reaction conditions are at temperatures ranging from 100 to 250°C, depending upon the thermal stability of the components, and using a sulfuric acid catalyst.

The following information describes many of the fatty acid–polyoxy–ethylene and polyethylene glycol esters that are used in fiber finishes.

POE and PEG Esters of Pelargonic Acid, CAS# 31621–91–7

Pelargonic, a C_9 fatty acid, is usually prepared either by the oxidation of oleic acid or through the oxo process (See Chapter 4, Section 4.1.3). This nine–carbon fatty acid is then esterified with either polyethylene glycol or by reaction with ethylene oxide. As with other low carbon chain acids and alcohols, the pelargonates are very effective cohesive agents, resulting in a high fiber to fiber friction and a high stick–slip. The ethoxylated fatty alcohols and PEG esters do tend to attack poylurethane parts on texturing equipment, probably because of the terminal primary hydroxyl group. The structure of this ester is:

$$CH_3(CH_2)_7-\overset{\overset{\textstyle O}{\|}}{C}-O-(CH_2CH_2O)_xH$$

The esters that are available in commercial quantities are those in which x is about 6.8, equivalent to PEG 300, or 9, which could be either PEG 400 or POE 9.

1. PEG (300) Pelargonate. If the polyethylene glycol has an average molecular weight of 300, and since the formula weight of ethylene oxide is 44, then about 6.8 moles of EO form the PEG 300 that is esterified with the acid. We must remember that this is an average molecular weight and the actual range of ethylene oxide units in PEG 300 is about two to ten. Chromatographic analyses show this distribution and that the average number of ethylene oxide units in solution are between six and seven [44]. PEG 300 pelargonate is manufactured by Henkel and sold as Emerest® 2634 [45][46]. The properties are given in Table 5–18.

Table 5–18

Typical Properties of PEG 300 Monopelargonate

Property	Value
Color, Gardner	2 max
pH, 5% in Distilled Water	4.5–6.5
Acid Value	5 max
Hydroxyl Value	115–135
Saponification Value	125–135
Water, %	1 max

2. PEG (400) Pelargonate and POE (9) Pelargonate. These two esters, synthesized by the two different reactions outlined on page 150, are essentially identical, although there might be small differences. The ethylene oxide content is about nine, but it can range from 3 to 17 [44]. Whatever the method of manufacturing, this distribution is about the same. Since the procedure of reacting polyethylene glycol produces water that must be removed, the combination of heat and vacuum or an azeotrope may remove some low molecular weight fractions, resulting in an increase in viscosity. The PEG (400) ester of pelargonic acid is a product of Henkel and supplied with the trade name Emerest® 2654 [45][47] and the POE (9) ester made by Ethox and sold as Ethox® 1122 [48][49]. General properties of PEG 400 pelargonate are listed in Table 5–19.

Table 5–19

Typical Properties of PEG 400 Monopelargonate

Property	Value
Color, APHA	100
pH, 5% in Distilled Water	4.5–6.5
Acid Value	1.5 max
Hydroxyl Value	95–105
Saponification Value	98–105
Viscosity, @ 25°C, cSt	60
Water, %	2 max

POE and PEG Esters of Lauric Acid, CAS# 9004–81–3

Dodecanoic acid, commonly called lauric acid, is the lowest molecular weight of the fatty acids found, as the glyceride, in coconut oil and palm kernel oil in amounts greater than 10%. These fats may be hydrolyzed with water, usually in the presence of an acid or alkaline catalyst, to yield glycerol and a mixture of fatty acids. The acids may be purified by distillation under vacuum [43]. This step usually removes any residual glycerides and improves the color and odor of the fatty acid.

Like the pelargonates, the laurate esters are excellent cohesive agents. On nylon, polyester, polypropylene, rayon, and acetate fibers, laurate esters show a medium level of fiber to fiber friction, but display a high stick–slip. When the ester is applied to acrylic fiber a medium level of friction and only a medium stick–slip is attained [25]. Laurate esters are available with about 5, 8, 9, 14, and 23 moles of ethylene oxide, either as the PEG or POE ester. The structure of the laurate esters is:

$$CH_3(CH_2)_{10}-\overset{\overset{\textstyle O}{\|}}{C}-O-(CH_2CH_2O)_x H$$

1. PEG (200) and POE (5) Monolaurate. In the structure above, if the number of ethylene oxide repeat units is approximately five, this compound is the result of the esterification. It must be remembered that the actual range of repeat units is a normal distribution and that only the average is about five. Normal properties of this ester are shown in Table 5–20.

Table 5–20

Typical Properties of PEG 200 Monolaurate

Property	Value
Color, Gardner	2 max
pH, 5% in Distilled Water	4.0–7.0
Acid Value	3 max
Hydroxyl Value	138–148
Saponification Value	138–145
Viscosity, @ 100°F, cSt	123

PEG (200) monolaurate is made by Henkel as Emerest® 2620 [45][50] and Stepan as Kessco® PEG 200 ML [51]. The polyethylene oxide analog, POE (5) monolaurate is a product of Ethox with the name Ethox® ML-5 [48] [52] and Witco as Witconol® 2620 [53].

2. PEG (400) and POE (9) Monolaurate. One of the most widely manufactured nonionic surfactants used as a fiber lubricant is PEG (400) monolaurate. It is another very effective cohesive agent but it also is results in an attack on the polyurethane parts of textile equipment. The PEG ester is made by Henkel and sold as Emerest® 2650 [45][54] and by Stepan as Kessco® PEG 400 ML [51][55]. The POE ester is a product from Ethox with the name Ethox® ML-9 [48][56], from ICI as Cirrasol® G-2109 [57], and from Witco as Witconol® 2650 [58]. Typical properties of this ester are presented in Table 5-21.

Table 5-21

Typical Properties of PEG 400 Monolaurate

Property	Value
Color, Gardner	2 max
pH, 5% in Distilled Water	4.0–6.0
Acid Value	3 max
Hydroxyl Value	85–100
Saponification Value	88–98
Iodine Value	2.5 max
Viscosity, @ 100°F, cSt	41
Water, %	2 max

3. PEG (600) and POE (14) Monolaurate. This ester is quite similar to the laurate with nine moles of ethylene oxide, but the five additional repeat units produce a more hydrophilic product. It is a liquid with a pour point of 14°C. This material, as with all of the other laurate esters, can be used as an emulsifier in addition to being a cohesive agent. PEG (600) monolaurate is sold by Henkel as Emerest® 2661 [45] and by Stepan as Kessco® PEG 600 ML [51][59]. The ester is made by Ethox by reacting with fourteen moles of ethylene oxide with lauric acid and sold as Ethox® ML-14 [48][60]. Usual properties are itemized by Table 5-22.

Table 5–22

Typical Properties of PEG 600 Monolaurate

Property	Value
Color, Gardner	2 max
pH, 5% in Distilled Water	4.0–7.0
Acid Value	5 max
Hydroxyl Value	65–75
Saponification Value	63–70
Viscosity, @ 100°F, cSt	60
Water, %	1 max

PEG and POE Esters of Oleic Acid, CAS# 9004–96–0

The major monounsaturated fatty acid is oleic, occurring in large amounts (greater than 30%) in beef tallow, corn oil, cottonseed oil, lard, olive oil, palm oil, and peanut oil. Since it is a liquid with a melting point of about 14°C, it is readily isolated from the mixtures of fatty acids that are products of hydrolyzed fats and oils. The effectiveness of ethylene oxide adducts of oleic acid as cohesive agents have been reported [61][62] and these studies showed that ethoxylated oleic acid had the highest level of static and dynamic coefficient of friction among all of the components investigated. This is again probably due to the *cis*– structure as:

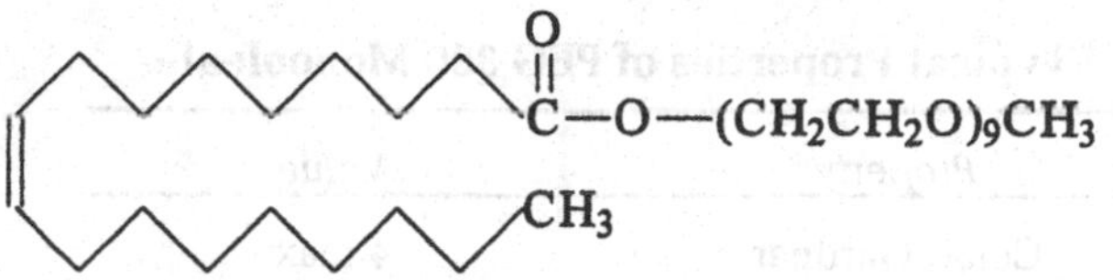

On hot surfaces unsaturated bonds can cross–link resulting in varnishes that seriously inhibit fiber processing. If heat is not involved, these materials should be considered as finish additives.

1. PEG (200) and POE (5) Monooleate. When PEG (200) or five moles of ethylene oxide are esterified with oleic acid, this component is formed. Because of the long alkyl hydrophobic chain and the short hydrophilic group, this component is only water dispersible, not water soluble, but it can be a co–emulsifier for mineral oil or ester lubricants. The PEG (200) ester is made by Henkel with the trade name Emerest® 2624 [45][63], and by Stepan with the name Kessco® 200 MO [51]. The ethylene oxide ester is made by Ethox as Ethox® MO–5 [48][64]. Various properties of this component are described in Table 5–23.

Table 5–23

Typical Properties of PEG 200 Monooleate

Property	Value
Color, Gardner	5 max
pH, 5% in Distilled Water	5.0–8.0
Acid Value	2 max
Hydroxyl Value	115–125
Saponification Value	115–121
Viscosity, @ 100°F, cSt	34
Water, %	1 max

2. PEG (300) and POE (7) Monooleate. This ester is a slightly more hydrophilic oleate, but it also is only water dispersible. It is made by Henkel and sold as Emerest® 2632 [45][65]. Some properties are listed in Table 5–24.

Table 5–24

Typical Properties of PEG 300 Monooleate

Property	Value
Color, Gardner	4 max
pH, 5% in Distilled Water	5.0–7.0
Acid Value	2 max
Hydroxyl Value	88–102
Saponification Value	93–99
Viscosity, @ 100°F, cSt	46
Water, %	1 max

3. PEG (400) and POE (9) Monooleate. Another in the homologous ethylene oxide oleate ester series, this ester does not contain enough ethylene oxide to make it water soluble. The PEG (400) ester is a product of Henkel and sold under the name Emerest® 2646 [45][66] and by Stepan as Kessco® PEG 400 MO [51]. Esters made with ethylene oxide are produced by Ethox as Ethox® MO-9 [48][67]. Table 5-25 gives the properties of this oleate ester.

Table 5-24

Typical Properties of PEG 400 Monooleate

Property	Value
Color, Gardner	4 max
pH, 5% in Distilled Water	5.0-7.5
Acid Value	2 max
Hydroxyl Value	77-90
Saponification Value	70-88
Viscosity, @ 100°F, cSt	52
Water, %	1 max

4. POE (10) Monooleate. This ester has been available for many years since it was originally made by Trylon Chemical, but is now sold by Henkel as Trydet® 2676 [45][68]. As expected, it is very similar to the preceding ester, POE (9) monooleate. The properties are shown in Table 5-25.

Table 5-25

Typical Properties of POE (10) Monooleate

Property	Value
Color, Gardner	3 max
Acid Value	2 max
Saponification Value	77-82
Viscosity, @ 100°F, cSt	54
Water, %	1.5 max

5. PEG (600) and POE (14) Monooleate. The reaction of the 600 average molecular weight polyethylene glycol or fourteen moles of ethylene oxide with oleic acid yields an ester with enough hydrophilic character to be water soluble. Besides being an effective cohesive agent, this ester is useful as a water soluble emulsifier and detergent. PEG (600) oleate is made by Henkel as Emerest® 2660 [45][69] and Stepan as Kessco® PEG 600 MO [51]. The ester made with nine moles of EO is a product of Ethox and sold as Ethox® MO–14 [48][70]. Characteristic properties are registered in Table 5–26.

Table 5–26

Typical Properties of PEG 400 Monooleate

Property	Value
Color, Gardner	4 max
pH, 5% in Distilled Water	5.0–7.0
Acid Value	2 max
Hydroxyl Value	62–73
Saponification Value	60–69
Viscosity, @ 100°F, cSt	75
Water, %	1 max

5.3.4 Derivatives of Fatty Acid Esters of Ethylene Oxide

As discussed in the section on ethoxylated alcohol ether derivatives, the terminal hydroxyl group of the fatty acid esters can be blocked to reduce the attack on polyurethane parts of textile processing equipment. Not only is this reactivity reduced, but foaming is decreased and stability to caustic materials is improved [71]. The reactions that may be used to prepare the blocked hydroxyl groups are described in Section 5.3.2. Individual hydroxyl capped components are described below.

POE (x) Pelargonate, Methoxy Capped, CAS# Proprietary

Esterification of nonanoic, or pelargonic, acid with ethylene oxide is accomplished by the procedures previously described [43]. The methyl end group is added, probably through a technique similar to the Williamson synthesis [23]. This component is made by Hoechst–Celanese and sold as Afilan® POM [72]. Usual properties of the material are recorded in Table 5–27 and the structure is:

$$CH_3(CH_2)_7-\overset{\overset{\displaystyle O}{\|}}{C}-O-(CH_2CH_2O)_xCH_3$$

Table 5-27

Typical Properties of POE (x) Pelargonate, Methoxy Capped

Property	Value
Color, Gardner	3 max
Acid Value	5 max
Hydroxyl Value	10 max
Saponification Value	90–110
Water, %	1 max

PEG (400) Pelargonate, Methoxy Capped, CAS# 109909-40-2

As proposed for the preceding material, which is very similar to this component, the capped pelargonates are very effective in building fiber cohesion. The PEG ester is made with typical procedures and the end group added through the Williamson synthesis [23]. The structure is:

$$CH_3(CH_2)_7-\overset{\overset{\displaystyle O}{\|}}{C}-O-(CH_2CH_2O)_9CH_3$$

This chemical is a product of Henkel and sold with the trade name Emery® 6724 [73]. Typical properties are shown in Table 5–28.

Table 5-28

Typical Properties of PEG (400) Pelargonate, Methoxy Capped

Property	Value
Color, Gardner	2 max
Viscosity, 100°F, cSt	20
Acid Value	3 max
Hydroxyl Value	8 max
Saponification Value	95–105

POE (9) Pelargonate, Ethoxy Capped, CAS# Proprietary

ICI Americas has reported the substitution of an ethoxy blocking group for a methoxy group results in some reduction in the tendency for materials to attack polyurethane while maintaining high fiber to fiber friction. This conclusion may be reached from Table 5–29 [74].

Table 5–29

Frictional Properties of Lubricants on Polyester Yarn (250/50) and Urethane Swelling

Product	Yarn to Yarn Friction, grams 1 cm/min	Yarn to Steel Friction, grams 50 m/min	Urethane Swelling, %
POE (9) Laurate	17 ± 9	300	5.7
Ethoxylated Glycerol Laurate	27 ± 7	388	4.9
Poe (9) Oleate, Methyl Ether	19 ± 11	325	9.7
POE (9) Laurate, Methyl Ether	18 ± 10	225	13.2
POE (9) Pelargonate, Ethyl Ether	20 ± 10	288	11.5

(Reference [74]. Reprinted by permission of ICI Americas)

The ethyl ether capped ethoxylated pelargonate ester is produced by ICI Americas as Cirrasol® G–3890 [75][76]. The structure is:

$$CH_3(CH_2)_7 - \overset{\overset{\textstyle O}{\textstyle \|}}{C} - O - (CH_2CH_2O)_9CH_2CH_3$$

Some properties of G–3890 are displayed in Table 5–30.

Table 5–30

Typical Properties of POE (9) Pelargonate, Ethoxy Capped

Property	Value
Viscosity, 25°C, cSt	45
pH, 5% in Distilled Water	5–7
Acid Value	10 max
Hydroxyl Value	15 max
Saponification Value	80–92
Water, %	1 max

POE (9) Laurate, Methoxy Capped, CAS# 9006–27–3

Fiber cohesion is certainly enhanced by using POE laurate esters in the finish composition, but the methyl ether revision does not reduce polyurethane swelling but actually increases that property (Table 5–29). This component is made by ICI with the trade name Cirrasol® G–1121 [75]. The structural formula is:

$$CH_3(CH_2)_{10}-\overset{\overset{\text{O}}{\|}}{C}-O-(CH_2CH_2O)_9CH_3$$

Several general properties of the POE (9) laurate ester, methyl ether capped are displayed in Table 5–31.

Table 5–31

Typical Properties of POE (9) Laurate, Methoxy Capped

Property	Value
Viscosity, 25°C, cSt	36
pH, 5% in Distilled Water	4.5–7.5
Acid Value	10 max
Hydroxyl Value	15 max
Saponification Value	80–110
Water, %	1 max

PEG (400) and POE (9) Oleate, Methoxy Capped, CAS# 34397–99–4

One might assume that the eighteen carbon oleic acid chain might reduce polyurethane swelling, but according to Table 5–29, this is certainly a wrong assumption. This, as with the cohesion induced upon use of oleates, is due to the *cis–* structure.

The compound is made by two finish component suppliers, Henkel who sells the PEG ester as Emery® 6799 [77] and ICI makes the POE ester as Cirrasol® G–1120 [75]. The properties are listed in Table 5–32.

Table 5–32

Typical Properties of PEG (400) Oleate, Methoxy Capped

Property	Value
Viscosity, 25°C, cSt	37
Color, Gardner	8
pH, 5% in Distilled Water	4.5–6.5
Acid Value	8 max
Hydroxyl Value	15 max
Saponification Value	80–90
Water, %	1 max

POE (20) Glycerol Monolaurate, CAS# 57107–95–6

Glycerol monolaurate can be prepared by the esterification of lauric acid with glycerol under controlled conditions. The esterification occurs on one of the primary hydroxyl groups, leaving two active groups available for additional reaction. Twenty moles of ethylene oxide are then added, probably at the other primary hydroxyl group, forming an ether ethoxylate. The resulting compound is:

$$\begin{array}{ccc}
\text{H} & & \text{O} \\
| & & \| \\
\text{H}-\text{C}-\text{O}-&\text{C}&-(\text{CH}_2)_{10}\text{CH}_3 \\
| & & \\
\text{H}-\text{C}-\text{OH} & & \\
| & & \\
\text{H}-\text{C}-\text{O}-&(\text{CH}_2\text{CH}_2\text{O})_{20}\text{H} \\
| & & \\
\text{H} & &
\end{array}$$

According to the data in Table 5–29, this material produces one of the highest levels of fiber to fiber friction and high stick–slip on nylon and polyester, but is not nearly as effective on other fibers [25]. It is made by ICI as Cirrasol® GM [78]. The properties are given by Table 5–33.

Table 5–33

Typical Properties of POE (20) Glycerol Monolaurate

Property	Value
Viscosity, 25°C, cSt	24
Color, Gardner	3
Acid Value	4 max
Hydroxyl Value	93 max
Saponification Value	41–55
Iodine Number	2 max

5.4 FIBER LUBRICANTS AND SOFTENERS

While the chemical components designed to be cohesive agents increase the level of fiber to fiber friction and the stick–slip, fiber lubricants and softeners reduce those properties. This is especially apparent with staple fiber. If a bundle of staple fiber is treated with a cohesive agent it will result in an increase in a property called scroop. This attribute is so named because of the sound that the staple bundle makes when it is squeezed. Scroop is caused by the resistance to fiber movement that results from the increase in friction between individual fibers. If a fabric is made from fiber and yarn that possess high scroop levels, that fabric has a harsh, undesirable feel, called the fabric handle, or hand. Fabrics with an improved hand may be obtained if a fiber lubricant, or softener, is applied at some stage in the manufacturing process [79].

In many instances, the fiber to fiber friction must be changed during yarn processing. As an example, consider the steps involved in converting staple fiber to yarn. A converter normally receives the staple fiber, either natural or man–made, in a bale. The bale is opened in a procedure immediately before carding. Sometimes, a lubricant is sprayed onto the bales to aid the mechanical separation and blending of the fiber. Loose fiber is then fed into the card input chute and carding begins. If the proper balance between cohesion and lubrication is not achieved, this operation will fail. The staple must have enough lubricant to assist the web in releasing from the card drum, but as the web is collected into a loose sliver, it must have enough cohesion to prevent the sliver from falling apart. During further processing of sliver to roving, roving to yarn, and yarn to fabric, a balance between cohesion and lubrication must be maintained.

Fiber lubricants, which we will define as ethoxylated long chain saturated fatty acids and their derivatives, and softeners are depicted in the following sections. Ethoxylated fatty alcohols will be defined as emulsifiers and described in the next chapter. As with the cohesive agents, fiber lubricants also can be useful in oil–in–water emulsion preparation.

5.4.1 Fatty Acid Ethoxylates as Fiber Lubricants

POE (7) Palmitate, CAS# 9004–94–8

Palmitic acid, isolated from the hydrolysis of palm oil or certain animal fats, is available in very pure grades [80][81]. When palmitic acid is esterified with seven moles of ethylene oxide, a product sold by Hoechst–Celanese as Afilan® 7PFA [82] is the resultant compound. Although it has some emulsification ability, the primary use is as a fiber lubricant. The structure is:

$$CH_3(CH_2)_{14} - \overset{\overset{\displaystyle O}{\|}}{C} - O - (CH_2CH_2O)_7H$$

Several properties of this palmitate ester are itemized in Table 5–34.

Table 5–34

Typical Properties of POE (7) Palmitate

Property	Value
pH, 5% in Distilled Water	4–7
Acid Value	<3
Density, g/cm³	0.91

*Ethylene Oxide–Propylene Oxide Ester of Coconut Fatty Acids,
Methoxy Capped, CAS# Proprietary*

Coconut fatty acids may be purchased in four grades, distilled, hydrogenated, stripped and hydrogenated stripped [81]. The fatty acid distribution of these grades is displayed in Table 5–35.

Table 5–35

Fatty Acid Distribution of Coconut Acids

Acid Description	Saturated						Unsat.
	C_8	C_{10}	C_{12}	C_{14}	C_{16}	C_{18}	C_{18}
Distilled	7	6	50	19	9	2	6
Hydrogenated	7	6	51	18	9	2	5
Stripped	—	—	55	24	12	1	4
Hydrogenated Stripped	1	1	55	23	12	8	—

(Reference [81]. Reprinted by permission of Witco Corporation)

If the amounts of the individual acids in each row of Table 5–35 are added, the total is not 100. There are very small amounts of other acids present which must be included for a material balance to be obtained.

An ethylene oxide–propylene oxide adduct of coconut acids has been synthesized by Hoechst–Celanese and that ester is sold as Afilan® 1080 [83]. The addition of propylene oxide improves the lubricating ability of the ester if it were to be compared to an ester formed only with ethylene oxide. Table 5–36 gives the typical properties of Afilan® 1080 and the probable structure is:

$$\underset{\substack{\|\\O}}{CH_3(CH_2)_{10\text{-}14}\ C}-O-(CH_2CH_2O)_x(CH_2\overset{CH_3}{\underset{|}{C}}\ O)_yCH_3$$

Table 5–36

Typical Properties of Ethoxylated–Propoxylated Coconut Fatty Acids, Methoxy Capped

Property	Value
Viscosity, 25°C, cSt	49
pH, 1% in Demineralized Water	4–6
Acid Value	2.5 max
Saponification Value	59–63

POE (x) C_{16}–C_{18} Fatty Acid, CAS# 68989–61–7

This compound is the ethylene oxide ester of a high stearic grade of a palmitic–stearic mixture. This particular raw material has very low unsaturation, and consequently a low iodine number. The ester is a paste to wax solid at room temperature and is an excellent fiber lubricant. The structure is:

$$\underset{\substack{\|\\O}}{CH_3(CH_2)_{14\text{-}16}\ C}-O-(CH_2CH_2O)_xH$$

This ester is made by Hoechst–Celanese and sold under the trade name Afilan® HSGR [84]. The specifications and some typical properties are listed in Table 5–37.

Table 5–37

Typical Properties of POE (x) C_{16}–C_{18} Fatty Acid

Property	Value
Melting Range, °C	28–32
pH, 1% in demineralized water	6.5–7.5
Acid Value	1 max
Iodine Number	1 max
Color, Gardner, Molten	1 max
Water, %	3 max
Saponification Value	73–83

POE and PEG Esters of Stearic Acid, CAS# 9004–99–3

Stearic acid–ethylene oxide condensates are widely recognized as effective fiber lubricants in which they are employed as a co–finish component with pigments, resins, and other auxiliaries [85]. As described in Section 4.1.2 of Chapter 4, stearic acid is obtained by the hydrolysis of animal fats, especially beef tallow. The POE and PEG esters are made using standard techniques.

1. PEG (200) and POE (4) Monostearate. Many of the stearic acid–EO adducts are hydrophobic and water dispersible, including this version of the series. It is made by Henkel as Trydet® 2670 [45][86] and by Stepan as Kessco® PEG 200 MS [51]. The structure of this compound is:

$$CH_3(CH_2)_{16}-\overset{\displaystyle O}{\overset{\displaystyle \|}{C}}-O-(CH_2CH_2O)_5H$$

Classical properties are recorded in Table 5–38.

Table 5–38

Typical Properties of PEG (200) and POE (5) Esters of Stearic Acid

Property	Value
pH, 5% in distilled water	5.0–8.0
Acid Value	3 max
Hydroxyl Value	100–115
Color, Gardner, Molten	1 max
Water, %	1.5 max
Saponification Value	105–115

2. POE (8) Monostearate. This stearate ester is a good wax and oil emulsifier and also a good fiber lubricant. It is commonly used in apparel fabric processing to improve the handle or drape of the fabric [85]. The structure is analogous to the POE (5) stearate, but with an increase in the ethylene oxide content. It is made by Ethox as Ethox® MS-8 [48][87], by Henkel as Trydet® 2671 [45], by ICI as Myrj® 45 [57], and by Witco as Witconol® 2640 [29]. Properties of this ester are given in Table 5–39.

Table 5–39

Typical Properties of POE (8) Esters of Stearic Acid

Property	Value
Melting Point, °C	30
pH, 5% in distilled water	5.0–7.0
Acid Value	2 max
Hydroxyl Value	80–97
Iodine Value	0.5 max
Saponification Value	105–115
Water, %	3 max

3. PEG (400) Monostearate. Very similar to the previously described ester, this compound contains approximately one additional ethylene oxide group and is made by esterification of stearic acid with PEG 400. The ester is produced by Henkel with the trade name Emerest® 2640 [45][88] and by

Stepan as Kessco® PEG 400 MS [51]. Properties are presented by Table 5–40.

Table 5–40

Typical Properties of PEG (400) Monostearate

Property	Value
Melting Point, °C	32
pH, 5% in Distilled Water	4.0–7.0
Acid Value	3 max
Hydroxyl Value	75–90
Saponification Value	82–90
Water, %	3 max

4. PEG (600) and POE (14) Monostearate. Another in the ethoxylated stearate esters, this version is slightly more hydrophilic than the PEG (400) ester, but remains only dispersible in water. The PEG (600) ester is a product of Henkel as Emerest® 2662 [45][89] and by Stepan as Kessco® PEG 600 MS [51]. The POE (14) stearate is made by Ethox with the designation Ethox® MS–14 [48]. Properties are listed in Table 5–41.

Table 5–41

Typical Properties of POE (14) and PEG (400) Monostearate

Property	Value
Melting Point, °C	40
pH, 5% in Distilled Water	4.0–6.0
Acid Value	2 max
Saponification Value	60–66
Water, %	3 max

5. PEG (1000) and POE (23) Monostearate. An increase from fourteen to twenty–three moles of ethylene oxide as the hydrophilic group for esterification with stearic acid is sufficient to make this compound water soluble. The long chains on both ends of the molecule result in a material that is valuable as a thickener in aqueous systems. It is also a good emulsifier, but most effective as a fiber lubricant. The ester synthesized by reaction with PEG 1000 is manufactured by Henkel as Emerest® 2610 [45]

[90] and by Stepan as Kessco® PEG 1000 MS [51]. Esterification with twenty–three moles of ethylene oxide is the procedure at Ethox and sold as Ethox® MS–23 [48][91]. Several properties are displayed in Table 5–42.

Table 5–42

Typical Properties of POE (23) and PEG (1000) Monostearate

Property	Value
Melting Point, °C	36
Color, Gardner, Molten	1.5 max
Hydroxyl Value	42–52
Acid Value	2 max
Saponification Value	41–49
Water, %	3 max

6. POE (40) Monostearate. Forty moles of ethylene oxide are equivalent to PEG (1800) in this ester. It may be purchased either as a solid or flaked and is effective as a stabilizer for aqueous dispersions. It is available from Ethox with the name Ethox® MS–40 [48][92], from Henkel as Emerest® 2715 [45] [93] and from ICI as Myrj® 52 [39]. With most fibers the fiber to fiber friction is low to medium, but it can cause a high stick–slip on rayon [25].

It is also recognized as a direct food additive in defoaming agents in the United States and is Generally Recognized as Safe (GRAS) by many national and international organizations. Extensive feeding studies have been conducted at dietary levels of 5 to 15% rats for ten weeks with little effect [94]. Some properties of this ester are listed in Table 5–43.

Table 5–43

Typical Properties of POE (40) Monostearate

Property	Value
Melting Point, °C	50
Color, Gardner, Molten	1 max
Hydroxyl Value	27–40
Acid Value	1 max
Saponification Value	25–35
Water, %	3 max

POE and PEG Esters of Isostearic Acid, CAS# 56002–14–3

Isostearic acid is a byproduct formed during the reaction of oleic acid in the presence of a neutral clay catalyst to yield a mixture of monomer acid, dimer acid, and trimer acid [95]. The reaction to form dimer acid is [96]:

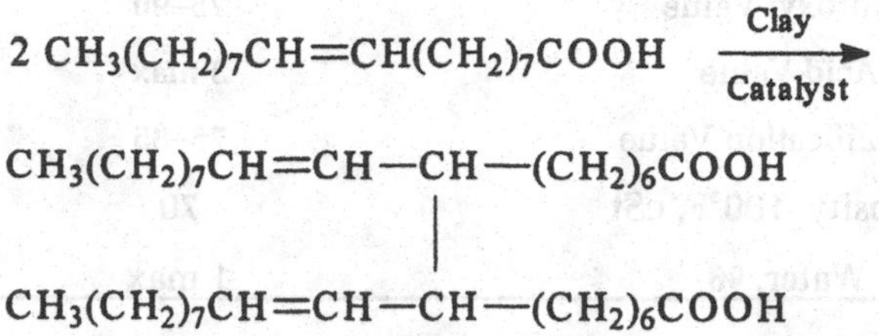

$$2 \ CH_3(CH_2)_7CH=CH(CH_2)_7COOH \xrightarrow[\text{Catalyst}]{\text{Clay}}$$

$$CH_3(CH_2)_7CH=CH-CH-(CH_2)_6COOH$$
$$|$$
$$CH_3(CH_2)_7CH=CH-CH-(CH_2)_6COOH$$

The bond between the two acids forms in many different positions and chemically several distinct structures are present although all contain 36 carbon atoms and two carboxyl groups. Monomer acid is a C_{18} acid that is also known as methyl heptadecanoic acid. The methyl group is randomly distributed along the carbon chain. When monomer acid is hydrogenated, the resultant product is isostearic acid. Isostearic acid is esterified with ethylene oxide or poylethylene glycol to produce this class of fiber lubricants. Although many structures are possible, one potential arrangement is:

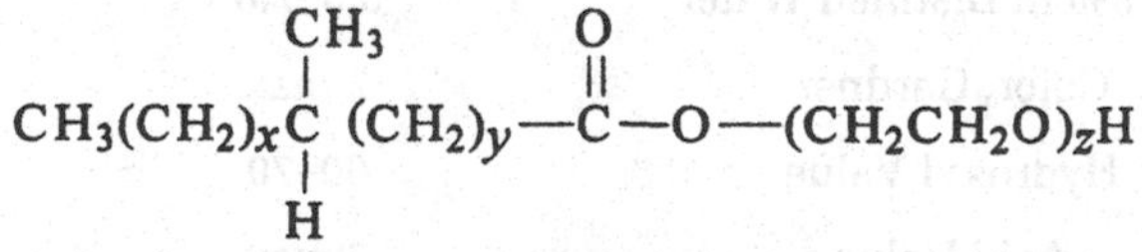

$$CH_3(CH_2)_x \overset{\overset{\displaystyle CH_3}{|}}{\underset{\underset{\displaystyle H}{|}}{C}} (CH_2)_y - \overset{\overset{\displaystyle O}{||}}{C} - O-(CH_2CH_2O)_zH$$

In the above structure x and y may range from zero to thirteen, but the total of $x + y$ always equals thirteen. In commercially available compounds, z is nine or fourteen.

1. PEG (400) and POE (9) Monoisostearate. In contrast to PEG (400) monostearate which is a solid with a melting point of 32°C, the isostearate ester is a liquid with a pour point of 10°C. Fiber to hard surface lubrication is improved with the reduction in the melting point, and the fiber to fiber friction is about the same for both ester types. PEG (400) monoisostearate is made by Henkel and sold as Emerest® 2644 [45][97] while the POE (9) ester is a product of Ethox with the name Ethox® MI–9 [48][98]. Typical properties are given in Table 5–44.

Table 5-44

Typical Properties of PEG (400) and POE (9) Monoisostearate

Property	Value
Pour Point, °C	10
Hydroxyl Value	75–90
Acid Value	3 max
Saponification Value	75–85
Viscosity, 100°F, cSt	70
Water, %	1 max

2. POE (14) Monoisostearate. A slightly more hydrophilic ester, the addition of five more moles of ethylene oxide raises the pour point to near room temperature so that in cool weather this compound is a semi–solid. It is produced by Ethox with the trade name of Ethox® MI–14 [48][99]. Some of the properties of POE (14) monoisostearate are recorded in Table 5–45.

Table 5-45

Typical Properties of POE (14) Monoisostearate

Property	Value
pH, 5% in Distilled Water	5.5–7.0
Color, Gardner	2 max
Hydroxyl Value	60–70
Acid Value	2 max
Saponification Value	58–65
Water, %	1 max

Ethylene Oxide–Propylene Oxide Ester of Oleic Acid, Methyl Capped, CAS# Proprietary

While most ethoxylated oleates are useful as cohesive agents, this ester containing both ethylene oxide and propylene oxide is effective as a fiber lubricant. This is due to the propylene oxide that reduces friction. It is a product of Hoechst–Celanese with the name Afilan® PTU [100]. A probable structure is:

Properties are shown in Table 5–45.

Table 5–45

Typical Properties of POE–POP Monooleate, Methyl Capped

Property	Value
pH, 1% in Distilled Water	3.5–5.5
Iodine Value	24–30
Viscosity, cSt @ 20°C	59–88
Acid Value	3 max
Saponification Value	57–60

5.4.2 Softeners and Softener Bases

Most of the finishes and components described earlier in this chapter and in previous chapters are recommended for use during the spinning of man–made fibers and yarns. Other finishes, called functional finishes, are chemical and mechanical processes that will impart specific performance properties to fabrics that are not normally present in those fabrics. The functional finishes are usually more durable than producer applied finishes, since a consumer will expect those additional properties to be retained throughout the life of the fabric [101]. Some of the most important functional finishes are softeners whose primary functions are to improve or enhance the hand, or feel, of the fabric and make it more appealing to the consumer. In addition, softeners may improve the abrasion resistance of the fabric where it serves as a lubricant, and they may also affect the absorbency and antistatic properties.

In order to assure the desired permanency, softeners are applied by fiber producers only to yarn that will not be processed through scouring and dyeing procedures. Examples of this type of yarn are pigmented polypropylene and nylon carpet fiber and producer dyed acrylic fiber. With other yarn types, softeners are applied at levels of about 1% during textile processing, either in the dyeing step or some later procedure.   These softeners are normally mixtures compounded by companies specializing in chemicals intended for use during fabric preparation and the formulations are proprietary to those organizations. They are all prepared from certain softener bases and can be categorized into three general classes: anionic, nonionic, and cationic. The softener bases for the three classes are described below.

1. Anionic. Anionic softeners, used primarily on cotton and rayon fabric, provide more lubricating properties than actual softening. They give fabric a smooth, slick surface and can be used to assist napping, shearing, sanforizing, and calendaring.   They also provide a finish with good absorbency and can be effective as towel softeners [102]. Generally anionic softeners have little affinity for textile substrates and many tend to develop odors during storage.

Anionic softener bases are sulfated glycerides with stearic acid soap as a dispersing agent. Because of the low cost of beef tallow, many anionics are based on that starting material and can include sulfated tallow alcohols. As shown in Table 4-3, tallow contains 25–30% C_{16} and 21–26% C_{18} saturated fatty acids and 39–42% monounsaturated C_{18} fatty acid. The reaction of tallow with sulfuric acid takes place at the double bond of the unsaturated acid. An approximate structure of sulfated tallow glyceride is:

$$
\begin{array}{c}
\qquad\quad \overset{\displaystyle O}{\overset{\|}{}} \\
CH_2-O-C-(CH_2)_7-CH_2-CH-(CH_2)_7-CH_3 \\
|\qquad\qquad\qquad\qquad\qquad\qquad | \\
|\qquad\quad O\qquad\qquad\qquad OSO_3Na \\
|\qquad\quad \| \\
CH-O-C-R \\
|\qquad\quad O \\
|\qquad\quad | \\
CH_2-O-C-R
\end{array}
$$

In this structure R can be either the C_{16} palmitic acid, the C_{18} stearic acid, or some of the other possible fatty acids present in much lower amounts. Sulfated tallow is treated with either sodium or potassium hydroxide to assist in emulsion preparation.

Another possible anionic softener is sulfated tallow alcohol, which has the advantage of low odor formation. These alcohols are prepared by hydrolysis of tallow to the acids and reduction to the alcohol followed by sulfonation. The resulting product is a mixture as:

$$CH_3(CH_2)_{14}CH_2-O-SO_3Na$$

or

$$CH_3(CH_2)_{16}CH_2-O-SO_3Na$$

and

$$CH_3(CH_2)_7-CH_2-CH-(CH_2)_7-CH_2-O-SO_3Na$$
$$|$$
$$OSO_3Na$$

2. *Nonionic.* While nonionic softeners are highly compatible with other finishing agents, such as resin finishes for permanent–press fabrics, they have low affinity for the fabric surface and must be applied by pad or spray application [103]. Nonionic softeners generally provide more lubrication then softness and often are blended with cationic components or silicones to give the desired hand.

Typical nonionic softeners are based on stearic acid esters of glycerol, ethylene glycol, or polyethylene glycol. At times, ethoxylated hydrogenated castor oil can be used as a softener for acrylic fabrics. The preparation of glycerol monostearate by the glycerolysis of tallow is shown in the following reaction [103]:

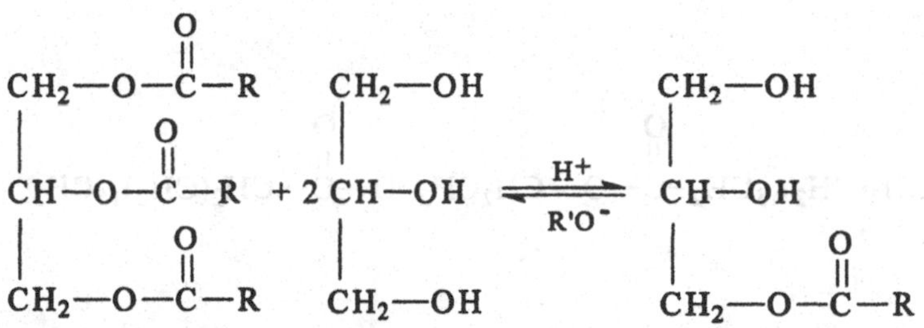

In this equation, R is the tallow fatty acid mixture described previously, but the resulting final produce is still called gylcerol monostearate (GMS). A typical softener blend with GMS might be 10% hydrogenated tallow, 10% GMS, 5% ethoxylated nonylphenol and 75% water. Other nonionic softener bases are:

Polyethylene glycol monostearate ($n= 9–25$):

$$CH_3(CH_2)_{16}-\overset{\overset{\displaystyle O}{\|}}{C}-O-(CH_2CH_2O)_nH$$

Polyethylene glycol distearate ($n=9–25$):

$$CH_3(CH_2)_{16}-\overset{\overset{\displaystyle O}{\|}}{C}-O-(CH_2CH_2O)_n-\overset{\overset{\displaystyle O}{\|}}{C}-(CH_2)_{16}CH_3$$

Ethylene glycol monostearate:

$$HO-\overset{\overset{\displaystyle H}{|}}{\underset{\underset{\displaystyle H}{|}}{C}}-\overset{\overset{\displaystyle H}{|}}{\underset{\underset{\displaystyle H}{|}}{C}}-O-\overset{\overset{\displaystyle O}{\|}}{C}-CH_2(CH_2)_{15}CH_3$$

Ethylene glycol distearate:

$$CH_3(CH_2)_{15}CH_2-\overset{\overset{\displaystyle O}{\|}}{C}-O-CH_2CH_2-O-\overset{\overset{\displaystyle O}{\|}}{C}-CH_2(CH_2)_{15}CH_3$$

Castor oil, which contains about 90% ricinoleic acid (12–hydroxy oleic acid) and lesser amounts of oleic and linoleic acids, can be hydrogenated to reduce the unsaturation and then ethoxylated to form another type of softener base. In general, the ethoxylation occurs on the hydroxyl group on the ricinoleic acid chain, but under extreme conditions polyethylene glycol can be inserted into the glyceride linkage. An approximate structure for this softener is:

$$
\begin{array}{l}
\quad\quad\quad\quad O \quad\quad\quad\quad O(CH_2CH_2)_xH \\
\quad\quad\quad\quad \| \quad\quad\quad\quad\quad | \\
CH_2-O-C-(CH_2)_{10}C-(CH_2)_5-CH_3 \\
| \quad\quad\quad\quad O \quad\quad\quad\quad O(CH_2CH_2)_yH \\
| \quad\quad\quad\quad \| \quad\quad\quad\quad\quad | \\
CH-O-C-(CH_2)_{10}C-(CH_2)_5-CH_3 \\
| \quad\quad\quad\quad O \quad\quad\quad\quad O(CH_2CH_2)_zH \\
| \quad\quad\quad\quad \| \quad\quad\quad\quad\quad | \\
CH_2-O-C-(CH_2)_{10}C-(CH_2)_5-CH_3
\end{array}
$$

The moles of ethylene oxide, $x + y + z$, are normally between 25 and 200 if the ethoxylate is to be used as a softener.

3. Cationic. Cationic softener bases are generally considered superior to the other classes of softeners because they are substantive to all types of fibers, natural and man–made. This unique property allows them to be applied by several techniques, including in the last cycle of the dye bath. The hydrophobic parts of the molecules are positively charged and consist of long–chain amides, alkanolamides, imidazolidines, and quaternary nitrogen compounds. Home laundry softeners are of the quaternary nitrogen types. Typically, quaternization is the preferred route because it reduces the tendency for yellowing. Representative cationic softeners are described below.

Stearic acid diamide, made from ethylenediamine and stearic acid, is a water–insoluble solid with a melting point of about 75°C that must be emulsified before use. The structure of this chemical, known as ethylenediamine–*bis*–stearamide, is:

$$
\begin{array}{l}
\quad\quad\quad\quad O \;\; H \quad\quad\quad\quad H \;\; O \\
\quad\quad\quad\quad \| \;\; | \quad\quad\quad\quad\; | \;\; \| \\
CH_3(CH_2)_{16}C-N-(CH_2)_2-N-C\,(CH_2)_{16}CH_3
\end{array}
$$

Fatty alkanolamides may be synthesized by reacting a fatty acid with an ethanolamine, especially diethanolamine, and perhaps followed by reaction with ethylene oxide, or by direct ethyoxylation of an amide. In the amide ethyoxlyation route, if no catalyst is used, the final product is identical with that formed with diethanolamine; however, by using a base catalyst the product is a monoalkanolamide. These reactions are:

$$RCOOH + HN(CH_2CH_2OH)_2 \xrightarrow{140\text{-}170°C} RCON(CH_2CH_2OH)_2 + H_2O$$

Diethanolamine   Diethanolamide

In commercial practice, if the diethanolamide is then reacted with ethylene oxide, a polyoxyethylene amide is formed [104]:

$$RCON\begin{smallmatrix}CH_2CH_2OH\\ \\CH_2CH_2OH\end{smallmatrix} + nH_2C\overset{O}{\overset{\diagdown}{-}}CH_2 \longrightarrow RCON\begin{smallmatrix}(CH_2CH_2O)_x + {}_1H\\ \\(CH_2CH_2O)_y + {}_1H\end{smallmatrix}$$

$$x + y = n$$

The other possible route includes the reaction of a primary or secondary amide with ethylene oxide [105], with and without a catalyst. When there is no catalyst, the hydrogen on the amide group is replaced:

$$RCONH_2 \xrightarrow{H_2C\overset{O}{\overset{\diagdown}{-}}CH_2} RCONH(CH_2CH_2OH) \xrightarrow{H_2C\overset{O}{\overset{\diagdown}{-}}CH_2} RCON(CH_2CH_2OH)_2$$

If a base catalyst is present, the ethylene oxide is added to the terminal hydroxyl in the ethylene oxide group and the reaction is:

$$RCONH_2 \xrightarrow{H_2C\overset{O}{\overset{\diagdown}{-}}CH_2} RCONH(CH_2CH_2OH)$$

$$\xrightarrow{H_2C\overset{O}{\overset{\diagdown}{-}}CH_2} RCONHCH_2CH_2OCH_2CH_2OH$$

Other cationic based softeners are typified by the imidazolines and quaternized imidazolines. Imidazoline is prepared by reacting hydrogenated tallow with a polyamine as [103]:

$$
\begin{array}{l}
CH_2-O-\overset{\overset{\displaystyle O}{\|}}{C}-R_1 \\[2pt]
CH-O-\overset{\overset{\displaystyle O}{\|}}{C}-R_1 + R_2(CH_2)_2\overset{\overset{\displaystyle H}{|}}{N}(CH_2)_2NH_2 \xrightarrow{\text{Heat}} R_2(CH_2)_2N(CH_2)_2NH \\[2pt]
CH_2-O-\overset{\overset{\displaystyle O}{\|}}{C}-R_1 \\
\end{array}
\qquad
\begin{array}{l}
O=C-R_1 \\
| \\
O=C-R_1
\end{array}
$$

| Hydrogenated Tallow | Polyamine | Amide Base |

$$
\begin{array}{c}
O=C-R_1 \\
| \\
R_2(CH_2)_2N\ (CH_2)_2NH \\
| \\
O=C-R_1
\end{array}
\xrightarrow{\text{Heat}}
\begin{array}{c}
N\diagdown\ R_1 \\
\diagup \\
N\diagdown R_2
\end{array}
$$

| Amide Base | | Imidazoline |

Imidazoline can be quaternized with dimethyl sulfate, or another salt, as:

$$
\begin{array}{c} N\diagdown R_1 \\ \diagup \\ N\diagdown R_2 \end{array}
+ \left[CH_3\right]_2 SO_4
\longrightarrow
\left[\begin{array}{c} N^+ \diagdown R_1 \\ CH_3 \\ N\diagdown R_2 \end{array} \right]
CH_3SO_4^- + H_2O
$$

| Imidazoline | Dimethyl Sulfate | Imidazoline Quaternary |

These cationic materials are not only effective as fabric softeners, but are also used in dishwashing detergents and in shampoos.

REFERENCES

1. Keesom, W. H. (1921). Van der Waals Attractive Force, *Phys. Z.*, Leipzig, Germany, **22**: 129
2. Debye, P. J. W.(1920). Van der Waals Cohesive Forces, *Phys. Z.*, Leipzig, Germany, **21**: 178
3. London, F. (1937). The General Theory of Molecular Forces, *Trans. Faraday Soc.*, **33**: 8

4. Good, R. L. (1967). Intermolecular and Interatomic Forces, in *Treatise on Adhesion and Adhesives*, (R. L. Patrick, ed.), Marcel Dekker, Inc., New York, p. 28

5. Rosen, M. J., *Surfactants and Interfacial Phenomena*, John Wiley & Sons, New York, 1978, pp. 1, 2–3,290–291

6. Hollen, N. and Sadler, J. (1973). *Textiles*, 4th Edition, Macmillian, New York, p. 12

7. Slade, P. E. and Hild, D. N. (1992). Surface Energies of Nonionic Surfactants Adsorbed onto Nylon Fiber and a Correlation with Their Solubility Parameters, *Textile Res. J.*, **62**: 535

8. Trappe, G. (1968). Polyurethane Elastomers, in *Advances in Polyurethane Technology* (Buist, J. M. and Gudgeon, H., eds), John Wiley and Sons, New York, pp. 25–61

9. Saunders, J. H. and Frisch, K. C. (1962). *Polyurethanes, Chemistry and Technology, Part 1. Chemistry*, Interscience Publishers, New York, pp. 302–303

10. Specialties from ICI for Fiber Producers (1987). *Laboratory Test Methods for Fiber Finishes: Effect of Finishes on Textile Belts*, ICI Americas, Wilmington, DE

11. Satkowski, W. B., Huang, S. K., and Liss, R. L. (1966). Polyoxyethylene Alcohols, in *Nonionic Surfactants*, (M. J. Schick, ed.), Marcel Dekker, Inc., New York, pp 86–141

12. Arné, M. (1984). Nonionic Surfactants, in *Process Economics Program, Report No. 168*, SRI International, Menlo Park, CA, pp. 29–38

13. Draves, C. Z. (1939). Evaluation of Wetting Agents–Official Methods, *American Dyestuff Rep.*, **28**: 425

14. Ethox Product Brochure (1994). *Ethoxylated Alcohols and Alkyl Phenols*, Ethox Chemicals, Inc., Greenville, SC

15. Ethox Specifications (1993). *Ethal® DA–4*, Ethox Chemicals, Inc., Greenville, SC

16. Textile Chemicals, Technical Bulletin 103A (1993). *Ethoxylated and Ethoxylated / Propoxylated Alcohols*, Henkel Corporation, Textile Chemicals, Charlotte, NC

17. Henkel Specifications (1989). *Trycol® 5950*, Henkel Corporation, Textile Chemicals, Charlotte, NC

18. Ethox Specifications (1993). *Ethal® DA–6*, Ethox Chemicals, Inc., Greenville, SC

19. Henkel Specifications (1989). *Trycol® 5952*, Henkel Corporation, Textile Chemicals, Charlotte, NC

20. Stepan Company (1995). Products for the Textile and Fiber Industries, Catalog on Computer Disk, *Stepantex® DA–6*, Stepan Company, Northfield, IL

21. Ethox Specifications (1993). *Ethal® DA–9*, Ethox Chemicals, Inc., Greenville, SC

22. Henkel Specifications (1989). *Trycol® 5956*, Henkel Corporation, Textile Chemicals, Charlotte, NC

23. Morrison, R. T. and Boyd, R. N. (1983). *Organic Chemistry, Fourth Edition*, Allyn and Bacon, Inc., Boston, pp. 537, 838, 1045

24. Ethox Specifications (1993). *Ethal® 326*, Ethox Chemicals, Inc., Greenville, SC

25. Specialties from ICI for Fiber Producers (1987). *Frictional Characteristics of ICI Products*, ICI Americas, Inc., Wilmington, DE

26. Ethox Specifications (1993). *Ethal® LA–4*, Ethox Chemicals, Inc., Greenville, SC

27. Henkel Specifications (1989). *Trycol® 5882*, Henkel Corporation, Textile Chemicals, Charlotte, NC

28. Specialties from ICI for Fiber Producers (1987). *Emulsifiers*, ICI Americas, Inc., Wilmington, DC

29. Fiber Production Auxiliaries (1994). *Fiber and Yarn Production. Oil Emulsifiers*, Witco Corporation, Greenwich, CT

30. Ethox Specifications (1993). *Ethal® LA–7*, Ethox Chemicals, Inc., Greenville, SC

31. Ethox Specifications (1993). *Ethal® 926*, Ethox Chemicals, Inc., Greenville, SC

32. Henkel Specifications (1989). *Trycol® 5964*, Henkel Corporation, Textile Chemicals, Charlotte, NC

33. Streitwieser, A. and Heathcock, C. H. (1985). *Introduction to Organic Chemistry*, Macmillian, New York, pp. 379, 510

34. Ethox Specifications (1993). *Ethal® OA–10*, Ethox Chemicals, Inc., Greenville, SC

35. Ethox Specifications (1993). *Ethal® OA–23*, Ethox Chemicals, Inc., Greenville, SC

36. Steinmiller, W. G., assignor to Fiber Industries, Inc., Charlotte, NC, (1976), Synthetic Fibers of Enhanced Processability, *U.S. Patent 3,997,450*

37. Kleber, R. and Billenstein, S., assignors to Hoechst Aktiengesellschaft, Frankfurt am Main, Federal Republic of Germany, (1979). Alkyl–Polyglycol Mixed Formals as Fiber Preparation Agents, *U.S. Patent 4,294,990*

38. Hoechst–Celanese Product Data (1993). *Afilan® GEV*, Hoechst–Celanese Corporation, Charlotte, NC

39. Specialties from ICI for Fiber Producers (1987). *Lubricants*, ICI Americas, Inc., Wilmington, DE

40. Coleman, J. H. letter to P. E. Slade (1981). *G–3886*, ICI Americas, Inc., Wilmington, DE

41. ICI Material Safety Data Sheet (1994). *G–3886*, ICI Americas, Inc., Wilmington, DE

42. Hoechst–Celanese Product Data (1994). *Afilan® ET*, Hoechst–Celanese Corporation, Charlotte, NC

43. Satowski, W. B., Huang, S. K., and Liss, R. L. (1966). Polyoxyethylene Esters of Fatty Acids, in *Nonionic Surfactants*, (M. J. Schick, ed.), Marcel Dekker,

Inc, New York, pp. 142–174

44. Bailey, F. E., Jr. and Koleske, J. V. (1966). Configuration and Hydrodynamic Properties of the Polyoxyethylene Chain in Solution, in *Nonionic Surfactants*, (M. J. Schick, ed.), Marcel Dekker, Inc, New York, pp. 794–822

45. Textile Chemicals, Technical Bulletin 103A (1993). *Ethoxylated Fatty Acids and Polyethylene Glycol Fatty Acid Esters*, Henkel Corporation, Textile Chemicals, Charlotte, NC

46. Henkel Specifications (1983). *Emerest® 2634*, Henkel Corporation, Textile Chemicals, Charlotte, NC

47. Henkel Specifications (1989). *Emerest® 2654*, Henkel Corporation, Textile Chemicals, Charlotte, NC

48. Ethox Product Brochure (1994). *Ethoxylated Fatty Acids*, Ethox Chemicals, Inc., Greenville, SC

49. Ethox Specifications (1989). *Ethox® 1122*, Ethox Chemicals, Inc., Greenville, SC

50. Henkel Specifications (1989). *Emerest® 2620*, Henkel Corporation, Textile Chemicals, Charlotte, NC

51. Stepan Textile Products, Technical Information (1995). *PEG Esters*, Stepan Company, Northfield, IL

52. Ethox Specifications (1987). *Ethox® ML-5*, Ethox Chemicals, Inc., Greenville, SC

53. Witco Fiber Production Auxiliaries (1994). *Desizing Agents*, Witco Corporation, Greenwich, CT

54. Henkel Specifications (1989). *Emerest® 2650*, Henkel Corporation, Textile Chemicals, Charlotte, NC

55. Stepan Company (1995). Products for the Textile and Fiber Industries, Catalog on Computer Disk, *Kessco® PEG 400 ML*, Stepan Company, Northfield, IL

56. Ethox Specifications (1986). *Ethox® ML-9*, Ethox Chemicals, Inc., Greenville, SC

57. Specialties from ICI for Fiber Producers (1987). *Lubricants*, ICI Americas, Inc., Wilmington, DE

58. Witco Fiber Production Auxiliaries (1994). *Lubrication*, Witco Corporation, Greenwich, CT

59. Stepan Company (1995). Products for the Textile and Fiber Industries, Catalog on Computer Disk, *Kessco® PEG 600 ML*, Stepan Company, Northfield, IL

60. Ethox Specifications (1988). *Ethox® ML-14*, Ethox Chemicals, Inc., Greenville, SC

61. Röder, H. L. (1953). Measurements of the Influence of Finishing Agents on the Friction of Fibres, *J. Textile Inst.*, **44**: T247

62. Schönfeld, N. (1969). *Surface Active Ethylene Oxide Adducts*, Pergamon Press, Oxford, U.K., p. 478

63. Henkel Specifications (1989). *Emerest® 2624*, Henkel Corporation, Textile Chemicals, Charlotte, NC

64. Ethox Specifications (1986). *Ethox® MO–5*, Ethox Chemicals, Inc., Greenville, SC

65. Henkel Specifications (1989). *Emerest® 2632*, Henkel Corporation, Textile Chemicals, Charlotte, NC

66. Henkel Specifications (1989). *Emerest® 2646*, Henkel Corporation, Textile Chemicals, Charlotte, NC

67. Ethox Specifications (1988). *Ethox® MO–9*, Ethox Chemicals, Inc., Greenville, SC

68. Henkel Specifications (1989). *Trydet® 2676*, Henkel Corporation, Textile Chemicals, Charlotte, NC

69. Henkel Specifications (1989). *Emerest® 2660*, Henkel Corporation, Textile Chemicals, Charlotte, NC

70. Ethox Specifications (1987). *Ethox® MO–14*, Ethox Chemicals, Inc., Greenville, SC

71. Burnett, L. W. (1966). Miscellaneous Nonionic Surfactants, in *Nonionic Surfactants* (M. J. Schick, ed.) Marcel Dekker, Inc., New York, pp 395–420

72. Hoechst–Celanese Product Data (1993). *Afilan® POM*, Hoechst–Celanese Corporation, Charlotte, NC

73. Henkel Product Data (1986). *Emery® 6724 Fiber Lubricant*, Henkel Corporation, Textile Chemicals, Charlotte, NC

74. Personal Communication, M. P. Broxterman, ICI Americas, to P. E. Slade, August 21, 1991

75. Specialties from ICI for Fiber Producers (1987). *Lubricants*, ICI Americas, Inc., Wilmington, DE

76. ICI Americas, Product Information Note (1991). *G–3890*, ICI Americas, Inc., Wilmington, DE

77. Henkel Product Data (1989). *Emery® 6779*, Henkel Corporation, Textile Chemicals, Charlotte, NC

78. Specialties from ICI for Fiber Producers (1987). *Antistats*, ICI Americas, Inc., Wilmington, DE

79. Hall, A. J. (1975). *The Standard Handbook of Textiles*, Butterworths, London, England, pp. 372–373

80. Perkins, G. P. (1978). Analysis and Standards of Fatty Acids, *Kirk–Othmer Encyclopedia of Chemical Technology*, Third Edition, Vol. 4, John Wiley & Sons, New York, pp. 845–853

81. Witco Corporation (1994). *Fatty Acids, Glycerine, Triglycerides*, Witco Corporation, Greenwich, CT, pp. 10, 9

82. Hoechst–Celanese Product Data (1993). *Afilan® 7PFA*, Hoechst–Celanese Corporation, Charlotte, NC

83. Hoechst–Celanese Product Data (1993). *Afilan® 1080*, Hoechst–Celanese Corporation, Charlotte, NC

84. Hoechst–Celanese Product Data (1991). *Afilan® HSGR*, Hoechst–Celanese Corporation, Charlotte, NC

85. Speel, H. C. and Schwarz, E. W. K. (1957). *Textile Chemicals and Auxiliaries*, Van Nostrand Reinhold, New York, pp. 121

86. Henkel Product Data (1989). *Trydet® 2670*, Henkel Corporation, Textile Chemicals, Charlotte, NC

87. Ethox Specifications (1990). *Ethox® MS–8*, Ethox Chemicals, Inc., Greenville, SC

88. Henkel Specifications (1989). *Emerest® 2640*, Henkel Corporation, Textile Chemicals, Charlotte, NC

89. Henkel Specifications (1989). *Emerest® 2662*, Henkel Corporation, Textile Chemicals, Charlotte, NC

90. Henkel Specifications (1989). *Emerest® 2610*, Henkel Corporation, Textile Chemicals, Charlotte, NC

91. Ethox Specifications (1986). *Ethox® MS–23*, Ethox Chemicals, Inc., Greenville, SC

92. Ethox Specifications (1986). *Ethox® MS–40*, Ethox Chemicals, Inc., Greenville, SC

93. Henkel Specifications (1989). *Emerest® 2715*, Henkel Corporation, Textile Chemicals, Charlotte, NC

94. Elworthy, P. H. and Treon, J. F. (1966). Physiological Activity of Nonionic Surfactants, in *Nonionic Surfactants* (M. J. Schick, ed.), Marcel Dekker, Inc., New York, pp. 923–970

95. Fulmer, R. W. (1968). Applications of Fatty Acids in Protective Coatings, in *Fatty Acids and Their Industrial Applications*, (E. S. Pattison, ed.), Marcel Dekker, Inc., New York, pp. 187–208

96. Wittcoff, H. A. and Reuben, B. G. (1980). *Industrial Organic Chemicals in Perspective. Part One: Raw Materials and Manufacture*, Wiley–Interscience, New York, pp. 138–140

97. Henkel Specifications (1989). *Emerest® 2644*, Henkel Corporation, Textile Chemicals, Charlotte, NC

98. Ethox Specifications (1990). *Ethox® MI–9*, Ethox Chemicals, Inc., Greenville, SC

99. Ethox Specifications (1990). *Ethox® MI–14*, Ethox Chemicals, Inc., Greenville, SC

100. Hoechst–Celanese Product Data (1994). *Afilan® PTU*, Hoechst–Celanese Corporation, Charlotte, NC

101. Smith, B. F. and Block, I. (1982). *Textiles in Perspective*, Prentice–Hall, Inc., Englewood Cliffs, NJ, pp. 290–292

102. Cohen, S. (1968). Fatty Acids and Their Derivatives in the Textile Industry, in *Fatty Acids and Their Industrial Applications* (E. S. Pattison, ed.), Marcel Dekker, Inc., New York, pp. 250–252

103. James, S.. (1994). *An Overview of Textile Softeners*, Presentation to Fabric Processors, Henkel Corporation, Textile Chemicals, Charlotte, NC

104. Jungermann, E. and Taber, D. (1966). Polyoxyethylene Alkylamides, in *Nonionic Surfactants* (M. J. Schick, ed.), Marcel Dekker, Inc., New York, pp. 208–246

105. Burnette, L. W. (1966). Miscellaneous Nonionic Surfactants, in *Nonionic Surfactants* (M. J. Schick, ed.), Marcel Dekker, Inc., New York, pp. 395–420

CHAPTER 6

EMULSIFIERS

An oil mixture, called a spin finish, is applied to man–made fibers during the spinning process at a fiber producer's plant to control tensions during internal operations and to enhance yarn performance at customers' mills. Although the choice of the lubricant allows the formulator to achieve the tension control that one must have, the additional components in the finish oil mixture will also affect the yarn processing ability. One of these components that will greatly influence tension and its control is the emulsifier. Some finishes are completely water soluble since the lubricant itself dissolves in water, but most lubricants are not miscible with water and must be diluted to insure uniform application. One technique that is widely used in the Far East and in Europe is to dissolve the lubricant in a solvent such as low–viscosity petroleum fraction and then apply this solution to the yarn. There are many safety hazards associated with solvent–based finishes, including inhalation of the solvent vapors and the ever–possible problem of fire.

Another option that can be used is the preparation of an emulsion of a proper concentration for good, uniform application [1]. An emulsion is described as a heterogeneous system, consisting of at least one immiscible liquid intimately dispersed in another liquid in the form of droplets, and generally whose diameters exceed 0.1 micrometers (μm). Such systems possess minimal stability, and that stability can be accentuated by such additives as surface active agents [2]. In discussing emulsions, one can distinguish between the two phases that are present. The phase that exists in the form of finely divided droplets is called either the *disperse, internal, or discontinuous* phase; the phase that forms the matrix in which the droplets are suspended is called the *continuous or external phase*. It should be noted that the inclusion of a solid in a liquid is called a *suspension* while gas particles mixed with a liquid are known as *foams*, so emulsions are liquid–liquid systems only. Spin finishes frequently are prepared as true emulsions with water forming the continuous phase and the finish oil mixture as the discontinuous phase. If any finish component is a solid, it is melted before addition to the water.

There are two distinct emulsion types, a classic example is one in which the continuous phase is water and an oil is present as the disperse phase, is known as an oil–in–water emulsion, abbreviated o/w. The other type is the situation in which oil is the continuous phase and water the discontinuous phase. The abbreviation for this water–in–oil system is w/o. This terminology is somewhat inexact for unusual emulsions, but holds for almost all circumstances. Spin finishes are found in both emulsion types. When a finish is prepared and initially applied to the yarn surface it is an oil–in–water emulsion. As the yarn moves along the machine during the spinning process, the water is absorbed into the fiber and/or evaporates into the air and the finish is transformed into a water–in–oil emulsion.

6.1 PARTICLE SIZE

Two immiscible, pure liquids cannot form an emulsion. For a suspension of one liquid in another to be stable enough to be called an emulsion, at least one other component must be present to stabilize the system. This third component, most likely a surface–active or other agent added to emulsions to increase stability by interfacial action, is called an emulsifier or emulsifying agent. A general rule for one to remember is that smaller particle sizes yield more stable emulsions. The effect of particle size on emulsion appearance is described in Table 6–1 [3]. Two types of emulsions based on the size of the suspended droplets are recognized: macroemulsions and microemulsions. The particles of the most common types of emulsions (which might be called a macroemulsion) range from 0.2 to 50 micrometers (1 micrometer = 10^{-6} meters = 1 micron) in average size and can be seen with a light microscope. The limit of the resolving power of a light microscope is about 0.25 μm, approximately the size of the smallest bacteria. The other type of emulsion, the microemulsion, has an average particle size of 0.01 to 0.20 μm [4]. It should be noted that the term "average particle size" has been used. All emulsions have a distribution in the size of the particles that generally follow a log–normal curve, as shown in Figure 6–1 in which a percentage of the number of particles is plotted against the particle diameter.

Table 6–1

Effect of the Particle Size of the Dispersed Phase on Emulsion Appearance

Particle Size	Appearance
Macroglobules	Two phases may be distinguished
Greater than 1 μm	Milky–white emulsion
1.0–0.1 μm	Blue–white emulsion
0.1–0.05 μm	Gray semitransparent, dries bright
0.05 μm and smaller	Transparent, dries bright

(Ref. [3]. Copyright 1978. Reprinted by permission of John Wiley & Sons, Inc.)

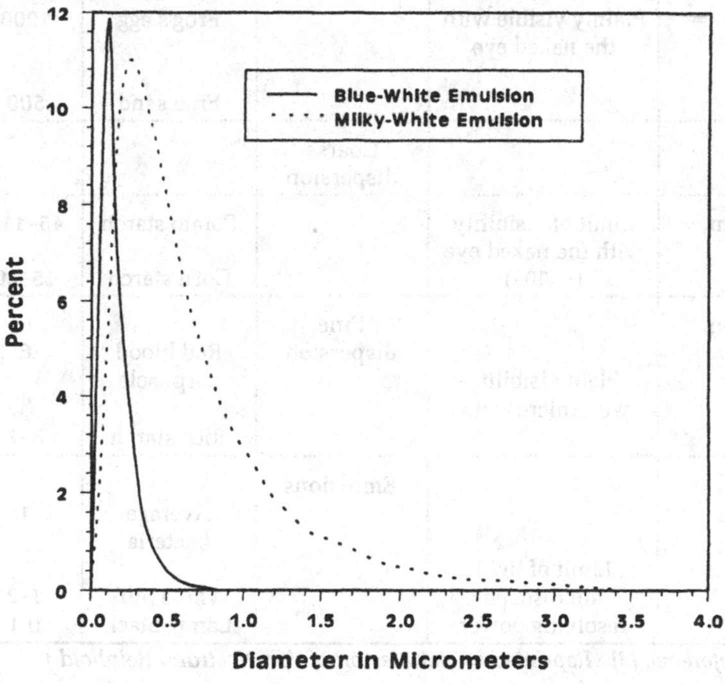

Figure 6–1. Comparison of Particle Size of Blue–White and Milky–White Finish Emulsions.

Most spin finish emulsions are considered satisfactory if they are blue–white or almost transparent. Two hypothetical emulsions were prepared and the particle size measured with a photon correlation spectrometer and the resulting data shown by Figure 6–1 [5]. The solid line represents the particle size distribution for a typical blue–white emulsion that has excellent stability. In contrast, the dashed line gives the particle size for a finish with low emulsifier content that is very milky and is unstable when held at ambient temperatures for an extended period.

Particle size values as stated above are hard to visualize without a comparison with other particles with which we are familiar. Table 6–2 lists particle sizes of some natural materials [2].

Table 6–2
Relative Particle Sizes

Range of Dimensions	Visibility	Description of State	Examples in Nature	
				μm
10^{-1} cm. = 1 mm	Plainly visible with the naked eye		Frog's egg	1000
			Fine sand	500
		Coarse dispersion		
10^{-2} cm.	Limit of visibility with the naked eye ($\sim 50\mu$)		Potato starch	45–110
			Corn starch	15–20
10^{-3} cm.	Plain visibility with microscope	Fine dispersion	Red blood corpuscle	8
			Rice starch	3–7
10^{-4} cm. = 1 μm	Limit of light microscope resolving power	Emulsions	Average bacteria	1
			Whole milk	1–2
			Lamp–black	0.1

(Reference [3]. Reprinted by permission of Van Nostrand Reinhold)

The particle size of emulsion droplets can be measured by several techniques, including field disruption, laser light scattering, and photon correlation spectroscopy. Several other methods used in determining the size of suspended solid materials (sedimentation, microscopic, and X–ray) are not normally the preferred techniques for emulsions.

In the field disruption procedure, a dilute emulsion in a conductive solvent is pumped through a capillary in which two electrodes have been placed. As the emulsion droplet passes through the electrical field generated between the electrodes they disrupt the field and are counted. One problem with this technique is the conductive solvent. If water is the solvent and a salt, such as potassium chloride, is added to enhance the conductivity, the salt might act to influence the emulsion particles and cause agglomeration. A representative instrument used for this technique is the Coulter® Multisizer IIe [6]. This device can measure particles ranging from 0.4 μm to 1,200 μm in diameter.

Laser diffraction is quite effective in determining the size of emulsion particles with diameters up to 3,500 μm. A dilute sample is placed in a small cell and stirred, and as a particle passes through the laser beam it is counted. Typical instruments utilizing this technique are made by Coulter Corporation and by Malvern Instruments. Both suppliers use Fraunhofer laser diffraction [7] and Mie particle scattering [8] to expand to a lower range of particle sizes. Fraunhofer theory assumes that the particles are opaque and larger than the wavelength of the light used, but Mie theory provides a better solution to the equations for the interaction of light with small particles, especially if they are transparent or translucent. The Coulter instruments are sold as the Coulter® LS Series [9]. The LS 100 measures a size range of 0.4–900 μm, the LS 200 a range from 0.4–2000 μm, and the LS 230 evaluates sizes from 0.04–2000 μm. Malvern also manufactures several particle size characterization [10] instruments. The Mastersizer X measures particles ranging from 0.1 to 2,000 μm, the Mastersizer S particles from 0.05 to 3,500 μm, the Mastersizer Micro particles from 0.3 to 300 μm, and the Mastersizer Microplus is useful in detecting sizes ranging between 0.05 to 550 μm.

Matec Applied Sciences has recently introduced two new techniques for characterizing particle size and size distribution. One of these, Capillary Hydrodynamic Fractionation [11], is founded on a size exclusion effect that occurs when a dispersion of particles flows through an open capillary tube [12]. As the fluid moves through the tube, the parabolic velocity signifies that the flow is faster in the center and slower near the wall. Since larger particles are unable to approach the wall as closely as smaller particles, the larger particles then travel at a higher velocity than the smaller particles [13]. The size and size distribution is determined by the amount of UV light scattered and absorbed by the particles. This apparatus is especially useful with suspensions having a multimodal distribution.

Another unique device developed by Matec is the AcoustoSizer® [14]. Although this instrument is designed primarily to measure particle size of concentrated suspensions of solids, it can also be applied to emulsions.

While almost all of the other techniques use light to determine the size, the AcoustoSizer® measures sound. The instrument applies an alternating voltage to a pair of parallel plate electrodes in contact with the suspension. Since essentially all particles have an electrical charge, the particles resonate at the frequency of the applied field. As they resonate, low amplitude sound waves are generated. This phenomenon is called the Electrokinetic Sonic Amplitude (ESA) effect [15]. ESA measurements are used to measure particle velocity in the electric field over a range of frequencies and the particle size information is extracted from the electrophoretic velocity spectrum. The particle size range extends from 0.1 to 10 μm [16].

One of the most important techniques used for emulsion particle size measurement is photon correlation spectroscopy. This type of instrument again uses a laser beam as the light source that passes through a cuvette containing the dilute emulsion. Brownian motion [17] causes the particles to diffuse through the beam and the photon digital correlator measures the speed at which the particle moves through that beam by determining the Doppler shift [18] of the particle. Since spherical particles diffuse through liquids at a rate that is a function of their diameters, the relationship between particle diameter and diffusion can be described by the Stokes–Einstein equation [19]:

$$D_T = \frac{k_B T}{3 \pi \eta d}$$
(6–1)

where:

D_T	=	diffusion coefficient,
k_B	=	Boltzmann's constant,
T	=	absolute temperature,
η	=	viscosity of the solution, and
d	=	particle diameter.

Photon correlation spectroscopy (PCS) is primarily designed to measure particle sizes below about 3 μm. Since the sizes are so small it is frequently more convenient to describe the sizes in nanometers (or 1×10^{-9} meters) instead of micrometers (or 1×10^{-6} meters). The typical size range for PCS instruments is 3 to 3000 nanometers. Representative PCS instruments for measuring particle size are supplied by Brookhaven Instruments [20], Coulter [21], and Malvern [22]. Brookhaven Instruments also makes a fiber optics PCS instrument for particle size analysis in concentrated solutions [23].

6.2 ZETA POTENTIAL

All particles in emulsions or dispersions are electrically charged to a lesser or greater degree. It has been suggested [2] that these charges can arise in three different ways, by ionization, adsorption, or frictional contact. For emulsion droplets, the difference between the first two mechanisms is quite small, and indeed may be equivalent. These particles generally have a negative charge, as Coehn's [24] rule proposes, *a substance having a high dielectric constant is positively charged when in contact with another substance having a lower dielectric constant.* Since water has a dielectric constant much higher than most of the substances that are likely to be in the other phase of the emulsion, it appears that the droplets in an o/w emulsion will probably have a negative charge. Regardless of how these charges arise, their presence on the emulsion droplet substantially contributes to the stability of the system since like charges repel each other and prevent close approach and the coagulation of the particles.

1. The electrical double layer. Early in the study of the charge on a colloidal particle, it was realized that the situation at the interface is not a simple one. In attempting to explain this phenomenon, Helmholtz [25] introduced the concept of the electrical double layer over a century ago. The theory of screening of the surface charge by the diffuse cloud was developed independently by Gouy [26] and Chapman [27]. The Gouy–Chapman theory was later modified by Stern [28]. According to Stern, the double layer is in two parts, one which is approximately a single ion thick, and remains fixed to the particle–liquid interface. In this phase there is a sharp drop in potential. The second part extends some distance into the liquid and is diffuse with a gradual fall in potential into the bulk of the solution. This charged electrical double layer is shown by Figure 6–2.

Charges on the particle at the surface affects the distribution of ions in the emulsion near the surface, causing an increase in the concentration of counter ions, that is, ions whose charge is opposite in sign to that of the particle in that region. The potential on the surface is therefore neutralized by the counter ions and decays as the distance from the surface increases, reaching zero in the bulk of the solution. The innermost portion of the double layer immediately adjacent to the particle surface includes adsorbed ions which are attached to the surface by electrostatic attraction or by van der Waals forces. This boundary around this surface in which these ions are located is known as the Stern plane or Stern layer. Beyond the Stern layer the excess counter ion concentration continues to fall, as does the electrical potential. This region is known as the diffuse layer and extends a distance from the particle determined by the surface charge on the particle, and the nature and concentration of the ions in the surrounding electrolyte.

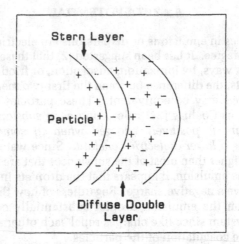

Figure 6–2. Stern Model of the Interface
for Charged Surfaces

Some ions are more or less rigidly attached to the particle, some others are loosely bound, while others in the bulk of the solution are not attached at all. The boundary between (1) the moving particle and counter ion units and (2) the surrounding solution constitutes a slipping plane, or surface of shear. While it is virtually impossible to make a direct measurement of the potential in the Stern layer, the potential of the slipping plane can be measured simply and accurately. The potential at this slipping plane is known as the zeta potential, and is one of the practical parameters in describing and predicting flocculation, adhesion, and especially finish emulsion stability. Generally, the zeta potential is the same sign as the surface potential, that is, negative, and with finishes the higher the negative potential, the smaller the emulsion particle and the more stable the emulsion.

2. Measurement of zeta potential. The measurement of zeta potential is usually straight forward. The most widely used and most convenient technique is that of particle electrophoresis measured by electrophoretic light scattering. Movement, or mobility, of a sphere of radius r in an electrical field is predicted by the Henry equation [18]:

$$U = \frac{Z e \chi (\kappa r)}{6 \pi r (1 + \kappa r)} \qquad (6\text{–}2)$$

where:

U	=	electrophoretic mobility,
Z	=	particle charge,
η	=	solution shear viscosity,
$\chi(\kappa r)$	=	Henry's function, and
r	=	particle radius.

Henry's function $\chi(\kappa r)$ is a sigmoid curve whose value increases from unity for κr less than 0.1 to a maximum of 1.5 for κr greater than 1000. Thus the electrophoretic mobility, or the velocity in an electrical field, is directly proportional to the particle charge but inversely proportional to the viscosity. When $\chi(\kappa r) < 1$, that is a small particle with a thick double layer, then $\chi(\kappa r)$ is taken to be 1 and the Henry equation in this limit is known as the Huckle equation. This applies only in very dilute solutions of small particles and in which the conductivity is low.

Frequently where we have an aqueous solution of an electrolyte and the thickness of the double layer is small when compared to the radius of curvature of the particle, then $\chi(\kappa r) = 1.5$ and the Henry equation reduces to that of Smoluchowski [29]:

$$U = \frac{\zeta \epsilon_0 D}{\eta} \qquad (6\text{-}3)$$

where:

U	=	electrophoretic mobility,
ζ	=	zeta potential,
ϵ_0	=	permittivity of free space,
D	=	dielectric constant, and
η	=	viscosity.

More detailed discussions of the theory of the zeta potential and its relationship to emulsion stability can be found in the work of Hunter [30], and Crow [31].

Commercially available instruments are sold by Brookhaven [20], Coulter [32], Malvern [22], and Matec [33].

6.3 HLB CONCEPT

6.3.1 The Hydrophile–Lipophile Balance (HLB) of Surfactants

The hydrophile–lipophile balance is a concept proposed to explain the attraction of an emulsifier for both the water phase and the oil phase of an emulsion. It is determined by the chemical composition, the distribution of the ethylene oxide content, and/or the extent of ionization of a given emulsifier. For example, propylene glycol monostearate,

$$CH_3CHOHCH_2COOC(CH_2)_{16}CH_3$$

has a low HLB (3.4) and thus is strongly lipophilic; polyoxyethylene (23) monostearate,

$$H(OCH_2CH_2)_{23}OOC(CH_2)_{16}CH_3$$

having a long polyoxyethylene chain, has a high HLB (15.7) and is hydrophilic; sodium stearate,

$$NaOOC(CH_2)_{16}CH_3$$

has a very high HLB (18.0) and is strongly hydrophilic, since it ionizes and thus provides an even stronger hydrophilic tendency.

Generally, emulsifiers with HLB values ranging from 3 to 6 are considered to be water–in–oil emulsifiers, products with HLB values of 7–9 are wetting agents, other materials whose HLB is 8–18 are oil–in–water emulsifiers. Other materials with an HLB of 13–15 are regarded as detergents and if the HLB is about 15–18 the component is called a solubilizer [34]. It should be noted that the HLB is only an indication of the type of emulsion that might be formed, not as a measure of emulsifier efficiency. The HLB can be a starting point for emulsifier selection; if the HLB of the oil is similar to the HLB of the emulsifier, an emulsion can be made. *Often it is necessary to blend emulsifiers to get the required HLB, for instance a blend of a low HLB and a higher HLB emulsifier is often superior to a single product with the same final HLB.*

6.3.2 Determination of HLB

Several different procedures can be used to determine the HLB of organic compounds: (1) estimation by water solubility, (2) calculation by one of the numerous techniques, or (3) preparation of an experimental emulsion. It is quite possible that one will obtain different HLB values with the different techniques, but they will be in the same general range. These procedures for HLB determination are detailed below.

1. Estimation by water solubility [34]. The following table, Table 6–3, lists the approximate HLB values of materials as determined by their solubility in water. There are exceptions to this general approximation, but

Table 6–3

Approximation of HLB by Water Solubility

Behavior When Added to Water	HLB Range
No dispersibility in water	1–4
Poor dispersion	3–6
Milky dispersion after vigorous agitation	6–8
Stable milky dispersion	8–10
Translucent to clear dispersion	10–13
Clear solution	> 13

(Reference [34])

it can be used as a rough guide for emulsifier HLB values.

2. *Calculation from saponification number [35].* The HLB for many materials, for which the saponification number is known, can be calculated by the equation:

$$HLB = 20 \left(1 - \frac{S}{A}\right) \qquad (6\text{–}4)$$

where S is the saponification number of the ester and A is the acid number of the fatty acid used to prepare the ester, not the acid value of the ester itself. This calculation, obviously, does not apply to ether alkoxylates made from fatty alcohols.

3. *Calculation from weight percent oxyethylene content [36].* A common method for calculating the HLB value is by finding the weight percent polyoxyethylene content, E, and the weight percent polyoyl content, P, adding the two, and dividing by five. The equation is:

$$HLB = \frac{E + P}{5} \qquad (6\text{–}5)$$

If the polyoxyethylene chain is the only hydrophilic group, this reduces to:

$$HLB = \frac{E}{5} \qquad (6\text{–}6)$$

where E is again the weight percent oxyethylene content.

4. *Calculation from the general formula [4]*. Another general formula is suggested for calculating the HLB for nonionics:

$$HLB = 20\left(\frac{M_H}{M_H + M_L}\right) \qquad (6\text{--}7)$$

where M_H is the formula weight of the hydrophilic portion of the molecule and M_L the formula weight of the lipophilic (hydrophobic) portion of the molecule.

5. *Structural calculation factors*. Davies and Rideal [37] have treated the HLB values as a summation of structural factors similar to the calculation of the parachlor. From this they attempted to correlate the various structural factors of the emulsifier and how each contributes, both positively and negatively, to the HLB value. For a given structure the HLB is calculated according to Equation 6–9, where

$$HLB = 7 + \sum (hydrophilic\ group\ numbers) - $$
$$\sum (lipophilic\ group\ numbers) \qquad (6\text{--}8)$$

The group assignments are in Table 6–5:

Table 6–5

HLB Group Numbers

Hydrophilic Group	
–SO₄Na	38.7
–COOK	21.1
–COONa	19.1
–N (tertiary amine)	9.4
Ester (sorbitan ring)	6.8
Ester (free)	2.4
–COOH	2.1
–OH (free)	1.9
–O–	1.3

Table 6–5, Continued

Hydrophilic Group	
–OH (sorbitan ring)	0.5

Lipophilic Group	
–CH–, –CH$_2$–, –CH$_3$, =CH–	–0.475

Derived Group	
–(CH$_2$–CH$_2$–O)–	0.33
–(CH$_2$–CHCH$_3$–O)–	–0.15

(Reference [37]. Reprinted by permission of Academic Press, Inc.)

This system gives a fair approximation for sorbitan based components, but the HLB values for ethoxylated alcohols and ethoxylated glyceride esters are lower than desired.

6.3.3 HLB of Cohesive Agents

The HLB values of the cohesive agents described in Chapter 5 have been obtained by their suppliers and are listed in Table 6–6.

Table 6–6

HLB of Cohesive Agents

Hydrophobe	Moles EO	Capped	HLB	Page in Chapter 5
Decyl Alcohol	4	No	10.5	138
	6	No	12.4	138
	9	No	14.3	139
Lauryl Alcohol	3	No	8	141
	4	No	9.2	141
	7	No	12.2	142
	9	No	13.3	142
	23	No	16.7	143
Oleyl Alcohol	10	No	12.6	144
	20	No	15.3	145

Table 6–6, Continued

Hydrophobe	Moles EO	Capped	HLB	Page in Chapter 5
Oleyl Alcohol	23	No	15.8	145
Lauryl Alcohol	5	Yes	11.4	147
	6	Yes	10.5	148
Pelargonic Acid	6.8	No	12.8	151
	9	No	14.3	152
Lauric Acid	5	No	9.3	153
	9	No	14.3	154
	14	No	14.9	154
Oleic Acid	5	No	8.3	155
	7	No	10.4	156
	9	No	12.0	157
	10	No	12.2	157
	14	No	13.5	157
Pelargonic Acid	9	Yes	13.9	160
	9	Yes	12.0	161
Oleic Acid	9	Yes	12.1	162
Palmitic Acid	7	No	10.9	164
Stearic Acid	4	No	9.2	167
	8	No	11.4	168
	9	No	12.0	168
	14	No	13.8	169
	23	No	15.7	169
	40	No	17.3	170
Isostearic Acid	9	No	11.3	171
	14	No	13.0	172

6.4 SURFACTANTS USED AS EMULSIFIERS

6.4.1 Soaps

Soap was discovered thousands of years ago when animal fats were boiled with wood ashes forming a product used for cleaning. Refined versions of these products were the earliest substances used for emulsifying lubricants in fiber finish compositions. Many of these soap emulsifiers were based on fatty acids isolated from natural fats, for example oleic and lauric acids. The acids were neutralized with sodium or potassium hydroxide, or with an amine, to produce the soap. Some natural fats were sulfated and neutralized to manufacture products that were more water soluble than the fatty acid soaps. In most instances, the use of soaps as finish emulsifiers was discontinued after fiber technology advanced from the old, slow speed flat yarn to the more modern high speed processes.

6.4.2 Alcohol Ether Ethoxylates

Although many ethoxylated alcohols are used in fiber finishes for increasing fiber cohesion and as co–emulsifiers, some others are included principally for their emulsification properties. These emulsifiers are based on alcohols with branched carbon chains or on chains containing sixteen or more carbon atoms. As with the alcohol cohesive agents, alcohol based emulsifiers are probably better when combined with a primary emulsifier and less effective when used alone.

(POE) Tridecyl Alcohols, CAS# 24938–91–8

Tridecyl alcohol is a branched 13–carbon component made by the condensation of propylene. The complete synthesis is described in Chapter 4, page 85. The alcohol is reacted with ethylene oxide in the presence of a base catalyst to yield the final product [38]. The structural formula for these emulsifiers is:

$$CH_3-\underset{\underset{CH_3}{|}}{\overset{\overset{H}{|}}{C}}-CH_2-\underset{\underset{CH_3}{|}}{\overset{\overset{H}{|}}{C}}-CH_2-\underset{\underset{CH_3}{|}}{\overset{\overset{H}{|}}{C}}-CH_2-\underset{\underset{CH_3}{|}}{\overset{\overset{H}{|}}{C}}-CH_2-O-(CH_2CH_2O)_xH$$

Tridecyl alcohol ethoxylates are not only valuable as emulsifiers, but also are very good wetting agents, scouring agents, and dispersants. They are useful in fiber finishes, detergent formulations, agricultural chemicals, and in fabric processing. Materials with a low ethylene oxide content are low foaming co–emulsifiers and those with higher levels of EO are effective in detergent products.

1. POE (3) Tridecyl Alcohol. The low level of ethylene oxide in this product causes it to be oil soluble and an excellent choice as a co–emulsifier in the preparation of micro–emulsions. It is one of the TDA ethers known for generating low levels of foam. This material is sold by Ethox as Ethal® TDA–3 [39][40], by Henkel as Trycol® 5993 [41][42], and by Stepan as Cedepal® TD–400 [43] . Typical properties of POE (3) tridecyl alcohol are listed in Table 6–7.

Table 6–7

Typical Properties of POE (3) Tridecyl Alcohol

Property	Value
HLB	7.9
Acid Value	1 max
Hydroxyl Value	168–180
Color, APHA	50 max
Water Solubility	Dispersible
Cloud Point, 1%, °C	<25
Draves Wetting, Seconds, (0.1% @ 25°C)	—

2. POE (6) Tridecyl Alcohol. This chemical is an excellent wetting agent, co–emulsifier, and low foaming cleaner. It finds extensive use in wool scouring and fabric scouring before dyeing. This alcohol may also be sulfated or phosphated to create specialty anionic surfactants.

The component is manufactured by Ethox using the trade name Ethal® TDA–6 [39][44], by Henkel as Trycol® 5940 [41][45], by ICI with the name Renex® 36 [46], by Stepan as Stepantex® TD–560 [43], and by Witco as Desonic® 6T [47]. It should be noted that Stepantex® TD–560 is reported to contain 5.5 moles of EO, but it is listed with the 6–mole products since the EO content is only an average value. Some features of the 6–EO–mole alcohol are shown in Table 6–8.

Table 6–8

Typical Properties of POE (6) Tridecyl Alcohol

Property	Value
HLB	11.4
Acid Value	0.5 max
Hydroxyl Value	115–130
Color, APHA	50 max
Water Solubility	Dispersible
Cloud Point, 1%, °C	<25
Draves Wetting, Seconds, (0.1% @ 25°C)	8

3. POE (8) Tridecyl Alcohol. The increase in ethylene oxide content from six to eight results in a product that is water soluble instead of being water dispersible. These chemicals are used in moderate to high foaming cleaners, as scouring agents, and as co–emulsifiers. Other properties are similar to the 6–mole component and are listed in Table 6–9. It is a product that can be obtained from Henkel as Trycol® 5949 [41][48] and by Stepan as Stepantex® TD–630 [43].

Table 6–9

Typical Properties of POE (8) Tridecyl Alcohol

Property	Value
HLB	12.5
Color, Gardner	1 max
Water Solubility	Soluble
Cloud Point, 1%, °C	43

4. POE (9) Tridecyl Alcohol. As expected, another mole of EO has very little effect on changing the properties of the 8–mole chemical. General characteristics are listed in Table 6–10.

Table 6–10

Typical Properties of POE (9) Tridecyl Alcohol

Property	Value
HLB	13.0
Acid Value	1 max
Color, APHA	100 max
Water Solubility	Soluble
Cloud Point, 1%, °C	53–60
Draves Wetting, Seconds, (0.1% @ 25°C)	7

5. POE (12) Tridecyl Alcohol. After reacting 12 moles of ethylene oxide with tridecyl alcohol the resulting product is a semi–solid with a melting point of about 16°C. Increasing water solubility makes this component valuable in higher concentrated detergent products and as a foam builder. It is a good dispersant and finish co–emulsifier. The material is made by Ethox as Ethal® TDA–12 [39][49], by Henkel with the designation Trycol® 5943 [41][50], and by ICI as Renex® 30 [46]. Some of the more important properties are presented by Table 6–11.

Table 6–11

Typical Properties of POE (12) Tridecyl Alcohol

Property	Value
HLB	14.5
Acid Value	1 max
Hydroxyl Value	72–84
Color, APHA	100 max
Water Solubility	Soluble
Cloud Point, 1%, °C	88–94
Draves Wetting, Seconds, (0.1% @ 25°C)	11

6. POE (15) and POE (18) Tridecyl Alcohol. These are similar low melting, water soluble solids that are high foaming detergents, dispersants, and emulsifiers. Typical attributes are shown in Table 6–12.

Table 6–12

Typical Properties of POE (18) Tridecyl Alcohol

Property	Value
HLB	16.0
Acid Value	1 max
Hydroxyl Value	57–69
Color, APHA	100 max
Draves Wetting, Seconds, (0.1% @ 25°C)	—

Fifteen–mole EO alcohol ethers are manufactured by Stepan and sold as Makon® TD–15 [43]. The 18–mole ether is made by Ethox as Ethal® TDA–18 [39][51], and by Henkel as Trycol® 5946 [41][52].

(POE) Cetyl Alcohols, CAS# 9004–95–9

Cetyl alcohol, the C_{16} saturated alcohol, may be synthesized by the reduction of naturally occurring fatty esters as shown in Chapter 5 on page 140. Cetyl acid esters are found in large amounts in beef tallow and palm oil. The general structure of this type of emulsifier is:

$$CH_3(CH_2)_{15}-O-(CH_2CH_2O)_xH$$

1. POE (10) Cetyl Alcohol. This is commercially available from ICI as Brij® 56 [46]. Some properties are displayed in Table 6–13.

Table 6–13

Typical Properties of POE (10) Cetyl Alcohol

Property	Value
HLB	12.9
Water Solubility	Dispersible
Pour Point, °C	23

2. POE (20) Cetyl Alcohol. The addition of more ethylene oxide makes this ether water soluble, as expected. It is made by ICI and sold as Brij® 58 [46]. Characteristics are listed in Table 6–14.

Table 6–14

Typical Properties of POE (20) Cetyl Alcohol

Property	Value
HLB	15.7
Water Solubility	Soluble
Pour Point, °C	38

(POE) Stearyl Alcohols, CAS# 9005–00–9

As with the cetyl alcohols, stearyl alcohol is made through reduction of the fatty acid esters. Ethoxylation then follows through standard techniques. The general structure is:

$$CH_3(CH_2)_{17}-O-(CH_2CH_2O)_xH$$

Adding two –CH$_2$– groups (from C$_{16}$ to C$_{18}$) to the hydrophobe chain results in lower water solubility and slightly lower HLB values.

1. POE (10) Stearyl Alcohol. This emulsifier has similar properties to the cetyl alcohol ethoxylate. It is also sold by ICI using the trade name Brij® 76 [46]. The properties are found in Table 6–15.

Table 6–15

Typical Properties of POE (10) Stearyl Alcohol

Property	Value
HLB	12.4
Water Solubility	Dispersible
Pour Point, °C	38

2. POE (20) Stearyl Alcohol. A homolog of the cetyl alcohol version, it is produced by ICI as Brij® 78 [46]. Some features appear in Table 6–16.

Table 6–16

Typical Properties of POE (20) Stearyl Alcohol

Property	Value
HLB	15.3
Water Solubility	Dispersible
Pour Point, °C	38

(POE) C_{12}–C_{18} Alcohols, CAS# 68213–23–0

When coconut oil is hydrolyzed to form glycerol and the free fatty acids present in that composition, further refinement is required to make the special grades needed for ethoxylation. One grade that is hydrogenated and the low molecular weight acids removed is available from Witco with the name Hystrene® 5012 [53]. Hydrogenation reduces the iodine value, an indication of the degree of unsaturation, from the normal 5–10 to less than one. The final composition is: 55% C_{12}, 23% C_{14}, 12% C_{16}, and 8% C_{18}. The acids are reduced to form the corresponding alcohols. A general structural formula of the ethoxylated alcohol with an EO content ranging from three to five moles is:

$$CH_3(CH_2)_{11-17}O(CH_2CH_2O)_xH$$

1. POE (3) C_{12}–C_{18} Alcohol. Because the fatty acids are mostly C_{12} and C_{14}, this component is a liquid. The low degree of ethoxylation yields a material that is water dispersible at room temperature. It is made by Ethox as Ethal® 3328 [39][54]. Normal properties are displayed in Table 6–17.

Table 6–17

Typical Properties of POE (3) C_{12}–C_{18} Alcohol

Property	Value
HLB	8.1
Color, APHA	50 max
Acid Value	1.0 max
Hydroxyl Number	147–162
Water, %	1 max
Viscosity, cSt, 25°C	31–38

2. POE (4) C_{12}–C_{18} Alcohol. Another in the series of oil–soluble emulsifiers, this product is valuable for mineral oil emulsification. It is produced by Hoechst–Celanese and called Afilan® 2299 [55]. Some of the attributes of this emulsifier are listed in Table 6–18.

Table 6–18

Typical Properties of POE (4) C_{12}–C_{18} Alcohol

Property	Value
HLB	9.7*
pH, 1%	5.0–8.0
Color, APHA	100 max
Hydroxyl Number	140–150
Water, %	1 max
Viscosity, cSt, 25°C	31–36

* Calculated by Equation (6–6)

3. POE (5) C_{12}–C_{18} Alcohol. Special manufacturing conditions were employed in synthesizing this emulsifier to insure that it contains a low level of free polyethylene glycol to insure improved emulsion quality. It is made by Hoechst–Celanese as Afilan® 5GC [56] and used as a lubricant/emulsifier in both staple and filament finishes. Properties are cited in Table 6–19.

Table 6–19

Typical Properties of POE (5) C_{12}–C_{18} Alcohol

Property	Value
HLB	10.7*
pH, 1%	6.0–8.0
Color, Gardner	2 max
Water, %	1 max
Viscosity, cSt, 25°C	40–45

* Calculated by Equation (6–6)

(POE) C_{16}–C_{18} Alcohols, CAS# 68439–49–6

The reduction of triple pressed stearic acid (51% C_{16}, 45% C_{18}, 4% other acids) to the alcohol forms the hydrophobic portion of these molecules.

A common name for the mixture is cetyl–stearyl alcohol. With the long carbon chains, they are pastes or solids at room temperature. The general structure is:

$$CH_3(CH_2)_{15-17}O(CH_2CH_2O)_xH$$

Long carbon chains also reduce the yarn friction, so these materials can be effective self–emulsifying lubricants.

1. POE (3) C_{16}–C_{18} Alcohol. An oil soluble emulsifier and lubricant, this material is a paste at room temperature. It is made by Ethox and called Ethal® 368 [57]. Some characteristics appear in Table 6–20.

Table 6–20

Typical Properties of POE (3) C_{16}–C_{18} Alcohol

Property	Value
HLB	6.9
Color, APHA	100 max
Water, %	1 max
Acid Value	0.1–1.1
Hydroxyl Number	132–140

2. POE (25) C_{16}–C_{18} Alcohol. This hard, waxy solid is a hydrophilic emulsifier that exhibits good detergency in hot aqueous solutions. It is manufactured by Ethox and sold as Ethal® CSA–25 [58]. Typical properties are given in Table 6–21.

Table 6–21

Typical Properties of POE (25) C_{16}–C_{18} Alcohol

Property	Value
HLB	16.3
Color, APHA	100 max
Water, %	1 max
Acid Value	1 max
Hydroxyl Number	40–55

6.4.3 Ethoxylated Alkylphenols

The reaction of ethylene oxide with an alkylphenol to yield a nonionic surfactant resulted from intense efforts in Germany to produce an improved product for textile auxiliaries and to replace fats and oils as sources for surfactants [59]. These products were made in both Germany and the United States before 1940. This type of surfactant is largely made from branched chain alkylphenols, but some linear polyoxyethylene alkyl–phenols are also available. Over 450 million pounds of alkylphenol ethoxylates were produced in the United States in 1990 [67]. The branched chain versions are more generally used in fiber finishes.

Nonylphenol is by far the most widely used base material (80%), with the combined volumes of octylphenol and dodecylphenol consumed at a much lower rate (20%) [60]. Octylphenol (isooctylphenol, diisobutylphenol) is made from the equilibrium dimer of isobutylene [60]:

$$\underset{\text{20\%}}{\underset{\underset{CH_3}{|}}{\overset{\overset{CH_3}{|}}{H_3CC\ CH{=}C\ CH_3}}} \rightleftharpoons \underset{\text{80\%}}{\underset{\underset{CH_3\ CH_3}{|\quad|}}{\overset{\overset{CH_3}{|}}{H_3CC\ CH_2C{=}CH_2}}} + \text{(benzene ring)}OH$$

$$\xrightarrow{BF_3} \underset{\underset{CH_3\ CH_3}{|\quad\ |}}{\overset{\overset{CH_3\ CH_3}{|\quad\ |}}{H_2CC\ CH_2C{-}}}\text{(benzene ring)}OH$$

Both isomers give the same *t*–carbonium ion and thus only a single structure is produced for the alkyl portion of the final product.

Side chains in nonyl and dodecyl phenols are derived from the propylene trimers and tetramers [61]. The reaction to form the trimers and tetramers is quite complex and the end products have varied isometric structures. As an example, propylene trimer preparation, using sulfuric acid catalyst might be:

$$3\ \underset{\underset{H}{|}}{\overset{\overset{H}{|}}{CH_3C{=}CH_2}} \xrightarrow{H_2SO_4} CH_3{-}\underset{\underset{CH_3}{|}}{\overset{\overset{H}{|}}{C}}{-}\underset{\underset{H}{|}}{\overset{\overset{CH_3}{|}}{C}}{-}\underset{\underset{CH_3}{|}}{\overset{\overset{H}{|}}{C}}{-}\underset{}{\overset{\overset{H}{|}}{C}}{=}CH_2$$

The structure shown for the trimer is not accurate, however, since there is a distribution of isomers. These are listed in Table 6–22.

Table 6–22

Olefin Types in Propylene Trimer

Structure	Weight %
$H_2C=CHR$	1
$RHC=CHR$	14
$H_2C=CR_2$	8
$RHC=CR_2$	35
$R_2C=CR_2$	42

(Reference [60])

The same types of isomers are also expected for the tetramer.

Ethoxylation of the alkylphenols is believed to proceed in two steps when the number of ethylene oxide molecules is greater than one. This general reaction sequence is:

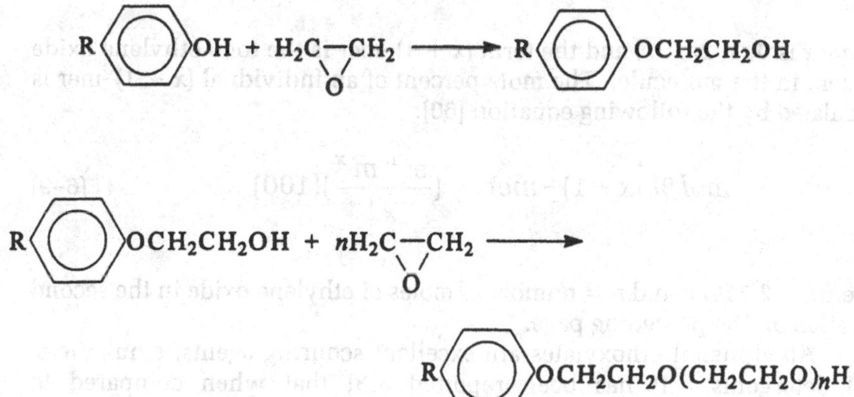

As with all other ethoxylated materials, the number of ethylene oxide units is distributed over a range of values, and is not a single number. Data reported by Enyeart [60] for Igepal® CO–210 (1.5 moles of ethylene oxide) confirm the conclusions of Miller, Bann, and Thrower [62] that the two step reaction is the appropriate representation for the preparation of ethoxylated alkylphenols. This data is displayed in Table 6–23.

Table 6–23

Composition of Igepal® CO–210

	Theory	Found
Molecular Weight	286	282
Nonylphenol		2.8%
NP/EO$_1$	55.99	52.0%
NP/EO$_2$	32.66	34.6%
NP/EO$_3$	9.33	7.9%
NP/EO$_4$	1.74	0.8%
NP/EO$_5$	0.25	Trace

(Reference [60])

We can then conclude that the average composition of polyoxyethylene alkylphenols is represented by:

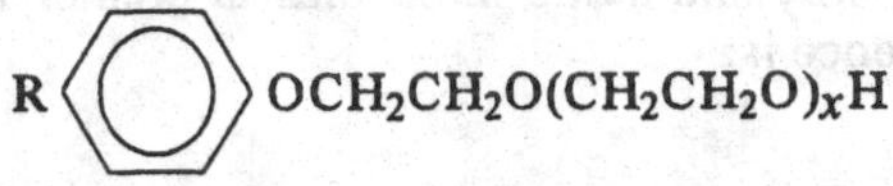

$$R \langle \bigcirc \rangle OCH_2CH_2O(CH_2CH_2O)_xH$$

where x is 0, 1, 2, . . . , and the term $(x + 1)$–mer is the total ethylene oxide content in the molecule. The mole percent of an individual $(x + 1)$–mer is calculated by the following equation [60]:

$$mol\,\%\,(x + 1) - mer = (\frac{e^{-n}m^x}{x!})(100) \qquad (6-9)$$

where $e = 2.71828$ and n = number of moles of ethylene oxide in the second equation on the preceding page.

Alkylphenol ethoxylates are excellent scouring agents, emulsifiers, and detergents. It has been reported [63] that when compared to polyoxyethylene alcohols, the presence of a phenoxy group retards microbial degradation in wastewater systems. Reports also suggest [64][65] that alkylphenol ethoxylates will be disallowed for washing and cleaning agents in Europe by the year 2000. Evaluations in the United States have shown that the biodegradation proceeds more readily than previously reported [66]–[69].

The polyoxyethylene alkylphenol components available commercially are listed in the following sections.

(POE) Octylphenols, CAS# 9036–19–5

1. POE (7) Octylphenol. The octyl phenol ethoxylates are quite similar to the nonyl phenol versions of this class of surfactants. The 7–mole component is produced by Henkel under the designation Hyonic® OP–7 [41][70]. Assorted properties are shown in Table 6–24.

Table 6–24

Typical Properties of POE (7) Octylphenol

Property	Value
HLB	12.0
pH, 2% in Distilled Water	6–8
Color, Gardner	1 max
Cloud Point, 1%, °C	23–27
Viscosity, cSt, 100°F	115

2. POE (10) Octylphenol. This is a product similar to the 8–mole component, but with a much higher cloud point. It is also made by Henkel as Hyonic® OP–10 [41][71]. Various features are listed in Table 6–25.

Table 6–25

Typical Properties of POE (10) Octylphenol

Property	Value
HLB	13.6
pH, 1% in Distilled Water	6–8
Color, Gardner	1 max
Cloud Point, 1%, °C	62–68
Viscosity, cSt, 100°F	117

(POE) Nonylphenols, CAS# 9016–45–9

The largest volume of all of the alkylphenols used commercially are the nonylphenols. The amount of ethylene oxide reacted with the base nonylphenol varies from 1 to 40 moles resulting in products having water solubility ranging from insoluble   liquids to soluble solids. These are described in the following classifications.

1. POE (1.5) Nonylphenol. The one mole alkoxylate is an oil–soluble liquid emulsifier that might be effective in the preparation of mineral oil dispersions. It is produced by Ethox as Ethal® NP–1.5 [39][72] and by Henkel as Trycol® 6960 [41][73]. Some properties are listed in Table 6–26.

Table 6–26

Typical Properties of POE (1.5) Nonylphenol

Property	Value
HLB	4.6
pH, 5%	5.5–7.5
Color, Gardner	1 max
Acid Value	1.0 max
Hydroxyl Number	190–210
Cloud Point, 1%, °C	<25
Viscosity, cSt, 100°F	150

2. POE (4) Nonylphenol. Another oil soluble nonylphenol ethoxylate, the 4–mole isomer can be used as an intermediate for anionic surfactants and as a defoamer in certain detergent and emulsion formulations. It is also an excellent product to blend with other surfactants to optimize emulsion formulations. This material is made by Ethox as Ethal® NP–4 [39][74], by Henkel as Trycol® 6961 [41][75], by Stepan as Makon® 4 [76], and by Witco as Witconol® NP–40 [77]. Some of the normal attributes of POE (4) nonylphenol are listed in Table 6–27.

Table 6–27

Typical Properties of POE (4) Nonylphenol

Property	Value
HLB	8.9
pH, 5%	5–8
Color, Gardner	2 max
Acid Value	1.0 max
Hydroxyl Number	135–150
Cloud Point, 1%, °C	<25
Viscosity, cSt, 100°F	472

3. POE (6) Nonylphenol. The change from four to six moles of ethylene oxide yields a product that is water dispersible. It is a low foaming scouring agent in cold water and effective as a mineral oil emulsifier. This component is produced by Ethox using the name Ethal® NP–6 [39][78], by Henkel as Trycol® 6962 [41][79], by Hoechst–Celanese called Afilan® 6NPR [80], by Stepan as Makon® 6 [81], and by Witco with the trade name Witconol® NP–60 [77]. Various properties are given in Table 6–28.

Table 6–28

Typical Properties of POE (6) Nonylphenol

Property	Value
HLB	10.9
pH, 5%	5–8
Color, Gardner	2 max
Water, %	1 max
Acid Value	1.0 max
Hydroxyl Number	111–121
Cloud Point, 1%, °C	<25
Viscosity, cSt, 100°F	100

4. POE (8) Nonylphenol. There are three nonylphenol ethoxylates that are very similar in composition with eight, nine, and ten moles of ethylene oxide. They are all soluble in cold water and the properties are similar. The 8–mole version is made by Stepan as Makon® 8 [82] and by Witco called Witconol® NP–80 [77]. Characteristics are shown in Table 6–29.

Table 6–29

Typical Properties of POE (8) Nonylphenol

Property	Value
HLB	12.0
pH, 5% IPA/Water	8
Color, APHA	100 max
Cloud Point, 1%, °C	23–26
Viscosity, cPs, 25°C	288

5. POE (9) Nonylphenol. The 9–ethylene–oxide–mole alkylphenol is an effective wetting agent, emulsifier, and general purpose detergent. It is produced by Ethox with the trade name Ethal® NP–9 [39][83], by Henkel where it is called Trycol® 6964[41][84], and also by Witco as Witconol® NP–90 [77]. Classical properties are displayed in Table 6–30.

Table 6–30

Typical Properties of POE (9) Nonylphenol

Property	Value
HLB	13.0
pH, 5%	5.5–8
Color, APHA	100 max
Acid Value	1.0 max
Hydroxyl Number	85–95
Draves Wetting, Seconds (0.1% @ 25°C)	9
Cloud Point, 1%, °C	52–56
Viscosity, cSt, 100°F	112

6. POE (10) Nonylphenol. This is one of the most commonly used nonyl–phenols as a detergent and emulsifier. In the textile industry it is employed as a dye dispersant, scour, cotton detergent, and lubricant emulsifier. It is made by Henkel as Trycol® 6974 [41][85], by Stepan as Makon® 10 [86] and by Witco as Witconol® NP–100 [77]. General features are listed in Table 6–31.

Table 6–31

Typical Properties of POE (10) Nonylphenol

Property	Value
HLB	13.5
pH, 5%	6–8
Color, APHA	50 max
Hydroxyl Number	80–100
Cloud Point, 1%, °C	61–65
Viscosity, cSt, 100°F	111

7. POE (14) and (15) Nonylphenol. The five additional moles of ethylene oxide on nonylphenol results in a product with a significant increase in viscosity. The 14–mole variant is made by Stepan as Makon® 14 [87] while Henkel sells the 15–mole isomer as Trycol® 6952 [41][88]. Table 6–32 displays some typical properties.

Table 6–32

Typical Properties of POE (15) Nonylphenol

Property	Value
HLB	15.0
pH, 5%	5–8
Color, Gardner	1 max
Cloud Point, 1%, °C	94–100
Viscosity, cSt, 100°F	139

8. POE (20) Nonylphenol. This alkoxylated nonylphenol is a soft solid that is a good high temperature detergent and it exhibits good electrolytic stability. Ethox makes this component where it is sold as Ethal® NP–20 [39] [89], and by Henkel who markets it as Trycol® 6967 [41][90]. The properties are shown in Table 6–33.

Table 6–33

Typical Properties of POE (20) Nonylphenol

Property	Value
HLB	16.0
pH, 5%	5.0–6.5
Color, Gardner	1 max
Moisture, %	1 max
Cloud Point, 5% Saline, °C	88
Hydroxyl Number	46–55
Melting Point, °C	34

9. POE (30) Nonylphenol. Continuing the list of nonylphenol ethoxylates, this is very similar to the 20–mole variant but with a slightly higher HLB and a higher melting point. It is produced by Henkel as Trycol® 6968 [41][91] and by Stepan as Makon® 30 [92]. Some specifications are in Table 6–34.

Table 6–34

Typical Properties of POE (30) Nonylphenol

Property	Value
HLB	17.1
pH, 5%	5–8
Color, Gardner	1 max
Hydroxyl Number	32–40
Cloud Point, 5% Saline, °C	93
Melting Point, °C	43

10. POE (40) Nonylphenol and POE (40) Nonylphenol/70% Solution. A high ethylene oxide content in this product makes it useful in emulsion polymerization and in blending with lower HLB products as a co–emulsifier. Since the 100% active alkyloxylate is a solid, it is also sold as a 70% solution for easier handling. The solid is made by Henkel as Trycol® 6957 [41][93]. Attributes of this solid are listed in Table 6–35.

Table 6–35

Typical Properties of POE (40) Nonylphenol

Property	Value
HLB	17.8
pH, 5%	5.5–7.5
Color @ Melt, Gardner	1 max
Hydroxyl Number	27–37
Cloud Point, 5% Saline, °C	90
Melting Point, °C	44

The solution is produced by Ethox and sold as Ethal® NP–407 [39] [94], and by Henkel as Trycol® 6970 [41]. Some of the properties of this solution are shown in Table 6–36.

Table 6-36

Typical Properties of POE (40)/70% Nonylphenol

Property	Value
HLB	17.8
pH, 5%	5.0–6.5
Color, Gardner	1 max
Moisture, %	28–32

(POE) Dinonylphenols, CAS# 9014–93–1

In the reaction between the nonene isomers and phenol to prepare the substituted phenol, if the ratio of nonene:phenol is 2:1, the yield is about 33% monoalkylphenol and 67% dialkylphenol [60]. The structural formula for dinonylphenol is:

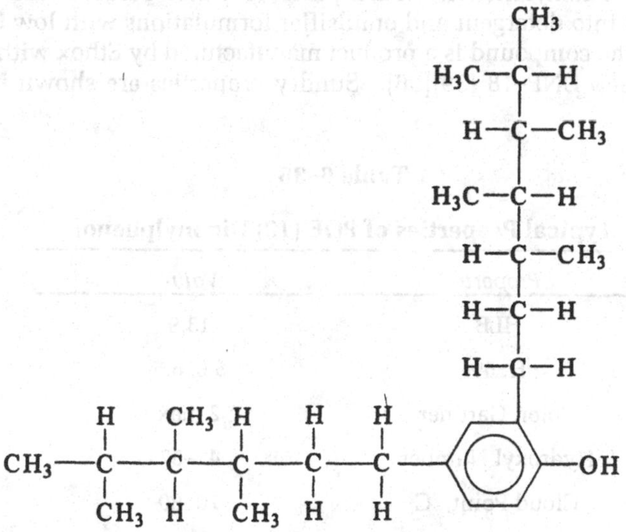

Ethoxylation proceeds through the normal reactions in the preparation of dialkyl ethoxylates. These materials are used as emulsifiers and detergents but are probably found in few spin finishes since the structure would probably result in high friction.

1. POE (8) Dinonylphenol. A compound that is useful as a low foaming emulsifier in solvent based systems since it is only water dispersible. It is produced by Ethox as Ethal® DNP–8 [39][95], and by Henkel as Trycol® 6985 [41]. Some of the properties are given in Table 6–37.

Table 6–37

Typical Properties of POE (8) Dinonylphenol

Property	Value
HLB	10.4
pH, 5%	5.0–6.5
Color, Gardner	2 max
Hydroxyl Number	77–84
Cloud Point, °C	<25
Viscosity, cSt, °F	173

2. POE (18) Dinonylphenol. This paste is water soluble and can be compounded into detergent and emulsifier formulations with low foaming properties. The compound is a product manufactured by Ethox with a trade name of Ethal® DNP–18 [39][96]. Sundry properties are shown in Table 6–38.

Table 6–38

Typical Properties of POE (18) Dinonylphenol

Property	Value
HLB	13.9
pH, 5%	5.0–6.5
Color, Gardner	2 max
Hydroxyl Number	41–57
Cloud Point, °C	70–80
Moisture, %	2.0 max

3. POE (150) Dinonylphenol and POE (150) Dinonylphenol, 50% Solution. The 150 mole dinonylphenol is a solid made by Henkel as Trycol® 6988 [41] [97]. Specifications are displayed in Table 6–39.

Table 6–39

Typical Properties of POE (150) Dinonylphenol

Property	Value
HLB	19.0
pH, 5%	5–8
Color, Gardner (Molten)	2 max
Hydroxyl Number	17–22
Cloud Point in 5% Saline, °C	82–88

Since this material is a solid, a 50% solution is available for ease of handling. The solution is used as a dyeing assistant, especially for acrylic, nylon, and wool fibers. It is also an excellent boarding lubricant in hosiery manufacturing and is also a component in hard surface cleaners. The solution is made by Ethox with the name Ethal® DNP–150/50% [39] and by Henkel as Trycol® 6989 [41][98]. Typical properties are listed in Table 6–40.

Table 6–40

Typical Properties of POE (150)/50% Dinonylphenol

Property	Value
HLB	19.0
pH, 5%	5–8
Color, Gardner (Molten)	1 max
Viscosity, cSt, 100°F	367
Moisture, %	51 max
Cloud Point, 2% in 5% Saline, °C	82–88

6.4.4 Ethoxylated Glycerides

Some of the most effective emulsifiers in fiber finish technology are the ethoxylated glycerides. Most commonly used are the derivatives of castor oil and hydrogenated castor oil. Castor oil is obtained from the seed of the castor plant, which grows in the tropics as a perennial plant and often becomes a small tree reaching a height of 30 feet or more. At one time this plant was grown extensively in the United States as an ornamental plant, but it is seen much less frequently now. Commercially, the major producer of castor seeds is Brazil. These seeds, called castor beans, weigh approximately 0.3 to 0.5 grams and are not edible (they are extremely poisonous).

Most of the oil is extracted from the seed by cold pressing, although some solvent extraction is used to prepare the oil for medicinal purposes. The residues from pressing or solvent extraction contain the toxin ricin, an albumin type chemical. The LDDo is 500 μg/kg by mouth and ingestion of as few as two seeds could be fatal [99]. Fortunately, the oil is non–toxic.

Castor oil is a triglyceride of a mixture of fatty acids with only about 2.5 percent saturated acids of all types, 7.4 percent oleic, 3.1 percent linoleic, and 87 percent ricinoleic. Ricinoleic acid is not commonly found in other plants and has an unusual molecular structure. The chemical name is 12–hydroxy–9–octadecenoic acid and the structure is:

$$
\underset{\underset{H}{|}}{\overset{\overset{OH}{|}}{CH_3(CH_2)_5C}}CH_2\overset{\overset{H}{|}}{C}=\overset{\overset{H}{|}}{C}(CH_2)_7COOH
$$

The hydroxyl group in the chain is the normal site for ethoxylation; however, ethylene oxide may be inserted into glyceride linkage under certain conditions containing water, heat, and alkaline catalyst [100]. The normal structure of ethoxylated castor oil is:

$$
\begin{array}{l}
CH_2O\overset{\overset{O}{\|}}{C}(CH_2)_7CH=CHCH_2\overset{\overset{O(CH_2CH_2O)_xH}{|}}{C}(CH_2)_5CH_3 \\
CHO\overset{\overset{O}{\|}}{C}(CH_2)_7CH=CHCH_2\overset{\overset{O(CH_2CH_2O)_yH}{|}}{C}(CH_2)_5CH_3 \\
CH_2O\overset{\overset{O}{\|}}{C}(CH_2)_7CH=CHCH_2\overset{\overset{O(CH_2CH_2O)_zH}{|}}{C}(CH_2)_5CH_3
\end{array}
$$

Hydrogenated castor oil has a similar structure, except the alkyl chain has been saturated. This hydrogenation step significantly reduces varnish deposits on hot surfaces. The level of ethylene oxide in the commercially available materials $(x + y + z)$ varies from 5 to 200.

(POE) Castor Oils, CAS# 61791–12–6

1. POE (5) Castor Oil. The five moles of ethylene oxide are so low that this molecule is insoluble in water but soluble in oils. It can be useful in emulsifying mineral oil, but the principal use is as a dispersant for pigments and clays. This compound is made by Abitec as Acconon® CA–5 [101], by Ethox as Ethox® CO–5 [102][103], and by Hoechst–Celanese and called Afilan® 5EL [104]. Features of POE (5) castor oil are listed in Table 6–41.

Table 6–41

Typical Properties of POE (5) Castor Oil

Property	Value
HLB	4.0
pH, 5%	6–8
Color, Gardner	4 max
Acid Value	1.5 max
Hydroxyl Number	128–140
Saponification Number	138–153
Iodine Value	63–73
Moisture, %	1 max

2. POE (16) Castor Oil. A water dispersible emulsifier, this component is one of the most valuable finish additives for emulsifying lubricant esters. As with the other castor oil products, the unsaturated linkage causes varnish formation on hot surfaces. A 15–mole material is produced by Abitec as Acconon® CA–15 [105] and the 16–mole version is available from Ethox as Ethox® CO–16 [102][106] and from Henkel with the name Trylox® 5902 [41] [107]. Classical properties are listed in Table 6–42.

Table 6-42

Typical Properties of POE (16) Castor Oil

Property	Value
HLB	8.6
pH, 5%	6–8
Color, Gardner	4 max
Acid Value	2 max
Hydroxyl Number	93–105
Saponification Number	95–105
Iodine Value	43–50
Viscosity, cSt, 100°F	546
Moisture, %	1 max

3. POE (25) Castor Oil. This material is widely used as an emulsifier and dispersant. When blended with other emulsifiers, it is especially valuable. Even with 25 moles of ethylene oxide, the compound remains water dispersible and not water soluble. Sundry properties are given in Table 6–43.

Table 6-43

Typical Properties of POE (25) Castor Oil

Property	Value
HLB	10.8
pH, 5%	5–7.5
Color, Gardner	5 max
Acid Value	2 max
Hydroxyl Number	70–85
Saponification Number	75–85
Viscosity, cSt, 100°F	396
Moisture, %	2 max

It is manufactured by Ethox as Ethox® CO–25 [102][108], by Henkel as Trylox® 5904 [41][109], and by Stepan as Stepantex® CO–25 [43][110].

4. POE (30) Castor Oil. The compound with 30 moles of EO is water soluble and manufactured by a large number of nonionic surfactant suppliers. One reason for this is that it is not only an emulsifier and dispersant, but it is a stabilizer for latex emulsions. The producers are Ethox with the designation Ethox® CO–30 [102][111], by Henkel and called Trylox® 5906 [41][112], by ICI as Cirrasol® G–1293 [113], by Stepan with the trade name Stepantex® CO–30 [110], and by Witco as Witconol® 5960 [47]. Table 6–44 lists the attributes of the compound.

Table 6–44

Typical Properties of POE (30) Castor Oil

Property	Value
HLB	11.8
pH, 5%	5–7.5
Color, Gardner	4 max
Acid Value	2 max
Hydroxyl Number	70–90
Saponification Number	68–80
Viscosity, cSt, 100° F	309

5. POE (36) Castor Oil. This material is a slightly higher HLB version in the castor oil ethoxylate series. It is produced by Ethox as Ethox® CO–36 [102] [114], by Henkel and called Trylox® 5907 [41][115], and by Stepan as Stepantex® CO–36 [110]. Properties are shown in Table 6–45.

Table 6–45

Typical Properties of POE (36) Castor Oil

Property	Value
HLB	12.6
pH, 5%	5.5–8
Color, Gardner	4 max
Acid Value	2 max
Hydroxyl Number	68–85
Saponification Number	57–72
Viscosity, cSt, 100° F	363

6. POE (40) Castor Oil. This cold–water–soluble surfactant is another in the homologous series of castor oil derivatives. The properties are very similar to the 36–mole compound but with a slightly higher HLB. It is also made by Ethox under the trade name Ethox® CO–40 [102][116], by Henkel as Trylox® 5909 [41][117], and also by Stepan as Stepantex® CO–40 [110]. Typical properties are listed in Table 6–46.

Table 6–46

Typical Properties of POE (40) Castor Oil

Property	Value
HLB	13.0
pH, 5%	5–7
Color, Gardner	4 max
Acid Value	2 max
Hydroxyl Number	77–89
Saponification Number	57–64
Iodine Number	24–30
Viscosity, cSt, 100° F	313

7. POE (80) Castor Oil. An ethoxylated castor oil that is only available as a 90% solution, this compound is generally utilized as a dyeing assistant and fabric lubricant. It can be blended with lower HLB surfactants for emulsification. It is made by Ethox as Ethox® CO–81 [102][118]. The characteristics are shown in Table 6–47.

Table 6–47

Typical Properties of POE (80) Castor Oil

Property	Value
HLB	15.9
pH, 5%	5.0–7.0
Color, Gardner	3 max
Acid Value	2 max
Hydroxyl Number	42–55
Saponification Number	28–38

8. POE (200) Castor Oil and POE (200)/50% Solution. Although the 200–mole EO castor oil is not an especially good emulsifier, it is included in this section to complete the range of ethoxylates available. The high level of ethylene oxide results in a product that is a waxy solid with a melting point of about 38°C. It is useful as a fabric softener and the high EO content causes the material to be hygroscopic and thus a good antistatic agent. The anhydrous wax is produced by Ethox as Ethox® CO–200 (A) [102][119], by Hoechst–Celanese as Afilan® 200EL [120], and by ICI as Cirrasol® G–1300 [113]. The solution is sold by Ethox as Ethox® CO–200/50% [101] and by Henkel as Trylox® 5918 [41]. Features of the anhydrous compound are depicted in Table 6–48.

Table 6–48

Typical Properties of POE (200) Castor Oil

Property	Value
HLB	18.1
pH, 5%	6.0–7.0
Color, Gardner	2 max
Acid Value	1 max
Hydroxyl Number	20–33
Saponification Number	14–20
Moisture, %	1 max

(POE) Hydrogenated Castor Oils, CAS# 61788–85–0

Castor oil is hydrogenated by the addition of hydrogen in the presence of a catalyst. The first reference to the use of catalytic hydrogenation appears to be that of Debus in 1863. In 1897 Paul Sabatier and Jean Baptiste Senderens at the University of Toulouse found that olefins passed over cobalt, iron, copper, or platinum gave the saturated versions of those gases. Sabatier worked in this area until his retirement in 1930. He was awarded the Nobel Prize in Chemistry, jointly with Victor Grignard, in 1912. Senderens went into industry and introduced contact catalysis into industrial organic chemistry. He was a very religious man and became an abbé and later a cannon in the Roman Catholic Church [121].

Hydrogenation of castor oil not only saturates the fatty acid chain and thus significantly reduces varnish formation, but the ethoxylated compound is a superior emulsifying agent. When finishes are prepared using either

ethoxylated castor oil or ethoxylated hydrogenated castor oil with the same
ethylene oxide content and with the same emulsifier content, the one with
ethoxylated hydrogenated castor oil is the better formulation. The particle
size is smaller and the finish is more stable during storage when compared
to the non–hydrogenated product.

1. POE (16) Hydrogenated Castor Oil. Although the HLB of the hydrogenated
component is the same as the non–hydrogenated version, this material is an
excellent emulsifier for esters. It is produced by both Ethox with the name
Ethox® HCO–16 [102][122] and by Henkel designated Trylox® 5921
[41][123]. The properties are listed in Table 6–49.

Table 6–49

Typical Properties of POE (16) Hydrogenated Castor Oil

Property	Value
HLB	8.6
pH, 5%	6.0–8.0
Color, Gardner	2 max
Acid Value	1.5 max
Hydroxyl Number	85–102
Saponification Number	95–105
Iodine Value	1.5 max
Moisture, %	1 max

2. POE (25) Hydrogenated Castor Oil. This compound is a versatile emulsifier
and lubricant for softeners in general textile processes and in fiber finishes.
It is available from Ethox as Ethox® HCO–25 [102][124], from Henkel as
Trylox® 5922 [41][125], from Hoechst–Celanese designated Afilan® 25HEL
[126], and from ICI as Cirrasol® G–1292 [113]. Representative attributes are
presented in Table 6–50.

Table 6–50

Typical Properties of POE (25) Hydrogenated Castor Oil

Property	Value
HLB	10.8
pH, 5%	5.0–7.0
Color, Gardner	2 max
Acid Value	2 max
Hydroxyl Number	70–84
Saponification Number	75–85
Iodine Value	1.0 max
Moisture, %	1 max

3. POE (200) Hydrogenated Castor Oil, 50% Solution. Even the 50% solution is a high viscosity liquid since the solid is difficult to handle. This liquid is useful as an emulsifier stabilizer and to increase the emulsion viscosity. It is produced by Ethox as Ethox® HCO–200/50% [102] and by Henkel as Trylox® 5925 [41]. Assorted properties are listed in Table 6–51.

Table 6–51

Typical Properties of POE (200) Hydrogenated Castor Oil 50% Solution

Property	Value
HLB	18.1
Color, Gardner	1 max
Viscosity, cSt, 100°F	5200
Moisture, %	55 max

POE (10) Coconut Glyceride, CAS# 68201–46–7

This ethoxylated glyceride is the only commercially available material in this classification not based on castor oil. It is prepared by reacting coconut fatty acids with glycerine to form the diglyceride followed by ethoxylation on the remaining hydroxyl group. A possible structure is:

$$H-\underset{\underset{H}{|}}{\overset{\overset{H}{|}}{C}}-O-\overset{\overset{O}{||}}{C}-(CH_2)_{10}-CH_3$$

$$H-\underset{\underset{|}{|}}{C}-O-\overset{\overset{O}{||}}{C}-(CH_2)_{10}-CH_3$$

$$H-\underset{\underset{H}{|}}{C}-O-(CH_2CH_2O)_{10}H$$

It is manufactured by Ethox where it is called Ethox® 1212 [102][127]. The typical properties are shown in Table 6–52.

Table 6–52

Typical Properties of POE (10) Coconut Oil

Property	Value
HLB	12.6
pH, 5%	5.6–7.0
Color, Gardner	1 max
Acid Value	2 max
Hydroxyl Number	100–105
Saponification Number	95–105
Moisture, %	2 max

6.4.5 Ethoxylated Fatty Acids

Although most of the ethoxylated fatty acids have already been described in Chapter 5, a few of these chemicals are not regarded as either cohesive agents or fiber lubricants. These are based on tall oil, or monomer acid, or are diesters.

(POE) Monomer Acids, CAS# 68876–85–7

Monomer acid is produced when oleic acid, probably derived from tall oil [128], is reacted in the presence of a clay catalyst. The resulting mixture consists of monomer acid, dimer acid, and trimer acid [129]. This fatty acid has an unsaturated, methyl substituted C_{18} carbon chain with both the double bond and the methyl group randomly distributed along that chain. Another name, based on the I.U.C. system, is methyl heptadecanoic acid. Since the structure is not clearly defined, it will not be displayed in the discussion.

1. POE (4) Monomerate. The cost of ethoxylated monomer acids is relatively low so they find use in surfactant formulations as well as emulsifiers. Viscosities are low, but monomerates are unsaturated and highly colored. The 4–mole EO monomerate is produced by Henkel as Trydet® 2691 [41] [130]. Distinguishing features are shown in Table 6–53.

Table 6–53

Typical Properties of POE (4) Monomerate

Property	Value
HLB	6.6
pH, 5%	4.5–6.0
Color, Gardner	6 max
Acid Value	2 max
Hydroxyl Number	115–125
Iodine Value	40–55
Saponification Number	110–130
Viscosity, cSt, 100°F	44
Moisture, %	1 max

2. POE (8) Monomerate. A higher HLB homolog, this surfactant is very cost effective in many formulations. It is water dispersible but soluble in mineral oil. The chemical is made by Ethox as Ethox® MA–8 [131][132] and by Henkel as Trydet® 2692 [41][133]. Properties are listed in Table 6–54.

Table 6–54

Typical Properties of POE (8) Monomerate

Property	Value
HLB	10.1
pH, 5%	6.0–7.0
Color, Gardner	9 max
Acid Value	2 max
Hydroxyl Number	70–94
Saponification Number	85–92
Viscosity, cSt, 100°F	59
Moisture, %	2 max

3. POE (15) Monomerate. This material is similar to the 8–mole monomerate but with a higher HLB. It is made by Ethox as Ethox® MA–15 [131][134]. Sundry properties are listed in Table 6–55.

Table 6–55

Typical Properties of POE (15) Monomerate

Property	Value
HLB	13.7
pH, 5%	6.0–7.0
Color, Gardner	7 max
Acid Value	2 max
Hydroxyl Number	53–65
Saponification Number	51–63
Moisture, %	2 max

(POE) Tall Oil, CAS# 61791–00–2

Tall oil is a by–product of the alkaline digestion of southern pine trees for the paper industry. Pine trees are converted to pulp by the Kraft process in which sodium hydroxide is used to separate the desired cellulosic fibers from the undesired lignins, rosins, fatty acids and other materials in the wood. The fatty acids, mainly oleic and linoleic, and rosin end up as their sodium salts in a smelly liquid, called black liquor. Acidification gives rosin and fatty acids, which may be separated by distillation [135]. The composition of crude tall oil is about 48% fatty acids and 42% rosin acids. A typical analysis of tall oil fatty acids reveals the distribution shown in Table 6–56 [136].

Table 6–56

Fatty Acid Content of Tall Oil

Acid	Percent
Stearic	2.1
Oleic	48.5
Linoleic	43.1
Pinolenic	3.5
Other	2.8

(Ref. [136]. Reprinted by permission of John Wiley & Sons, Inc.)

1. POE (8) Tallate and PEG (400) Tall Oil Fatty Acid Monoester. These products are high rosin acid base surfactants that are good dispersants and low–foaming detergents. As indicated in Table 6–56, they are highly unsaturated and they find limited use as spin finish emulsifiers. They can be useful, however, in textile processing applications. The polyoxyethylene ester is a product from Ethox with the name Ethox® TO–8 [131][137] and the polyethylene glycol ester is made by Stepan as Stepantex® TMO–9 [43]. Properties representative of these products are presented by Table 6–57.

Table 6–57

Typical Properties of POE (8) Tallate

Property	Value
HLB	9.8
Color, Gardner	14 max
Acid Value	2 max
Hydroxyl Number	80–92
Saponification Number	46–55
Moisture, %	1 max

2. POE (16) Tallate and PEG (600) Tall Oil Fatty Acid Monoester. These
products are water soluble detergents and co–emulsifiers for various solvents.
The POE ester is made by Ethox as Ethox® TO–16 [131][138] and by Henkel
with the name Trydet® 2682 [41][139] while the PEG ester is produced by
Stepan as Stepantex® TMO–14 [43]. Typical properties are listed in Table
6–58.

Table 6–58

Typical Properties of POE (16) Tallate

Property	Value
HLB	13.4
pH, 5%	6.5–7.5
Color, Gardner	12 max
Acid Value	2 max
Hydroxyl Number	50–65
Iodine Value	20–35
Saponification Number	25–40
Moisture, %	2 max

PEG (400) Sesquioleate, CAS# 61791–06–8

The term sesqui– means one and one–half so this compound is a 400
molecular weight polymer of ethylene glycol esterified with 1½ moles of
oleic acid, leaving some unreacted hydroxyl end groups. It is made by Ethox
as Ethox® 2966 [131] and by Henkel as Emerest® 2647 [41]. Some features
are described in Table 6–59.

Table 6–59

Typical Properties of PEG (400) Sesquioleate

Property	Value
HLB	9.4
Color, Gardner	3 max
Viscosity, cSt, 100°F	50

(PEG) Dilaurates, CAS# 9005–02–1

      Polyethylene glycol dilaurate esters are low foaming emulsifiers that are also good lubricants in textile systems. The chemical structure is:

$$CH_3(CH_2)_{10}\overset{\overset{\displaystyle O}{\|}}{C}-O-(CH_2CH_2O)_{\overline{x}}-\overset{\overset{\displaystyle O}{\|}}{C}(CH_2)_{10}CH_3$$

1. PEG (200) Dilaurate. This water dispersible lubricant–emulsifier is the lowest molecular weight dilaurate commercially available. It is made by Ethox as Ethox® DL–5 [140][141] and by Stepan as Kessco® PEG 200 DL [43] [142]. Assorted properties are shown in Table 6–60.

Table 6–60

Typical Properties of PEG (200) Dilaurate

Property	Value
HLB	6.1
pH, 5%	5.5–7.5
Color, Gardner	3 max
Acid Value	10 max
Hydroxyl Number	35–45
Saponification Number	168–195
Moisture, %	2 max

2. PEG (400) Dilaurate. Even with the increase in the amount of poly–ethylene glycol, this compound is still only water dispersible. It is also a good lubricant and provides low foaming emulsification. It is produced by Ethox as Ethox® DL–9 [140][143] and by Stepan as Kessco® PEG 400 DL [43]. Representative characteristics are presented in Table 6–61.

Table 6–61

Typical Properties of PEG (400) Dilaurate

Property	Value
HLB	10.8
Color, Gardner	3 max
Acid Value	7.5 max
Hydroxyl Number	35 max
Saponification Number	131–141
Moisture, %	1 max

3. PEG (600) Dilaurate. A water soluble lubricant–emulsifier, this soft–solid homolog is made by Stepan as Kessco® PEG 600 DL [43][144]. Various properties are listed in Table 6–62.

Table 6–62

Typical Properties of PEG (600) Dilaurate

Property	Value
HLB	12.0
Color, Gardner	1 max
Pour Point, °F	75

(PEG) Dioleates, CAS# 9005–07–6

If a formulator allows for the unsaturation of oleic acid and the problems that this causes when in contact with hot surfaces, the dioleates are excellent oil and fat emulsifiers and low–viscosity lubricants. The structure of this class of chemicals is:

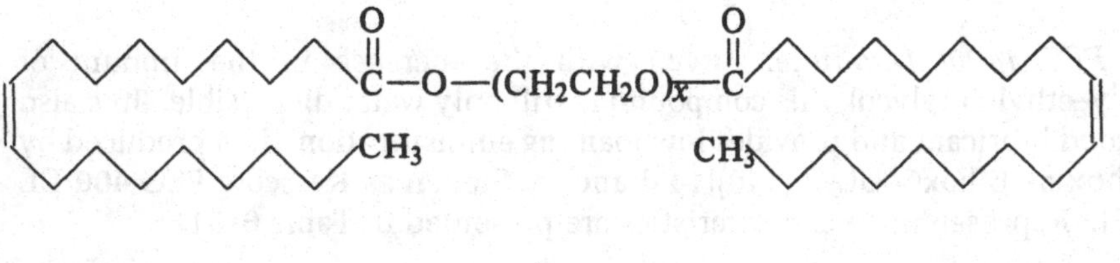

1. PEG (400) Dioleate. This water dispersible emulsifier is especially valuable in dispersing solvents and other industrial process fluids. It is manufactured by Ethox with the designation Ethox® DO–9 [140][145], by Henkel as Emerest® 2648 [41][146], and by Stepan as Kessco® PEG 400 DO [43]. Typical properties are displayed in Table 6–63.

Table–63

Typical Properties of PEG (400) Dioleate

Property	Value
HLB	8.8
pH, 5%	5.0–7.0
Color, Gardner	5 max
Acid Value	10 max
Hydroxyl Number	40 max
Saponification Number	105–123
Iodine Value	55 max
Moisture, %	1 max

2. PEG (600) Dioleate. A compound that is similar to the previous material, it is made by Ethox as Ethox® DO–14 [140][147], by Henkel as Dyafac® PEG 6DO [41][148], and by Stepan as Kessco® PEG 600 DO [43]. Some attributes are listed in Table 6–64.

Table 6–64

Typical Properties of PEG (600) Dioleate

Property	Value
HLB	10.0
Color, Gardner	5 max
Acid Value	10 max
Hydroxyl Number	30 max
Saponification Number	92–102
Moisture, %	1 max

PEG (6000) Distearate, CAS# 9005–08–7

When removal of lubricants and hydrophilic oils from fabric surfaces is wanted, this solid component can be effectively utilized. Many aqueous solutions have high viscosities and it can be a thickener for emulsions. It is produced by Stepan as Kessco® PEG 6000 DS [43][149]. Some properties are listed in Table 6–65.

Table 6–65

Typical Properties of PEG (6000) Distearate

Property	Value
HLB	18.0
Acid Value	9 max
Color, Gardner	1 max
Pour Point, °F	135

6.4.6 Ethoxylated Sorbitol Esters

Sorbitol occurs naturally in many fruits, particularly in the berries of the mountain ash and in certain types of seaweed. It is commercially synthesized by the catalytic reduction of glucose.

$$
\begin{array}{ccc}
\text{CHO} & & \text{CH}_2\text{OH} \\
| & & | \\
\text{H—C—OH} & & \text{H—C—OH} \\
| & & | \\
\text{HO—C—H} & \xrightarrow[\text{Catalyst}]{\text{H}_2} & \text{HO—C—H} \\
| & & | \\
\text{H—C—OH} & & \text{H—C—OH} \\
| & & | \\
\text{H—C—OH} & & \text{H—C—OH} \\
| & & | \\
\text{CH}_2\text{OH} & & \text{CH}_2\text{OH}
\end{array}
$$

Ethoxylated sorbitol esters are first ethoxylated and then esterified with a fatty acid. A probable structure of these hexa–esters is:

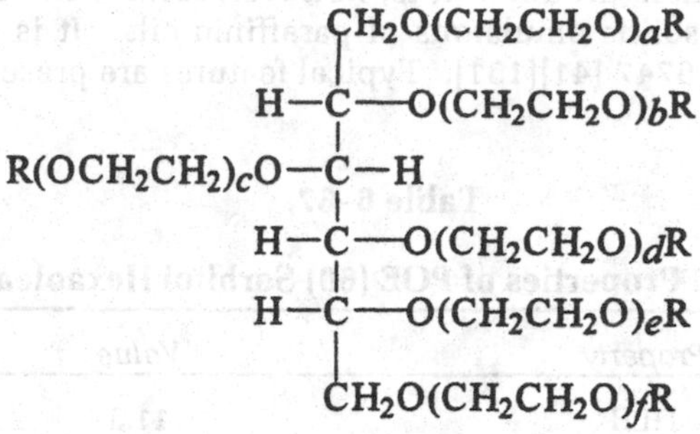

$$CH_2O(CH_2CH_2O)_aR$$
$$H-C-O(CH_2CH_2O)_bR$$
$$R(OCH_2CH_2)_cO-C-H$$
$$H-C-O(CH_2CH_2O)_dR$$
$$H-C-O(CH_2CH_2O)_eR$$
$$CH_2O(CH_2CH_2O)_fR$$

(POE) Sorbitol Hexaoleates, CAS# 57171–56–9

Ethoxylated sorbitol esters are nonionic, hydrophilic surfactants that are relatively low foaming emulsifiers for petroleum oils, vegetable oils, and organic solvents. They are not only emulsifiers but act as dispersants, wetting agents, lubricants, and solubilizers in the textile industry. The esters that are commercially available contain six moles of oleic acid.

1. POE (40) Sorbitol Hexaoleate. In this substance, the total number of moles of ethylene oxide $(a + b + c + d + e + f)$ in the above structure equals 40, with an unpredictable distribution of EO along the molecule. The fatty acid, R, is oleic. It is made by Henkel under the trade name Trylox® 6746 [41] [150]. Normal properties of this material are shown in Table 6–66.

Table 6–66

Typical Properties of POE (40) Sorbitol Hexaoleate

Property	Value
HLB	10.4
Acid Value	10 max
Saponification Value	90–100
Color, Gardner	6 max
Moisture, %	0.5 max
Viscosity, cSt, 100°F	120

2. POE (60) Sorbitol Hexaoleate. After 20 additional moles of ethylene oxide have been reacted with sorbitol, this compound has only a slightly higher

HLB than the 40–mole product. It is, however, soluble in mineral oil and gives stable, high solids emulsions of paraffinic oils. It is also made by Henkel as Trylox® 6747 [41][151]. Typical features are presented by Table 6–67.

Table 6–67

Typical Properties of POE (60) Sorbitol Hexaoleate

Property	Value
HLB	11.3
Acid Value	8 max
Saponification Value	82–92
Color, Gardner	6 max
Moisture, %	0.5 max
Viscosity, cSt, 100°F	110

6.4.7 Sorbitan Esters and Ethoxylated Sorbitan Esters

The starting material for the synthesis of sorbitan esters, also called anhydrohexitol esters, is sorbitol, a simple sugar, which is prepared by the high pressure hydrogenation of glucose [152]. Sorbitol is converted into 1,4–sorbitan by dehydration with sulfuric acid [153]. The reaction may be described as:

In commercial practice sorbitan molecules are normally reacted with one or three molecules of fatty acid to yield products used as emulsifiers for many applications. Since most of the sorbitan esters and ethoxylated sorbitan esters are recognized as being low in toxicity [154], these applications include food products, pharmaceuticals, cosmetics, textiles, agricultural chemicals, and detergents.

Mono–substituted sorbitan esters have slightly more than one fatty acid unit per molecule, approaching 1.3 acid molecules on each sorbitan molecule [155]. Although the reaction of the fatty acid on the primary hydroxyl group is kinetically favored, the esterification occurs on all of the available hydroxyls. The structure of the monosubstituted esters is:

$$
\begin{array}{c}
\text{H} \\
| \\
\text{H}-\text{C} \\
| \\
\text{H}-\text{C}-\text{OOCR}_a \qquad \text{O} \\
| \\
\text{R}_b\text{COO}-\text{C}-\text{H} \\
| \\
\text{H}-\text{C} \\
| \\
\text{H}-\text{C}-\text{OOCR}_c \\
| \\
\text{H}-\text{C}-\text{OOCR}_d \\
| \\
\text{H}
\end{array}
$$

where $(a + b + c + d)$ equals 1.3. For the sesquiester the sum is 1.5 and for the triesters, which are very similar in structure but the total of $(a + b + c + d)$ equals about three.

Ethoxylated sorbitan esters are synthesized by first preparing the sorbitan ester and then reacting the esters with ethylene oxide. When the ester has already been formed, a transesterification occurs and the ethylene oxide chain is attached to each hydroxyl group and the ester group rearranges to the end of the ethylene oxide moiety [154]. The structure of the polyoxyethylene sorbitan esters is [156]:

$$
\begin{array}{c}
\text{H} \\
| \\
\text{H}-\text{C} \\
| \\
\text{H}-\text{C}-\text{O}(\text{CH}_2\text{CH}_2\text{O})_w\text{OCR}_a \quad \text{O} \\
| \\
\text{R}_b\text{CO}(\text{OCH}_2\text{CH}_2)_x\text{O}-\text{C}-\text{H} \\
| \\
\text{H}-\text{C} \\
| \\
\text{H}-\text{C}-\text{O}(\text{CH}_2\text{CH}_2\text{O})_y\text{OCR}_c \\
| \\
\text{H}-\text{C}-\text{O}(\text{CH}_2\text{CH}_2\text{O})_z\text{OCR}_d \\
| \\
\text{H}
\end{array}
$$

In this structure $(w + x + y + z)$ equals the number of moles of ethylene oxide and as in the previous structure $(a + b + c + d)$ indicates how many moles of fatty acid are present. A description of the commercially available sorbitan esters is presented in the sections below.

Sorbitan Monolaurate, CAS# 1338–39–2

If the fatty acid is laurate, R in the previous structures is $CH_3(CH_2)_{11}-$. This ester is widely used as an emulsifier, especially in conjunction with ethoxylated sorbitan esters. It is, however, used alone as a dispersant in certain pharmaceuticals, and in metal working fluids. The ester is produced by Henkel as Emsorb® 2515 [41][157], by ICI where it is called Span® 20 [46], and by Witco as Kemester® S–20 [47]. Several properties are shown in Table 6–68.

Table 6–68

Typical Properties of Sorbitan Monolaurate

Property	Value
HLB	8.0
Acid Value	7 max
Saponification Value	158–170
Hydroxyl Value	330–358
Color, Gardner	6 max
Viscosity, cSt, 100°F	1000

Sorbitan Monopalmitate, CAS# 26266–57–9

If about 1.5 moles of palmitic acid (R is $CH_3(CH_2)_{14}-$) is reacted with one mole of sorbitan, the monopalmitate ester is the result. It is produced by ICI as Span® 40 [46] and by Witco as Kemester® S–40 [158]. Some attributes are presented in Table 6–69.

Table 6–68

Typical Properties of Sorbitan Monopalmitate

Property	Value
HLB	6.7
Water Solubility	Dispersible
Pour Point, °C	48

Sorbitan Monostearate, CAS# 1338–41–6

Stearic acid, $R = CH_3(CH_2)_{16}-$, forms the monoester in this product. It is made by ICI and called Span® 60 [46] and by Witco as Kemester® S–60 [158]. The properties are in Table 6–69.

Table 6–69

Typical Properties of Sorbitan Monostearate

Property	Value
HLB	4.7
Acid Number	5–10
Hydroxyl Number	235–260
Saponification value	147–157
Pour Point, °C	53

Sorbitan Monooleate, CAS# 1338–43–8

Oleic acid monoesters of sorbitan have been used in many applications, including acting as a water–in–oil emulsifier in reverse emulsion polymerization. Since it has a low HLB, it is useful in incorporating water in non–aqueous systems. About 1.3 moles of oleic acid is incorporated into 1 mole of sorbitan to form the ester. It is a product of

Henkel where it is called Emsorb® 2500 [41][159], by ICI and known as Span® 80 [46], and by Witco as Kemester® S–80 [47]. A table of typical properties is found in Table 6–70.

Table 6–70

Typical Properties of Sorbitan Monooleate

Property	Value
HLB	4.6
Acid Value	8 max
Saponification Value	149–160
Hydroxyl Value	193–209
Color, Gardner	9 max
Viscosity, cSt, 100°F	360

Sorbitan Sesquioleate, CAS# 8007–43–0

Instead of 1.3 moles of oleic acid, the sesquioleate contains 1.5 moles of that acid. The characteristics are very similar, as shown in Table 6–71. It is made by Henkel as Emsorb® 2502 [41][160].

Table 6–71

Typical Properties of Sorbitan Sesquioleate

Property	Value
HLB	4.5
Acid Value	13 max
Saponification Value	145–160
Hydroxyl Value	185–215
Color, Gardner	8 max
Viscosity, cSt, 100°F	475

Sorbitan Tristearate, CAS# 26658–19–5

The tristearate ester is formed when 3 moles of stearic acid are reacted with sorbitan. When incorporated in fiber finishes, it significantly increases fiber–to fiber friction in acrylic yarn [161]. It is manufactured by ICI with the name Span® 65 [46]. Some properties are listed in Table 6–72.

Table 6–72

Typical Properties of Sorbitan Tristearate

Property	Value
HLB	2.1
Water Solubility	Insoluble
Pour Point, °C	53

Sorbitan Trioleate, CAS# 26266–58–0

This ester is valuable as a dye solubilizer in textile processing and acts as a wetting agent in explosive ammonium nitrate preparations in preventing deterioration during storage [153]. It is also made by ICI as Span® 85 [46] and by Witco as Kemester® S–85 [158]. Features are shown in Table 6–73.

Table 6–73

Typical Properties of Sorbitan Trioleate

Property	Value
HLB	1.8
Water Solubility	Insoluble
Viscosity, cSt, 25°C	210

(POE) Sorbitan Monolaurates, CAS# 9005–64–5

These emulsifiers also exhibit very low toxicity, both by ingestion and dermal application, and provide detergent action in nonirritating shampoos [153]. In other emulsification end uses, these products are often blended with the corresponding sorbitan esters to effect optimum emulsion quality. They are widely used as dispersants for agricultural chemicals and in foods, where they are listed as polysorbates. In one interesting application, POE (20) sorbitan monolaurate was added to soil in experiments to evaluate the effect of surfactants on the growth and yield of vegetables. The result was an increase in plant growth rate and the number of pea pods per plant [162].

1. POE (4) Sorbitan Monolaurate. As with most of the other polyoxyethylene sorbitan monoesters, this product in spin finishes causes high fiber to metal friction on nylon and polyester fibers [161]. It is dispersible in water, soluble in isopropanol, but insoluble in mineral oil. This material is made by ICI as Tween® 21 [46]. Several properties are shown in Table 6–74.

Table 6–74

Typical Properties of POE (4) Sorbitan Monolaurate

Property	Value
HLB	13.3
Water Solubility	Dispersible
Viscosity, cSt, 25°C	600

2. POE (20) Sorbitan Monolaurate. This compound is probably one of the most widely used sorbitan–based emulsifiers and in many instances it is mixed with sorbitan monolaurate to increase its effectiveness. It is a water–soluble emulsifier for oils and fats and a good lubricant for hydrophilic fibers as cotton and rayon. The compound is manufactured by Ethox as Ethsorbox® L–20 [163][164], by Henkel as Emsorb® 6915 [41][165], by ICI with the name Tween® 20 [46], and by Witco as Witconol® 6915 [47]. A list of classical properties is shown by Table 6–75.

Table 6–75

Typical Properties of POE (20) Sorbitan Monolaurate

Property	Value
HLB	16.5
Color, Gardner	6 max
Acid Value	2.0 max
Hydroxyl Number	96–108
Saponification Number	40–50
Moisture, %	3.0 max
Viscosity, cSt, 25°C	400

POE (20) Sorbitan Monopalmitate, CAS# 9005–66–7

The alkyl chain length was increased from 14 carbons in the monolaurate to 16 in this monopalmitate. It is also water soluble and useful as an emulsifier with a slightly lower HLB. It is made by ICI as Tween® 40 [46]. Properties are listed in Table 6–76.

Table 6–76

Typical Properties of POE (20) Sorbitan Monopalmitate

Property	Value
HLB	15.6
Water Solubility	Soluble
Viscosity, cSt, 25°C	500

(POE) Sorbitan Monostearates, CAS# 9005–67–8

As with many other stearate esters, the ethoxylated sorbitan stearates are lubricants and provide some softening properties for hydrophilic fibers. They tend to be soft waxy solids near room temperature.

1. POE (4) Sorbitan Monostearate. This material is water dispersible, soluble in isopropanol, but insoluble in mineral oil. It can be used as a low HLB emulsifier for oils and waxes. Produced by ICI, it is sold under the name Tween® 61 [46]. The published attributes are shown in Table 6–77.

Table 6–77

Typical Properties of POE (4) Sorbitan Monostearate

Property	Value
HLB	9.6
Water Solubility	Dispersible
Pour Point, °C	38

2. POE (20) Sorbitan Monostearate. A water soluble version of sorbitan monostearate alkoxylate is available with the addition of twenty moles of ethylene oxide to the sorbitan stearate monoester. This compound is used in whipped toppings, as a fat emulsifier in pharmaceuticals, and in lotions and hand creams. It is also an effective emulsifier in textile processing as a lubricant and softener [153]. The material is made by Ethox as Ethsorbox® S–20 [163][166] and by ICI as Tween® 60 [46]. Typical properties of the compound are displayed in Table 6–78.

Table 6–78

Typical Properties of POE (20) Sorbitan Monostearate

Property	Value
HLB	15.0
Color, Gardner	5 max
Acid Value	2.0 max
Hydroxyl Number	81–96
Saponification Number	45–55
Moisture, %	3.0 max
Viscosity, cSt, 25°C	550

(POE) Sorbitan Monooleates, CAS# 9005–65–6

The uses of the monooleate esters are similar to the stearates, but they do not exhibit the softening properties of the saturated carbon chains. They are, however, frequently present as emulsifiers in textile processing since they result in low fiber to fiber friction on nylon and polyester [161].

1. POE (5) Sorbitan Monooleate. As one might expect, the low amount of ethylene oxide results in a product that is water and mineral oil dispersible and one that is useful as an emulsifier for industrial lubricants. It is manufactured by Henkel as Emsorb® 6901 [41][167] and by ICI as Tween® 81 [46]. Features are shown in Table 6–79.

Table 6–79

Typical Properties of POE (5) Sorbitan Monooleate

Property	Value
HLB	10.0
Color, Gardner	6
Viscosity, cSt, 25°C	450

2. POE (20) Sorbitan Monooleate. This widely manufactured chemical is a valuable emulsifier in textile and fiber applications, as well as in foods and cosmetics. It is made by Ethox as Ethsorbox® O–20 [163][168], by Henkel as Emsorb® 6900 [46][169], by ICI as Tween® 80 [46], and by Witco as Witconol® 2722 [47]. Various properties are listed in Table 6–80.

Table 6–80

Typical Properties of POE (20) Sorbitan Monooleate

Property	Value
HLB	15.0
Acid Value	2.0 max
Hydroxyl Value	65–80
Saponification Number	45–55
Color, Gardner	7 max
Moisture, %	3 max
Viscosity, cSt, 25°C	425

POE (20) Sorbitan Tristearate, CAS# 9005–71–4

This component is also a good emulsifier for fats and oils and has many applications in textile processing. It is produced by Ethox as Ethsorbox® TS–20 [163][170] and by ICI as Tween® 65 [46]. Typical attributes are presented by Table 6–81.

Table 6–81

Typical Properties of POE (20) Sorbitan Tristearate

Property	Value
HLB	11.1
Acid Value	2.0 max
Hydroxyl Number	44–60
Saponification Number	88–98
Color, Gardner	5.0 max
Moisture, %	3.0 max

POE (20) Sorbitan Trioleate, CAS# 9005–70–3

Trioleate esters of ethoxylated sorbitan compounds are excellent lubricants and emulsifiers for industrial process fluids, including those in the textile industry. The POE (20) ester is made by Ethox as Ethsorbox® TO–20 [163][171], by Henkel as Emsorb® 6903 [41][172], by ICI as Tween® 85 [46], and by Witco as Witconol® 6903 [47]. Sundry properties are shown by Table 6–82.

Table 6–82

Typical Properties of POE (20) Sorbitan Trioleate

Property	Value
HLB	11.1
Acid Value	2.0 max
Hydroxyl Number	39–52
Saponification Number	82–95
Color, Gardner	6 max
Moisture, %	30.
Viscosity, cSt, 25°C	315

6.4.8 Alkyl Polyglycosides

Alkyl polyglycosides are new classes of nonionic surfactants based on glucose and fatty alcohols. The structure is:

In this structure x is predominantly zero to three carbohydrate units and R is an alkyl chain between 8 and 16 carbon atoms. All of the raw materials are based on renewable sources, glucose from corn and the fatty alcohols from coconut and palm kernel oils. They are excellent detergents and wetting agents and also exhibit mildness to the skin and eyes. Although these compounds were initially formulated for formulation into cleaning products, they are also useful in textile processing. Information about these products is listed in Table 6–83 and they are made by Henkel as Glucopon® surfactants [41][173].

Table 6–83

Properties of Alkyl Polyglycosides

Glucopon	225	425	600	625
HLB	13.6	13.1	11.6	12.1
CAS#	68515–73–1	68515–73–1	110615–47–9	110615–47–9
Alkyl Chains Present	8,10	8,10,12,14,16	12,14,16	12,14,16
Average Alkyl Chain	9.1	10.3	12.8	12.8
% Active	70	50	50	50
Ross Miles Foam Height, mm, 0.1%, 49°C	160	160	115	150
Draves Wetting, sec, 0.1% @ 25°C	120	32	20	23
Viscosity, cPs@ 35°C	2,150	300	4,230	6,250

6.4.9 Ethoxylated Alkylamines

A class of surfactants prepared by reacting amines with ethylene oxide was first described by Schöller and Wittwer in 1934 [174]. A large variety of commercial products prepared by the ethoxylation of alkylamines are available from a number of suppliers. These surfactants are mildly cationic, but when the length of the ethylene oxide chain is increased they progressively acquire the properties of nonionic surfactants.

The first step in the reaction is the addition of two molecules of ethylene oxide to a primary amine at about 150°C [175] followed by polyoxyethylation using a basic catalyst at about the same temperature or at a slightly higher one.

$$RNH_2 \ + \ 2 \ H_2C \underset{O}{\overset{}{\diagup\!\!\!\!\diagdown}} CH_2 \xrightarrow{150°C} RN \overset{CH_2CH_2OH}{\underset{CH_2CH_2OH}{\diagup}}$$

$$RN \overset{CH_2CH_2OH}{\underset{CH_2CH_2OH}{\diagup}} + \ n \ H_2C \underset{O}{\overset{}{\diagup\!\!\!\!\diagdown}} CH_2 \xrightarrow[150°C]{OH^-} RN \overset{(CH_2CH_2O)_xH}{\underset{(CH_2CH_2O)_yH}{\diagup}}$$

Polyoxyethylene alkylamines are used for emulsifiers in neutral and acidic solutions, foaming agents, corrosion inhibitors, wetting agents, dye leveling agents, and in textile finishes [175]. The primary amine in the first reaction is a product derived from the fatty acid amide made by the thermal decomposition of the acid ammonium salt. The amine is synthesized by the Hoffman Rearrangement [121]:

$$RCONH_2 + NaOX + 2NaOH \longrightarrow RNH_2 + Na_2CO_3 + NaX + H_2O$$

The fatty acids in the amides are coconut, stearic, oleic, and tallow, with tallow being predominant.

(POE) Cocoamines, CAS# 61791–14–8

Coconut oil is a mixture of fatty acids with the distribution shown in Table 4–3. The largest percentages are lauric (C_{12}) at 44–51% and myristic (C_{14}) at 13–18% with lesser amounts of other saturated acids and only a small amount (less than 10%) of unsaturated fatty acids. These products can be employed when a lower level of unsaturation is required.

1. POE (2) Cocoamine. The low ethylene oxide content results in a product that is an oil–soluble emulsifier and an intermediate for the synthesis of amphoteric antistatic agents. It is made by Ethox as Ethox® CAM–2 [176][177]. Some properties of this product are shown by Table 6–84.

Table 6–84
Typical Properties of POE (2) Cocoamine

Property	Value
HLB	6.0
Neutralization Equivalent	280–303
Amine Value	185–200
% Tertiary Amine	95 min
Color, Gardner	6 max
Moisture, %	1.0

2. POE (15) Cocoamine. As would be expected, this amine is water soluble and its cationic nature produces some antistatic activity. It is a hydrophilic emulsifier and penetrating agent in several textile processes. In viscose rayon production, this special surfactant is used to keep spinnerets clean and to extend coagulation time that allows the production of high tenacity rayon. It is manufactured by Ethox as Ethox® CAM–15 [176][178] and by Witco with the designation Varonic® K–215–LC [179]. Classical features are given in Table 6–85.

Table 6–85
Typical Properties of POE (15) Cocoaimine

Property	Value
HLB	15.2
pH, 5%	9.0–10.5
Neutralization Equivalent	825–905
Amine Value	62–68
% Tertiary Amine	96 min
Color, Gardner	12 max
Moisture, %	1.0

(POE) Stearyl Amines, CAS# 26635–92–7

Triple pressed stearic acid is neutralized with ammonium hydroxide to prepare the stearamide that is heated to make the primary amine. Ethoxylation follows in the preparation of the alkoxylated stearyl amines. The commercial products have a wide range of HLB values and thus several applications.

1. POE (2) Stearyl Amine. This chemical product is a good emulsifier for waxes and is also used as a mold release agent. As expected, it is not soluble in water. Manufactured by Ethox, it is commercially available as Ethox® SAM–2 [176][180]. The properties are listed in Table 6–86.

Table 6–86

Typical Properties of POE (2) Stearyl Amine

Property	Value
HLB	4.9
Neutralization Equivalent	355–372
Amine Value	151–158
% Tertiary Amine	95 min
Color, Gardner	7 max
Moisture, %	1.0

2. POE (10) Stearyl Amine. This is a water soluble emulsifier that is also compounded into corrosion inhibitor formulations. Most surfaces, either metallic, polymeric, or mineral, have a negative charge that can assume importance when using a cationic surface active agent. When polyoxyethylene amines are adsorbed onto a surface they form a protective and tightly bound film, which will effectively protect the metal surface against corrosion. This property is especially important in petroleum refining operations. The continuous injection of polyoxyethylated amines into many refinery processes will protect metal surfaces against the attack of water, hydrogen sulfide, hydrogen chloride, or sulfur oxides. This adsorptive and protective property is also useful in greases, oils, and cutting oils [175]. POE (10) stearyl amine is also made by Ethox as Ethox® SAM–10 [176][181]. Some characteristics are shown in Table 6–87.

Table 6–87

Typical Properties of POE (10) Stearyl Amine

Property	Value
HLB	12.3
Neutralization Equivalent	685–709
Amine Value	79–82
% Tertiary Amine	97 min
Color, Gardner	14 max
Moisture, %	1.0

3. POE (50) Stearyl Amine. The high level of ethylene oxide in this molecule results in it being a solid and the highest HLB of all commercially available amine ethoxylates. This chemical is a leveling and dispersing agent for dyeing with acid, disperse, and cationic dyes. It is made by Ethox as Ethox® SAM–50 [176][182] and by Henkel as Trymeen® 6617 [41][183]. Properties are displayed in Table 6–88.

Table 6–88

Typical Properties of POE (50) Stearyl Amine

Property	Value
HLB	18.0
Neutralization Equivalent	2337–2672
Amine Value	21–24
pH, 5%	8.0–10.5
Pour Point, °C	35
Color, Gardner	10 max
Moisture, %	1.0

(POE) Tallow Amines, CAS# 61791–26–2

Since tallow is a by–product of the meat–packing industry, it is a low cost starting material for the synthesis of a number of fatty acids. The distribution of acids in tallow is:

C_{14} 2–3 % Saturated
C_{16} 25–30 % Saturated
C_{18} 21–26 % Saturated
C_{16} 2–3 % Unsaturated
C_{18} 39–42 % Unsaturated

The unrefined fatty acids are used to prepare the ethoxylated tallow amine emulsifiers and textile processing aids.

1. POE (2) Tallow Amine. A low ethylene content in many emulsifiers, including tallow amines, results in water insoluble compounds that are useful in preparing wax dispersions. This material, POE (2) tallow amine, is also an effective lubricant. It is produced by Ethox as Ethox® TAM–2 [176] [184] and also by Stepan as Stepantex® TA–2 [43]. Some of the features of this compound are listed in Table 6–89.

Table 6–89

Typical Properties of POE (2) Tallow Amine

Property	Value
HLB	4.2
Neutralization Equivalent	351–365
Amine Value	154–160
% Tertiary Amine	96 min
Color, Gardner	8 max
Moisture, %	1.0

2. POE (5) Tallow Amine. This substance is also an oil–soluble emulsifier and a coating agent that is substantive to many inorganic surfaces. It can be a lubricating agent in some applications. The amine is made by Ethox where the designation is Ethox® TAM–5 [176][185] and by Stepan as Stepantex® TA–5 [43]. Common properties are shown in Table 6–90.

Table 6–90

Typical Properties of POE (5) Tallow Amine

Property	Value
HLB	8.8
Neutralization Equivalent	470–510
Amine Value	110–120
pH, 5%	9.0–10.5
% Tertiary Amine	96 min
Color, Gardner	12 max
Moisture, %	1.0

3. POE (10) Tallow Amine. A sufficient amount of ethylene oxide is reacted with this tallow amine base to make a water–soluble emulsifier. This compound is also an intermediate in the preparation of amphoteric antistats. It is also made by Ethox as Ethox® TAM–10 [176][186] and by Stepan as Stepantex® TA–9 [43]. Attributes are presented in Table 6–91.

Table 6–91

Typical Properties of POE (10) Tallow Amine

Property	Value
HLB	12.5
Neutralization Equivalent	668–738
Amine Value	76–84
pH, 5%	9.0–10.5
Color, Gardner	10 max
Moisture, %	1.0

4. POE (15) Tallow Amine. In addition to being a water–soluble emulsifier, this amine is a mildly cationic surfactant that is used in the application of dyes to fibers, yarns, fabrics, and carpeting containing both anionic and cationic dyeable yarns. It greatly aids dye solubility, promoting effective dye exhaustion of both anionic and cationic dyes in the same bath. This compound is an excellent dye antiprecipitant that contributes to dye leveling and shows no adverse effect on dye fastness. The product is supplied by Ethox as Ethox® TAM–15 [176][187], by Henkel as Trymeen® 6606

[41][188], by Stepan as Stepantex® TA–15 [43], and by Witco as Varonic® T–215 [189]. Typical properties are shown in Table 6–92.

Table 6–92

Typical Properties of POE (15) Tallow Amine

Property	Value
HLB	14.3
Neutralization Equivalent	890–951
Amine Value	59–66
pH, 5%	9.0–10.5
Color, Gardner	8 max
Viscosity, cSt, 100°F	96
Moisture, %	1.0

5. POE (20) Tallow Amine. The 20–mole ethoxylated tallow amine has applications similar to the 15–mole product, but with a higher HLB. It is made by Ethox as Ethox® TAM–20 [176][190], by Henkel as Trymeen® 6607 [41][191], and by Witco as Varonic® T–220 [189]. Characteristics are described in Table 6–93.

Table 6–93

Typical Properties of POE (20) Tallow Amine

Property	Value
HLB	15.4
Neutralization Equivalent	1100–1220
Amine Value	44–51
Color, Gardner	8 max
Moisture, %	5.0

6. POE (25) Tallow Amine. This product is another dye leveling agent, especially important for dyeing nylon. Although the material is a liquid at room temperature, the pour point is 16°C so some long–chain saturated fatty amines may solidify and the product must be stirred before use. It is made by Ethox as Ethox® TAM–25 [176][192] and by Henkel as Trymeen® 6609 [41][193]. Diverse properties are shown in Table 6–94.

Table 6–94

Typical Properties of POE (25) Tallow Amine

Property	Value
HLB	16.0
Neutralization Equivalent	1305–1438
Amine Value	39–43
% Tertiary Amine	96.0 min
Viscosity, cSt, °F	128
Color, Gardner	8 max
Moisture, %	5.0

7. POE (40) Tallow Amine. This product is a solid with the same application in dye leveling as the other products. It is made by Henkel as Trymeen® 6610 [41][194]. The features are listed in Table 6–95.

Table 6–95

Typical Properties of POE (40) Tallow Amine

Property	Value
HLB	17.4
pH, 5%	8.0–10.0
Amine Value	27–31
Pour Point, °C	37
Color, Gardner	6 max
Moisture, %	5.0

POE (30) Oleyl Amine, 80% Active, CAS# 26635–93–8

Ethoxylated (30 moles EO) oleyl amine is dissolved in water to form an 80% solution for ease of handling. It is a cold–water–soluble hydrophilic emulsifier and textile dyeing assistant. The material functions as an anti–precipitant in cross dyeing, as a mild stripping agent, and as a leveler for acid dyes. The solution is supplied by Ethox as Ethox® OAM–308 [176][195] and by Henkel as Trymeen® 6620 [41][196]. Typical properties are presented by Table 6–96.

Table 6–96

Typical Properties of POE (30) Oleyl Amine, 80%

Property	Value
HLB	16.0
Neutralization Equivalent	1870–2150
Amine Value	26–30
Cloud Point, 1%, °C	>100
% Solids	79–81
Viscosity, cSt, °F	445
Color, Gardner	4 max
Moisture, %	19–21

Table 6–97

HLB Values of Emulsifiers

Emulsifier	HLB
Sorbitan Trioleate	1.8
Sorbitan Tristearate	2.1
POE (5) Castor Oil	4.0
POE (2) Tallow Amine	4.2
Sorbitan Sesquioleate	4.5
Sorbitan Monooleate	4.6
POE (1.5) Nonylphenol	4.6
Sorbitan Monostearate	4.7
POE (2) Stearyl Amine	4.9
POE (2) Cocoamine	6.0
PEG (200) Dilaurate	6.1
POE (4) Monomerate	6.6
Sorbitan Monopalmitate	6.7
POE (3) C_{10}–C_{18} Alcohol	6.9

Table 6–97, Continued

Emulsifier	HLB
POE (3) Tridecyl Alcohol	7.9
Sorbitan Monolaurate	8.0
POE (3) C_{12}–C_{18} Alcohol	8.1
POE (16) Castor Oil	8.6
POE (16) Hydrogenated Castor Oil	8.6
POE (5) Tallow Amine	8.8
PEG (400) Dioleate	8.8
POE (4) Nonylphenol	8.9
PEG (400) Sesquioleate	9.4
POE (20) Sorbitan Monostearate	9.6
POE (4) Sorbitan Monostearate	9.6
POE (4) C_{12}–C_{18} Alcohol	9.7
POE (8) Tallate	9.8
PEG (600) Dioleate	10.0
POE (5) Sorbitan Monooleate	10.0
POE (8) Monomerate	10.1
POE (8) Dinonylphenol	10.4
POE (40) Sorbitol Hexaoleate	10.4
POE (5) C_{12}–C_{18} Alcohol	10.7
POE (25) Castor Oil	10.8
PEG (400) Dilaurate	10.8
POE (25) Hydrogenated Castor Oil	10.8
POE (6) Nonylphenol	10.9
POE (20) Sorbitan Tristearate	11.1
POE (20) Sorbitan Trioleate	11.1
POE (60) Sorbitol Hexaoleate	11.3

Table 6–97, Continued

Emulsifier	HLB
POE (6) Tridecyl Alcohol	11.4
POE (30) Castor Oil	11.8
POE (8) Nonylphenol	12.0
POE (7) Octylphenol	12.0
PEG (600) Dilaurate	12.0
POE (10) Stearyl Amine	12.3
POE (10) Stearyl Alcohol	12.4
POE (10) Tallow Amine	12.5
POE (8) Tridecyl Alcohol	12.5
POE (10) Coconut Oil	12.6
POE (36) Castor Oil	12.6
POE (10) Cetyl Alcohol	12.9
POE (40) Castor Oil	13.0
POE (9) Nonylphenol	13.0
POE (9) Tridecyl Alcohol	13.0
POE (4) Sorbitan Monolaurate	13.3
POE (16) Tallate	13.4
POE (10) Nonylphenol	13.5
POE (10) Octylphenol	13.6
POE (15) Monomerate	13.7
POE (18) Dinonylphenol	13.9
POE (15) Tallow Amine	14.3
POE (12) Tridecyl Alcohol	14.5
POE (15) Nonylphenol	15.0
POE (20) Sorbitan Monooleate	15.0
POE (15) Cocoamine	15.2

Table 6-97, Continued

Emulsifier	HLB
POE (20) Stearyl Alcohol	15.3
POE (20) Tallow Amine	15.4
POE (20) Sorbitan Monopalmitate	15.6
POE (20) Cetyl Alcohol	15.7
POE (80) Castor Oil	15.9
POE (18) Tridecyl Alcohol	16.0
POE (25) Tallow Amine	16.0
POE (30) Oleyl Amine	16.0
POE (20) Nonylphenol	16.0
POE (25) C_{16}–C_{18} Alcohol	16.3
POE (20) Sorbitan Monolaurate	16.5
POE (30) Nonylphenol	17.1
POE (40) Tallow Amine	17.4
POE (40) Nonylphenol	17.8
PEG (6000) Distearate	18.0
POE (50) Stearyl Amine	18.0
POE (200) Castor Oil	18.1
POE (150) Dinonylphenol	19.0

REFERENCES

1. Bernholz, W. F., Redston, J. P., and Schlatter, C. (1984). Spin Finish Usage and Compounding for Man–Made Fibers, in *Emulsions and Emulsion Technology*, (K. J. Lissant, ed.), Marcel Dekker, Inc., New York, pp. 215–239
2. Becher, P. (1965). *Emulsions: Theory and Practice, Second Edition, ACS Monograph Series No. 162*, Van Nostrand Reinhold, New York, pp. 2, 50, 118–140
3. Griffin, W. C. (1979). Emulsions in *Kirk–Othmer Encyclopedia of Chemical Technology, Third Edition*, Volume 8, John Wiley & Sons, New York, pp. 900–930

4. Rosen, M. J. (1978). *Surfactants and Interfacial Phenomena*, John Wiley & Sons, New York, pp. 224, 244

5 Slade, P. E., Unpublished data

6. Coulter Corporation (1994). *The Coulter System, Multisizer IIe,* Coulter Corporation, Miami, FL

7. Hecht, E. and Zajac, A. (1979). *Optics*, Addison–Wesley, Reading, MA, pp. 336–364

8. van de Hulst, H. C. (1957). *Light Scattering by Small Particles*, John Wiley & Sons, New York, pp. 397–400

9. Coulter Corporation (1994). *The Coulter LS Series*, Coulter Corporation, Miami, FL

10. Malvern Instruments (1995). *The Mastersizer Family of Particle Characterization Systems*, Malvern Instruments, Malvern, Worcestershire, U. K.

11. Matec (1993). *CHFD–1100 Capillary Hydrodynamic Fraction*, Matec Applied Sciences, Hopkinton, MA

12. Dos Ramos, J. G. and Silebi, C. A. (1990). The Determination of Particle Size Distribution of Submicrometer Particles by Capillary Hydrodynamic Fractionation, *J. Colloid Interface Science*, **135:** 165

13. Silebi, C. A. and Dos Ramos, J. G. (1989). Separation of Submicrometer Particles by Capillary Hydrodynamic Fractionation, *J. Colloid Interface Science*, **130:** 14

14. Cannon, D. W. and O'Brien, R. W. (1994). The AcoustoSizer®: A New Instrument for the Characterization of Concentrated Colloidal Suspensions, Matec Applied Sciences, Hopkinton, MA

15. Oja, T, Petersen, G. L., and Cannon, D. W., assingors to Matec, Inc., Warwick, RI (1985). Measurement of Electrokinetic Properties of Solutions, *U.S. Patent 4,497,208*

16. Matec Applied Sciences (1994). *AcoustoSizer®*, Matec Applied Sciences, Hopkinton, MA

17. Lavenda, B. H. (1985). Brownian Motion, *Scientific American*, **252:** 70

18. Ware, B. R. and Haas, D. D. (1983). Electrophoretic Light Scattering, in *Fast Methods in Physical Biochemistry and Cell Biology*, (R. I Sha'afi and S. M. Fernandez, eds.), Elsevier Science Publishers, Amsterdam, The Netherlands, pp. 173–220

19. Bott, S. E. (1984). *Submicron Particle Sizing by Photon Correlation Spectroscopy: Use of Multiple Angle Detection*, Langley Ford Instruments, Division of Coulter Electronics, Amherst, MA

20. Brookhaven Instruments (1994). *Brookhaven ZetaPlus®*, Brookhaven Instruments Corporation, Holtsville, NY

21. Coulter Corporation (1987). *Coulter® Model N4 Series Analyzers*, Coulter Corporation, Miami, FL

22. Malvern Instruments (1995). *Zetasizer® 4 Zeta Potential and Submicron Particle Size Distribution Analyzers*, Malvern Instruments Limited, Malvern, Worcestershire, U.K.

23. Brookhaven Instruments (1994). *Brookhaven FOQELS®*, Brookhaven Instruments Corporation, Holtsville, NY

24. Coehn, A. (1898). Über ein Gesetz der Elektrizitätserrrgung, *Ann. Physik*, Leipzig, Germany, **64:** 217

25. von Helmholz, H. (1879). Stüdien über Electrische Grenschichten, *Ann. der Physik und Chemie*, Leipzig, Germany, **7:** 337

26. Gouy, G., (1909). Constitution of the Electric Charge at the Surface of an Electrolyte, *Compt. Rend.*, Paris, France, **149:** 654

27. Chapman, D. L. (1913). A Contribution to the Theory of Electrocencapillarity, *Phil. Mag.*, **25:** 475

28. Stern, O. Z. (1924). The Theory of the Electrolytic Double–layer, *Elektrochem.*, Halle an der Saale, Germany, **30:** 508

29. von Smoluchowski, M.(1903). Contribution à la Théorie de L'endosmose Électrique et de Quelques Phenomènes Corrélatifs, *Bulletin International de l'Academie des Sciences de Cracouie*, Cracow, Poland, **8:** 182

30. Hunter, R. J. (1981). *Zeta Potential in Colloid Science*, Academic Press, New York, pp. 1–386

31. Crow, D. R. (1974). *Principles and Applications of Electrochemistry*, Chapman and Hall, London, England, pp. 1–260

32. Oja, T., Bott, S., and Sugrue, S. (1988). *Doppler Electrophoretic Light Scattering Analysis Using the Coulter® DELSA 440*, Coulter Corporation, Miami, FL

33. Matec Applied Sciences (1994). *Electrokinetic Sonic Analysis ESA®–8000 Zeta Potential Determination in Concentrated Colloidal Suspensions*, Matec Applied Sciences, Hopkinton, MA

34. Becher, P. (1966). Emulsification, in *Nonionic Surfactants* (M. J. Schick, ed.), Marcel Dekker, Inc., New York, pp. 604–626

35. Myers, D. (1991). *Surfaces, Interfaces, and Colloids: Principles and Applications*, VCH Publishers, Inc., New York, pp. 237–240

36. Kao Corporation (1983). *Surfactants: A Comprehensive Guide*, Kao Corporation, Tokyo, Japan, pp. 58–66

37. Davies, J. T. and Rideal, E. K. (1963). *Interfacial Phenomena*, Academic Press, New York, pp. 372–374

38. Schönfeld, N. (1969). *Surface Active Ethylene Oxide Adducts*, Pergamon Press, Oxford, U.K., pp. 24–45

39. Ethox Product Brochure (1994). *Ethoxylated Alcohols and Alkyl Phenols*, Ethox Chemicals, Inc., Greenville, SC

40. Ethox Specifications (1993). *Ethal® TDA–3*, Ethox Chemicals, Inc., Greenville, SC

41. Henkel Corporation (1993). *Textile Chemicals: Technical Bulletin 103A*, Henkel Corporation, Textile Chemicals, Charlotte, NC

42. Henkel Specifications (1988). *Trycol® 5993*, Henkel Corporation, Textile Chemicals, Charlotte, NC

43. Stepan Company (1995). *Textile Products, Technical Information*, Stepan Company, Northfield, IL

44. Ethox Specifications (1993). *Ethal® TDA–6*, Ethox Chemicals, Inc., Greenville, SC
45. Henkel Specifications (1988). *Trycol® 5940*, Henkel Corporation, Textile Chemicals, Charlotte, NC
46. Specialties from ICI for Fiber Producers (1987). *Emulsifiers*, ICI Americas Inc., Wilmington, DE
47. Witco Fiber Production Auxiliaries (1994). *Oil Emulsifiers*, Witco Corporation, Greenwich, CT
48. Henkel Specifications (1988). *Trycol® 5949*, Henkel Corporation, Textile Chemicals, Charlotte, NC
49. Ethox Specifications (1988). *Ethal® TDA-12*, Ethox Chemicals, Inc., Greenville, SC
50. Henkel Specifications (1988). *Trycol® 5943*, Henkel Corporation, Textile Chemicals, Charlotte, NC
51. Ethox Specifications (1993). *Ethal® TDA-18*, Ethox Chemicals, Inc., Greenville, SC
52. Henkel Specifications (1989). *Trycol® 5946*, Henkel Corporation, Textile Chemicals, Charlotte, NC
53. Witco (1994). *Fatty Acids, Glycerine, Triglycerides*, Witco Corporation, Greenwich, CT
54. Ethox Specifications (1993). *Ethal® 3328*, Ethox Chemicals, Inc., Greenville, SC
55. Hoechst–Celanese Product Data (1993). *Afilan® 2299*, Hoechst–Celanese Corporation, Charlotte, NC
56. Hoechst–Celanese Product Data (1993). *Afilan® 5GC*, Hoechst–Celanese Corporation, Charlotte, NC
57. Ethox Specifications (1993). *Ethal® 368*, Ethox Chemicals, Inc., Greenville, SC
58. Ethox Specifications (1993). *Ethal® CSA-25*, Ethox Chemicals, Inc., Greenville, SC
59. Steindorff, A., Balle, K., Horst, K., and Michel, R., assignors to General Aniline and Film Corporation (1940). Glycol and Polyglycol Ethers of Isocyclic Hydroxyl Compounds, *U.S. Patent 2,213,477*
60. Enyeart, C. R. (1966). Polyoxyethylene Alkylphenols, in *Nonionic Surfactants* (M. J. Schick, ed.), Marcel Dekker, Inc., New York, pp. 44–85
61. Demianiw, D. G. (1978). Olefins, Higher, *Kirk–Othmer Encyclopedia of Chemical Technology*, Third Edition, Vol. 16, John Wiley & Sons, New York, pp. 480–499
62. Miller, S. A., Bann, B., and Thrower, R. D. (1950). The Reaction Between Phenol and Ethylene Oxide, *J. Chem. Soc.*, **1950**: 3623
63. Schick, M. J. (1966). Biodegradation, in *Nonionic Surfactants* (M. J. Schick, ed.), Marcel Dekker, Inc., New York, pp. 971–996
64. Balekjian, J., Hoechst–Celanese Corporation (1992). *Waste–Water Regulations Affecting the Textile & Fiber Industries in Europe*, Presentation to Monsanto Company, August 14, 1992

65. Oslo and Paris Conventions for the Prevention of Marine Pollution (1992). *PARACOM Recommendations on Nonylphenol–Ethoxylates*, Oslo, Norway, June 22–26, 1992

66. Melnikoff, A., letter to Mr. Stig Borgvang (1993). *Comments on the PARACOM Recommendations on Nonylphenol Ethoxylates*, Union Carbide Chemicals and Plastics (Europe) S.A., January 11, 1993

67. Alkylphenol & Ethoxylates Panel (1994). *Alkylphenol Ethoxylates in the Environment: An Overview*, Chemical Manufacturers Association, Washington, D.C., August, 1994

68. Naylor, C. G. (1992). Environmental Fate of Alkylphenol Ethoxylates, *Soap/Cosmetics/Chemical Specialties*, **68**(8): 27

69. Naylor, C. G., Mieure, J. P., Adams, W. J., Weeks, J. A., Castaldi, F. J., Ogle, L. D., and Romano, R. R. (1992). Alkylphenol Ethoxylates in the Environment, *J. Am. Oil. Chem. Soc.* **69**: 695

70. Henkel Specifications (1991). *Hyonic® OP-7*, Henkel Corporation, Textile Chemicals, Charlotte, NC

71. Henkel Specifications (1991). *Hyonic® OP-10*, Henkel Corporation, Textile Chemicals, Charlotte, NC

72. Ethox Specifications (1993). *Ethal® NP-1.5*, Ethox Chemicals, Inc. Green–ville, SC

73. Henkel MSDS (1993). *Trycol® 6960*, Henkel Corporation, Textile Chemicals, Charlotte, NC

74. Ethox Specifications (1991). *Ethal® NP-4*, Ethox Chemicals, Inc., Greenville, SC

75. Henkel Specifications (1989). *Trycol® 6961*, Henkel Corporation, Textile Chemicals, Charlotte, NC

76. Stepan Company (1991). Products for the Textile Industry, *Makon® 4*, Stepan Company, Northfield, IL

77. Witco Fiber Production Auxiliaries (1994). , *Desizing Agents*, Witco Corporation, Greenwich, CT

78. Ethox Specifications (1991). *Ethal® NP-6*, Ethox Chemicals, Inc., Greenville, SC

79. Henkel Specifications (1989). *Trycol® 6962*, Henkel Corporation, Textile Chemicals, Charlotte, NC

80. Hoechst–Celanese Product Data (1993). *Afilan® 6NPR*, Hoechst–Celanese Corporation, Charlotte, NC

81. Stepan Company (1991). Products for the Textile Industry, *Makon® 6*, Stepan Company, Northfield, IL

82. Stepan Company (1991). Products for the Textile Industry, *Makon® 8*, Stepan Company, Northfield, IL

83. Ethox Specifications (1986). *Ethal® NP-9*, Ethox Chemicals, Inc., Greenville, SC

84. Henkel Specifications (1989). *Trycol® 6964*, Henkel Corporation, Textile Chemicals, Charlotte, NC

85. Henkel Specifications (1989). *Trycol® 6974*, Henkel Corporation, Textile Chemicals, Charlotte, NC

86. Stepan Company (1991). Products for the Textile Industry, *Makon® 10*, Stepan Company, Northfield, IL

87. Stepan Company (1991). Products for the Textile Industry, *Makon® 14*, Stepan Company, Northfield, IL

88. Henkel Specifications (1983). *Trycol® 6952*, Henkel Corporation, Textile Chemicals, Charlotte, NC

89. Ethox Specifications (1993). *Ethal® NP-20*, Ethox Chemicals, Inc., Greenville, SC

90. Henkel Specifications (1989). *Trycol® 6967*, Henkel Corporation, Textile Chemicals, Charlotte, NC

91. Henkel Specifications (1989). *Trycol® 6968*, Henkel Corporation, Textile Chemicals, Charlotte, NC

92. Stepan Company (1991). Products for the Textile Industry, *Makon® 30*, Stepan Company, Northfield, IL

93. Henkel Specifications (1989). *Trycol® 6957*, Henkel Corporation, Textile Chemicals, Charlotte, NC

94. Ethox Specifications (1993). *Ethal® NP-407*, Ethox Chemicals, Inc., Greenville, SC

95. Ethox Specifications (1993). *Ethal® DNP-8*, Ethox Chemicals, Inc., Greenville, SC

96. Ethox Specifications (1993). *Ethal® DNP-18*, Ethox Chemicals, Inc., Greenville, SC

97. Henkel Specifications (1983). *Trycol® 6988*, Henkel Corporation, Textile Chemicals, Charlotte, NC

98. Henkel Specifications (1983). *Trycol® 6989*, Henkel Corporation, Textile Chemicals, Charlotte, NC

99. Sax, N. I. and Lewis, R. J. (1987). *Dangerous Properties of Industrial Materials, Seventh Edition, Volume 2*, van Nostrand Reinhold, New York, p. 730

100. Burnette, L. W. (1966). Miscellaneous Nonionic Surfactants, in *Nonionic Surfactants* (M. J. Schick, ed.), Marcel Dekker, Inc., New York, pp. 395–440

101. Abitec Performance Products (1994). *Acconon® CA-5*, Abitec Corporation, Columbus, OH

102. Ethox Product Brochure (1994). *Ethoxylated Glycerides*, Ethox Chemicals, Inc., Greenville, SC

103. Ethox Specifications (1986). *Ethox® CO-5*, Ethox Chemicals, Inc., Greenville, SC

104. Hoechst–Celanese Product Data (1993). *Afilan® 5EL*, Hoechst–Celanese Corporation, Charlotte, NC

105. Abitec Performance Products (1994). *Acconon® CA-15*, Abitec Corporation, Columbus, OH

106. Ethox Specifications (1987). *Ethox® CO-16*, Ethox Chemicals, Inc., Greenville, SC

107. Henkel Specifications (1989). *Trylox® 5902*, Henkel Corporation, Textile Chemicals, Charlotte, NC

108. Ethox Specifications (1990). *Ethox® CO–25*, Ethox Chemicals, Inc., Greenville, SC

109. Henkel Specifications (1989). *Trylox® 5904*, Henkel Corporation, Textile Chemicals, Charlotte, NC

110. Stepan Product Bulletin (1994). *Stepantex® CO Series*, Stepan Company, Northfield, IL

111. Ethox Specifications (1986). *Ethox® CO–30*, Ethox Chemicals, Inc., Greenville, SC

112. Henkel Specifications (1989). *Trylox® 5906*, Henkel Corporation, Textile Chemicals, Charlotte, NC

113. ICI Specialty Chemicals (1987). *Lubricants*, ICI Americas, Inc., Wilmington, DE

114. Ethox Specifications (1986). *Ethox® CO–36*, Ethox Chemicals, Inc., Greenville, SC

115. Henkel Specifications (1994). *Trylox® 5907*, Henkel Corporation, Textile Chemicals, Charlotte, NC

116. Ethox Specifications (1993). *Ethox® CO–40*, Ethox Chemicals, Inc., Greenville, SC

117. Henkel Specifications (1989). *Trylox® 5909*, Henkel Corporation, Textile Chemicals, Charlotte, NC

118. Ethox Specifications (1989). *Ethox® CO–81*, Ethox Chemicals, Inc., Greenville, SC

119. Ethox Specifications (1986). *Ethox® CO–200 (A)*, Ethox Chemicals, Inc., Greenville, SC

120. Hoechst–Celanese Product Data (1993). *Afilan® 200EL*, Hoechst–Celanese Corporation, Charlotte, NC

121. Noller, C. R. (1958). *Textbook of Organic Chemistry, Second Edition*, W. B. Saunders, Philadelphia, p. 40, 180

122. Ethox Specifications (1986). *Ethox® HCO–16*, Ethox Chemicals, Inc., Greenville, SC

123. Henkel Specifications (1989). *Trylox® 5921*, Henkel Corporation, Textile Chemicals, Charlotte, NC

124. Ethox Specifications (1986). *Ethox® HCO–25*, Ethox Chemicals, Inc., Greenville, SC

125. Henkel Specifications (1989). *Trylox® 5922*, Henkel Corporation, Textile Chemicals, Charlotte, NC

126. Hoechst–Celanese Product Data (1993). *Afilan® 25HEL*, Hoechst–Celanese Corporation, Charlotte, NC

127. Ethox Specifications (1985). *Ethox® 1212*, Ethox Chemicals, Inc., Greenville, SC

128. Peters, R. M., assignor to Emery Industries, Inc., Greenville, SC (1957). Hydrogenation of Structurally Modified Fatty Acids, *U.S. Patent 2,812,342*

129. Fulmer, R. W. (1968). Applications of Fatty Acids in Protective Coatings in *Fatty Acids and Their Industrial Applications* (E. S. Pattison, ed.), Marcel Dekker, Inc., New York, pp. 187–208

130. Henkel Specifications (1991). *Trydet® 2691*, Henkel Corporation, Textile Chemicals, Charlotte, NC

131. Ethox Product Brochure (1944). *Ethoxylated Fatty Acids*, Ethox Chemicals, Inc., Greenville, SC

132. Ethox Specifications (1986). *Ethox® MA–8*, Ethox Chemicals, Inc., Greenville, SC

133. Henkel Specifications (1989). *Trydet® 2692*, Henkel Corporation, Textile Chemicals, Charlotte, NC

134. Ethox Specifications (1987). *Ethox® MA–15*, Ethox Chemicals, Inc., Greenville, SC

135. Wittcoff, H. A. and Reuben, B. G. (1980). *Industrial Organic Chemicals in Perspective Part One: Raw Materials and Manufacture*, John Wiley & Sons, New York, pp. 134–135

136. Arlt, H. G. (1983). Tall Oil, *Kirk–Othmer Encyclopedia of Chemical Technology*, Third Edition, Volume 22, John Wiley & Sons, New York, pp. 531–541

137. Ethox Specifications (1986). *Ethox® TO–8*, Ethox Chemicals, Inc., Greenville, SC

138. Ethox Specifications (1987). *Ethox® TO–16*, Ethox Chemicals, Inc., Greenville, SC

139. Henkel Specifications (1989). *Trydet® 2682*, Henkel Corporation, Textile Chemicals, Charlotte, NC

140. Ethox Product Brochure (1994). *Esters*, Ethox Chemicals, Inc., Greenville, SC

141. Ethox Specifications (1986). *Ethox® DL–5*, Ethox Chemicals, Inc., Greenville, SC

142. Stepan Typical Properties (1990). *Kessco® PEG 200 DL*, Stepan Company, Northfield, IL

143. Ethox Specifications (1992). *Ethox® DL–9*, Ethox Chemicals, Inc., Greenville, SC

144. Stepan Typical Properties (1990). *Kessco® PEG 600 DL*, Stepan Company, Northfield, IL

145. Ethox Specifications (1994). *Ethox® DO–9*, Ethox Chemicals, Inc., Greenville, SC

146. Henkel Specifications (1982). *Emerest® 2648*, Henkel Corporation, Textile Chemicals, Charlotte, NC

147. Ethox Specifications (1993). *Ethox® DO–14*, Ethox Chemicals, Inc. Greenville, SC

148. Henkel Specifications (1982). *Dyafac® PEG 6DO*, Henkel Corporation, Textile Chemicals, Charlotte, NC

149. Stepan Typical Properties (1990). *Kessco® PEG 6000 DS*, Stepan Company, Northfield, IL

150. Henkel Specifications (1989). *Trylox® 6746*, Henkel Corporation, Textile Chemicals, Charlotte, NC

151. Henkel Specifications (1989). *Trylox® 6747*, Henkel Corporation, Textile Chemicals, Charlotte, NC

152. Morrison, R. T. and Boyd, R. N. (1983). *Organic Chemistry, Fourth Edition,*, Allyn and Bacon, Inc., Boston, pp. 1056–1057

153. Benson, F. R. (1966). Polyol Surfactants, in *Nonionic Surfactants* (M. J. Schick, ed.), Marcel Dekker, Inc., New York, pp. 247–299

154. Elworthy, P. H. and Treon, J. F. (1966). Physiological Activity of Nonionic Surfactants, in *Nonionic Surfactants* (M. J. Schick, ed.), Marcel Dekker, Inc., New York, pp. 923–970

155. F. Norman Tuller, Henkel Corporation (1996). Personal communication

156. Elworthy, P. H., Florence, A. T., and Macfarlane, C. B. (1968). *Solubilization by Surface–Active Agents*, Chapman and Hall Ltd., London, England, pp. 320–321

157. Henkel Specifications (1989). *Emsorb® 2515*, Henkel Corporation, Textile Chemicals, Charlotte, NC

158. Humko Chemical Technical Information (1994). *Humko Chemical Oleo Esters for Textile and Fiber Industries*, Humko Chemical Division, Witco Corporation, Memphis, TN

159. Henkel Specifications (1989). *Emsorb® 2500*, Henkel Corporation, Textile Chemicals, Charlotte, NC

160. Henkel Specifications (1986). *Emsorb® 2502*, Henkel Corporation, Textile Chemicals, Charlotte, NC

161. Specialties from ICI for Fiber Producers (1987). *Laboratory Test Methods for Fiber Finishes: Friction Test*, ICI Americas, Wilmington, DE

162. Slade, K. N. (1989). *Effect of Surfactants on Pisum Sativum*, Thirty Fourth Florida State Science and Engineering Fair, West Palm Beach, FL

163. Ethox Product Brochure (1994). *Ethoxylated Sorbitan Esters*, Ethox Chemicals, Inc., Greenville, SC

164. Ethox Specifications (1987). *Ethsorbox® L–20*, Ethox Chemicals, Inc., Greenville, SC

165. Henkel Specifications (1989). *Emsorb® 6915*, Henkel Corporation, Textile Chemicals, Charlotte, NC

166. Ethox Specifications (1986). *Ethsorbox® S–20*, Ethox Chemicals, Inc., Greenville, SC

167. Henkel Specifications (1989). *Emsorb® 6901*, Henkel Corporation, Textile Chemicals, Charlotte, NC

168. Ethox Specifications (1986). *Ethsorbox® O–20*, Ethox Chemicals, Inc., Greenville, SC

169. Henkel Specifications (1989). *Emsorb® 6900*, Henkel Corporation, Textile Chemicals, Charlotte, NC

170. Ethox Specifications (1986). *Ethsorbox® TS–20*, Ethox Chemicals, Inc., Greenville, SC

171. Ethox Specifications (1992). *Ethsorbox® TO–20*, Ethox Chemicals, Inc., Greenville, SC

172. Henkel Specifications (1989). *Emsorb® 6903*, Henkel Corporation, Textile Chemicals, Charlotte, NC

173. Henkel Technical Data Sheet (1993). *Glucopon® Alkyl Polyglycoside Surfactants*, Henkel Corporation, Textile Chemicals, Charlotte, NC

174. Schöller, C. and Wittwer, M. (1934), assignors to I. G. Farbenindustrie A. G., Wetting and Dispersing Agents for use in the Textile and Other Industries, *U. S. Patent 1,970,578*

175. Reck, R. A. (1966). Polyoxyethylene Alkylamines, in *Nonionic Surfactants* (M. J. Schick, ed.), Marcel Dekker, Inc., New York, pp. 187–207

176. Ethox Product Brochure (1994). *Ethoxylated Fatty Amines*, Ethox Chemicals, Inc., Greenville, SC

177. Ethox Specifications (1992). *Ethox® CAM–2*, Ethox Chemicals, Inc., Greenville, SC

178. Ethox Specifications (1992). *Ethox® CAM–15*, Ethox Chemicals, Inc., Greenville, SC

179. Witco Fiber Production Auxiliaries (1994). *Spinning Auxiliaries*, Witco Corporation, Greenwich, CT

180. Ethox Specifications (1989). *Ethox® SAM–2*, Ethox Chemicals, Inc., Greenville, SC

181. Ethox Specifications (1988). *Ethox® SAM–10*, Ethox Chemicals, Inc., Greenville, SC

182. Ethox Specifications (1989). *Ethox® SAM–50*, Ethox Chemicals, Inc., Greenville, SC

183. Henkel Specifications (1989). *Trymeen® 6617*, Henkel Corporation, Textile Chemicals, Charlotte, NC

184. Ethox Specifications (1992). *Ethox® TAM–2*, Ethox Chemicals, Inc., Greenville, SC

185. Ethox Specifications (1992). *Ethox® TAM–5*, Ethox Chemicals, Inc., Greenville, SC

186. Ethox Specifications (1988). *Ethox® TAM–10*, Ethox Chemicals, Inc., Greenville, SC

187. Ethox Specifications (1995). *Ethox® TAM–15*, Ethox Chemicals, Inc., Greenville, SC

188. Henkel Specifications (1989). *Trymeen® 6606*, Henkel Corporation, Textile Chemicals, Charlotte, NC

189. Witco Fiber Production Auxiliaries (1994). *Antistats*, Witco Corporation, Greenwich, CT

190. Ethox Specifications (1987). *Ethox® TAM–20*, Ethox Chemicals, Inc., Greenville, SC

191. Henkel Specifications (1989). *Trymeen® 6607*, Henkel Corporation, Textile Chemicals, Charlotte, NC

192. Ethox Specifications (1992). *Ethox® TAM–25*, Ethox Chemicals, Inc., Greenville, SC

193. Henkel Specifications (1979). *Trymeen® 6609*, Henkel Corporation, Textile Chemicals, Charlotte, NC
194. Henkel Specifications (1989). *Trymeen® 6610*, Henkel Corporation, Textile Chemicals, Charlotte, NC
195. Ethox Specifications (1986). *Ethox® OAM–308*, Ethox Chemicals, Inc., Greenville, SC
196. Henkel Specifications (1989). *Trymeen® 6620*, Henkel Corporation, Textile Chemicals, Charlotte, NC

CHAPTER 7

ANTISTATS

It would be very difficult for us to live in today's technology–based culture without electrical energy. Although we accept electricity as a basic requirement for our existence, some aspects of electricity can cause problems. Those of us who have worked in the fiber and textile industries have seen many examples of *static* electricity causing filaments to be repelled from each other, thus causing monumental processing problems. The textile consumer has also experienced problems with static in clinging of apparel and the electrical shocks experienced after walking on carpets or sliding over upholstery. All static charges do not result in problems, however, because we can take advantage of this phenomena in the construction of pile fabrics, in the lay–down of some non–woven fabric, as well as in electrostatic copying and even in the clinging of plastic wrap on leftover food containers. Because of the advantages and disadvantages of this type of energy, it is important to know some of the factors influencing the generation and dissipation of static electricity on fibers and how these changes affect the behavior of textile materials.

7.1 TECHNICAL BACKGROUND

The generation of static electricity by rubbing one object with another has been recognized since ancient times. It has been said [1] that the Greek philosopher, Thales of Miletus, in about 600 B.C. rubbed amber with animal

273

fur and observed that after rubbing, the amber attracted light particles such as dust or ash. From the Greek word for amber, "elektron," came the word electricity, and from the Greek word "tribein," to rub, comes the term triboelectrification, the generation of electricity by rubbing. William Gilbert (1540–1603), an English pioneer researcher into magnetism, discovered that many other materials behave like amber when rubbed [2]. About one hundred years later, Dufay recognized that there were two "kinds" of electricity, which he identified as "vitreous" (glass) and "resinous" (amber). He discovered that objects charged with vitreous electricity repel each other but attract objects charged with resinous electricity. Thirteen years later Benjamin Franklin introduced the terms "positive" and "negative" electricity and identified them with the "vitreous" and "resinous" electricity of Dufay. Franklin also proposed the single–fluid theory, which infers that negative electricity results from an excess of negative charge (electrons) and positive electricity results from a deficiency of negative charge. After the existence of positive and negative "electricity" was recognized, the way was open to establishing a *triboelectric* series, that is, a list of materials so ordered that after rubbing two together, the one listed first would acquire a positive charge and the one listed below a negative charge [3]. The first such list was published by Wilche [4] in 1757. Several other lists have been published and are presented in Table 7–1. It must be remembered, however, that it is difficult to predict the sign of the charge on materials that are far apart in the triboelectric series when they are rubbed together. Kinetic effects are also important in charge generation. Many other factors such as differences in the molecular surface structure of the materials, surface impurities, intensity of contact, and other effects may be responsible for inconsistencies in the triboelectric series.

Our conventional concept of elementary particles and atomic structure form the basic premises to describe a theory of the origin of static electricity. Throughout the time span from the mid 1700s until today we have followed Franklin's arbitrary assignment of negative and positive signs to denote electrical charges. It is generally agreed that atoms are composed of tightly bound protons in the nucleus of the atom while negative electrons surround this nucleus. Although the electrons are attracted to this positive nucleus, they can easily be excited and removed from the electron cloud. The loss of electrons results in a net positive charge, while the gain in electrons will impart a negative charge to the body.

Although many industrial processes include the rubbing of one substance against another, this mechanism is not necessary to generate a static charge. When two surfaces are brought into contact, electrons continuously pass through the interface in both directions. Generally, that transfer of electrons in not necessarily in equilibrium, and even with identical substances, by some happenstance one material may get an excess of electrons at the expense of the other. As long as the contact exists, there

Table 7–1

Triboelectric Series

Ballou [5]	Lehmicke [6]	Hersh and Montgomery [7]	Henry [8]
		MORE POSITIVE	
Wool	Glass	Wool	Platinum
Nylon	Human hair	Nylon	Formvar
Silk	Nylon yarn	Viscose	Filter paper
Viscose	Nylon polymer	Cotton	Cellulose Acetate
Cordura®	Wool	Silk	Cellulose Triacetate
Human skin	Silk	Acetate	Polyethylene
Fiber glass	Viscose	Lucite®	Aluminum
Cotton	Cotton	Poly(vinyl alcohol)	Polystyrene
Glass	Paper	Dacron®	Copper
Ramie	Ramie	Dynel®	Rubber (natural)
Dacron®	Steel	Velon®	
Chromium	Hard rubber	Polyethylene	
Orlon®	Acetate	Teflon®	
Polyethylene	Synthetic rubber		
	Orlon®		
	Polyethylene		
	MORE NEGATIVE		

are no further repercussions; upon separation, however, one of the two
substances retains the excess of electrons and the other is left with a shortage
of electrons. These two substances are then charged with an electrical charge
of the same magnitude but of opposite sign [9]. If these substances are
electrically conductive the electrical charge acquired upon separation will be
insignificant, because electrons will flow from the internal body of the
substance to achieve equilibrium, or any residual charge will be dispersed by

leakage. If these substances are *insulators*, the charge will linger for some time and the phenomena of static electricity is exhibited. In the ancient experiments of Thales, both the amber and the fur could be considered as insulators. The specific resistance (the reciprocal value of the conductance) of both materials expressed in the usual units is roughly 10^{16} Ω-cm. By contrast, the specific resistance of metals is about 10^{-6} Ω-cm. Therefore, there is a ratio of 10^{22} between the conductances of typical insulators and electrical conductors [5].

After the separation of two bodies an electric field is produced that exerts a force that attempts to drive back the charged particles. A part of these charges will return so that the net charge remaining after the bodies are separated is reduced. This leakage through the gap produced during and after the separation can take place by means of several mechanisms:

1. Quantum–mechanical tunneling across very small gaps occurring during the initial stages of separation.
2. Conduction through the atmosphere.
3. Conduction through the contacting bodies.
4. Conduction along the surface of the contacting bodies [3].

Harper [10] has calculated that for conductive materials electronic tunneling effectively stops after a separation of about 25 Å, with about half the charge flowing back. Unfortunately, the tunneling during the separation of insulator–metal and insulator–insulator contacts has not been studied in detail. It has been suggested that the fraction of charge remaining will be reduced from about one–half in the metal–metal case to less than one–tenth if insulators are separated [11].

When the separation distance between the two bodies has reached several times that of the mean free path of the gas molecules in the air, dissipation through the atmosphere can occur if the electric field produced by the act of separation is greater than the ionization potential, or dielectric strength, of that surrounding atmosphere. The magnitude of the charge density, σ, in this type of separation is $4\pi\sigma$, if all of the lines of force are in one direction. If this value of $4\pi\sigma$ reached the ordinary dielectric strength of air, σ would be 8 e.s.u./cm^2, or approximately 2400 kV/cm [12]. This is true, however, only if there are no local concentrations of lines of force. If there are any aspirates, or if the surface has a small radius of curvature, the resulting charge on the surface could be much larger.

If one has a surface with a small radius of curvature, such as a fiber, the emerging lines of force diverge rapidly, and the field strength falls off quickly with the distance from the surface. This reduces the ability of the surface to ionize the air and the charge will be abnormally large before ionization will occur. The dielectric strength of air at atmospheric pressure is given by:

$$F_{max} = \frac{30 + 9}{\sqrt{r}} \, kV/cm \qquad\qquad (7-1)$$

for a thin cylinder of radius r cm. The highest charge can thus be obtained on fibers with the lowest denier per filament, and it has been shown that for a 20μ diameter nylon filament the charge might build up to 50 e. s. u./cm^2, or 15,000 kV/cm. The experiments of Medley [12] have shown that:

1. Gaseous discharge is easily produced by a limited amount of rubbing, and is obtained by rubbing small objects (e.g., single fibers).
2. Discharge over the distances roughly 1 mm or more consists partly or wholly of discrete pulses of charge rather than a continuous slow leakage.
3. The discharge is quenched when the charge falls only slightly below the value needed for its initiation.

The belief that moist air conducts electricity and that this is the reason that experiments will fail if performed in high humidity conditions is erroneous. Air at any relative humidity has about the same dielectric strength [12].

One technique that can be used to increase the discharge through the air is by using static eliminators that ionize the air. These static eliminators are divided into two main groups: (1) corona discharge eliminators, such as inductive eliminators or high voltage eliminators that ionize the air by applying a strong electrostatic field; and (2) radioactive eliminators that provide a multitude of ions from independent ion sources [2].

Although some static is undoubtedly dissipated through the air, most of the static developed is dispersed by an inductive mechanism along the filament to an electrical ground. This can be accomplished by either incorporating a conductive material into the polymer itself, or in the application of an antistatic coating to the yarn. One typical type of conductive materials is a hygroscopic wax, such as a highly ethoxylated hydrogenated castor oil 13] that was incorporated as micro–fibrils into both nylon apparel and carpet fibers. Unfortunately, this wax is water soluble and after repeated washing the antistatic properties were lost. A more permanent solution is to coat fibers with metals or by inserting a filament containing a carbon stripe along one side of the filament [14]. One of the successful carbon stripe yarns is produced by Monsanto as No–Shock® conductive nylon [15]. One or two conductive filaments are inserted into the whole bundle of carpet fibers to provide the resistance to static development.

Figure 7–1 presents a graph of static buildup after walking on staple or bulked continuous filament (BCF) carpet [16].

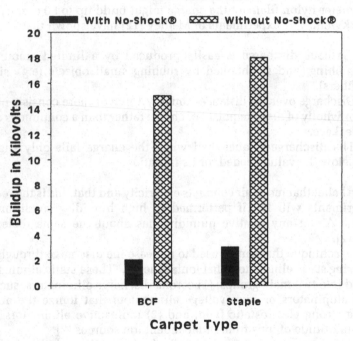

Figure 7–1. Carpets With and Without No–Shock® Yarn.
(Ref. [16]. Reprinted by permission of the Monsanto Company)

One problem with any conductive filament containing carbon black is color pollution. The term *color pollution* means that carpets that have been dyed to a very light shade will look off–color if they contain the conductive filament. This problem has not yet been solved, but fiber producers are actively investigating conductive materials to replace carbon black. The cross–section of this yarn, showing the carbon stripe causing the color problem, is displayed in Figure 7–2 [16]. Antistatic finish additives are also effective in reducing the electrical resistance, thus improving the conductivity of fibers and fabrics.

Figure 7–2. No–Shock®
Conductive Nylon
Cross–Section

The electrical resistance of a material is an important factor in determining whether that substance will show large changes in static after being separated from another material. If the resistance of the material is low, current flows back through the material near the point of separation and prevents large charges from being left on the material. If, however, the resistance is higher than a critical value, this value changing depending upon the conditions at separation, the reduction of current flow through the material is negligible. The charges then remain behind and can cause problems in yarn processing [17].

7.1.1 Conductivity of Polymers and Fibers

The conductance of organic polymers is in the order of 10^{16} Ω–cm; therefore, they are considered to be insulators. When exposed to an atmosphere of high relative humidity, many polymeric materials absorb significant amounts of water. It has been shown [18] that the conductivity of yarn and other textile materials increases exponentially with a decrease in the relative humidity with which it is in equilibrium. The sorption isotherm varies with different polymers, depending on the type of polymer or fiber, depending on the affinity of the polymer for water and the accessibility of the polymeric molecules to the water molecules. Whereas most natural fibers, such as cotton, silk, and wool, are composed of strongly hydrogen–bonding groups and are quite hygroscopic, the synthetic polymers such as polyolefins, polyamides, polyesters, and polyacrylonitrile are less hygroscopic. The presence of water in these polymeric bodies increases the electrical conductivity, so we may conclude that, generally, synthetic fibers are better insulators than natural fibers. There are many deviations from this generality, however: regenerated cellulose (rayon) is hygroscopic, and wool, in spite of being highly hygroscopic, is an insulator at medium relative humidity.

The effect of absorbed water is not difficult to explain. Pure water has a specific resistance of 10^8 Ω–cm, or 10^8 times lower than the pure polymer. In addition, water that contains dissolved electrolytes (such as tap water) has a conductivity as much as 10^3 times greater than pure water. It is thus very reasonable that polymers containing absorbed water have much higher conductivity than the original pure polymer. There are certain assumptions that one must make in relating moisture content to an increase in conductivity. We must presume that the water is present in a continuous channel extending from surface to surface. Thus, the role of dissolved electrolytes would be the same as in the conductivity of an aqueous solution; namely, the transport of electrical charges by migration of hydrated ions. There are many observations supporting this concept as an approximation, but there are also others indicating that the situation is more complicated due to the discontinuity of the channels, and limitations in the availability of the absorbed water for acting as the solvent [5].

For ionic conduction, the resistance will be determined by the number of ions available for conduction and their mobility. It is well known that a reduction in the total salt content of a fiber increases its electrical resistance; but there is also strong evidence that the tendency of the ions to associate into neutral ion–pairs, reducing the number of charged ions available for conduction, is the most important factor in the determination of the electrical resistance of fibers. The lower the dielectric constant of the material, the stronger the forces attracting the oppositely charged particles in the material. Thus, a reduction of dielectric constant results in an increase in the number of neutral ion–pairs, and consequent increases in the resistance. An approximate quantitative analysis of this concept by Hearle [17] has led to the equation:

$$\log R = \frac{\Psi}{\epsilon} + \chi \tag{7-2}$$

where:

R	=	resistance of the material,
ϵ	=	dielectric constant, and
Ψ and χ	=	constants.

The constant χ depends on the dimensions of the sample, the nature and amount of the total ion content and the mobility of the ions. Hearle [17] has also shown that:

$$\Psi = \frac{U_o \log e}{2\,k\,T} \tag{7-3}$$

and in this equation, U_o is the energy needed to separate ions in a vacuum, k is the Boltzmann's constant, and T the absolute temperature.

Hearle [19] has plotted experimental data, shown in Figure 7-3, relating the moisture content of fibers to the specific resistance of those fibers. Thus the fundamental difference between electrically insulating and

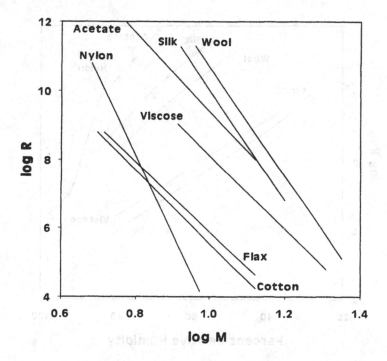

Figure 7-3. Relationship Between the Log Mass Specific Resistance, R, and the Log Percent Moisture Content, M.
(Reference [19]. Reprinted by permission of the Textile Institute)

non–insulating fibers is the actual moisture content under the conditions at which they are used. Figure 7–3 shows that the range of resistance, even in this limited range of moisture content, includes eight orders of magnitude. The reality that at the same moisture content hydrophobic fibers tend to exhibit higher resistance than hydrophilic fibers indicates a reduced conducting efficiency of water in the hydrophobic groups. This could be related to the lower dielectric constant and consequently lower degree of ionization in those polymers. The unusual position of wool might suggest that water is absorbed by the interior (cortex) rather than the exterior (cuticle and epicuticle) of the fibers so that the surface presents a higher resistance. It should also be noted in Figure 7–4 that the resistance of nylon does not decrease until the relative humidity has increased to greater than 80%, and

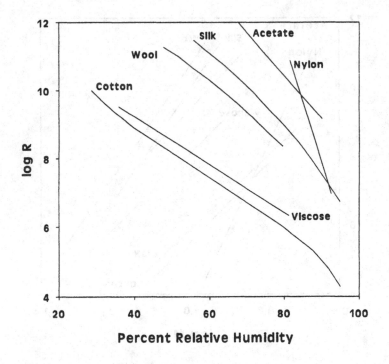

Figure 7–4. Relationship Between the Log Mass Specific
Resistance, R, and Percent Relative Humidity.
(Reference [19]. Reprinted by permission of the Textile Institute)

then the decrease is very sharp. This is possibly related to a potential barrier for hydrogen bonding of the water molecules, and this barrier cannot be overcome until the humidity has reached a large value.

A study has been made [20] on the carding of various fibers, including nylon. The authors found that for nylon, acetate, and wool the charge measured on card sliver decreases as the moisture regain increases, until at moisture levels corresponding roughly to a regain of 60–70% R.H., the static charge reaches a steady value, which is approximately the same for all these fibers. Thus it could be expected that electrification would be present in carding up to 60–70% R. H., but that at higher humidity levels the charge would decrease rapidly.

Attempts have been made to correlate the charge decay rate of polymers with their chemical structure [21]. It is not surprising that groups such as $-Cl$, $-CN$, or $-COOCH_3$ in the polymer contribute less to antistatic properties than the presence of other functional groups such as $-COONa$, $-SO_3Na$, or $-CH_2CH_2O-$.

7.1.2 Conductivity Through Antistatic Agents

Antistatic agents are chemicals that are applied to the surface of yarn or fabric to control the propensity of those materials to accumulate static electricity. Although these agents can function either by reducing the generation of the electrical charge, by increasing the conductance of the materials to which they are added, or by both mechanisms, most antistatic chemicals probably act through the conductivity mechanism.

Chemicals of high electrical conductivity include solutions of electrolytes, liquids of high dielectric constant, and metals. The liquids of high dielectric constant are normally volatile and thus their effectiveness as antistatic agents are temporary. Water is outstanding among these liquids because of its low cost, absence of toxic or environmental effects, and the complete lack of flammability. It is even more important that water, although very volatile, can be replenished from the moisture content of the atmosphere. Most of the available chemical antistatic agents use the conductivity of water as the mechanism of their action and actually are present only to assure the presence of water. Another way to state this observation is that antistatic agents are hygroscopic and are frequently electrolytic. In spin finishes, many of the compounds in a formulation are nonionic surfactants and are hygroscopic, but these compounds alone are quite poor antistatic agents. For effective reduction of static electricity, it is almost necessary to add some ionic species to the spin finish, preferably one that is also hygroscopic.

Water has a high electrical conductivity in comparison to dry yarn or fabric and it follows that this conductivity is increased by the presence of dissolved ions. In order to contribute its conductivity to the system, the water must be present in a continuous phase; therefore, for highest efficiency the antistatic agent must be present on the surface of the fiber rather than in the body of that substance. This surface concentration of antistatic agent then increases the surface conductivity rather than the volume conductivity. One of the problems with the antistatic behavior of yarn in some stages of processing is that the antistat is absorbed into the yarn and is no longer on the surface [22].

Since a spin finish is probably not a uniform coating on the entire perimeter of a filament nor continuous along each fiber length, we must assume that the conduction is at a point of contact, either at a place on the processing equipment or at some other filament. Let us consider two curved fiber surfaces A and B in contact at point O with a small wedge of slightly conducting liquid located between the surfaces around O. If one surface then moves relative to the other, as in rolling, and a charge is developed, the lines of force will pass through the liquid almost perpendicular to those surfaces. A point on one surface moving away from the point of contact will remain opposite the oil drop for a time OP/v, where OP is the length of the oil filled region and v the speed of the moving surface. The charge density at that point will fall according to the exponential law [12]:

$$\sigma = \sigma_o \exp(-4\,\pi\,k\,t/D) \tag{7-4}$$

where D is the dielectric constant, k is the conductivity of the oil, and t is time. This equation expresses the dependance of the time constant of a condenser on the conductivity and the dielectric constant of the material of which it is composed, and the fact that it is independent of the size and shape. The relationship between the initial and final charge will be:

$$\sigma = \sigma_o \exp(-4\,\pi\,k\,[OP]/Dv) \tag{7-5}$$

Electrification will fall very rapidly when:

$$Dv \approx 4\,\pi\,k\,[OP] \tag{7-6}$$

Taking the logarithms and converting, we have:

$$\frac{\sigma}{\sigma_o} = 1.13 \times 10^{13} \exp(-k[OP]/Dv) \tag{7-7}$$

where k is in ohm^{-1} cm^{-1} [12].

The essential assumption in the above theory is that the liquid must wet both of the solid surfaces so that no separation charging occurs at either liquid–solid interface. The only separation is a liquid–liquid one where all lines of force tend to vanish as the result of conduction, leaving the separated portions uncharged. Despite a high electrical charge, the static will fall off rapidly once the conductivity rises beyond the critical value [12]. Additional derivations for electrical charge dissipation can be found in other sources [23]–[25].

As these authors assume, a spin finish forms a continuous layer between the filament and the surface, and the conductance is proportional to its volume or weight and the specific conductance of the finish. If it is assumed that the fiber is a cylindrical filament of l cm length, a radius of r cm, and denoting the thickness of the finish layer as Δr and the specific conductance of the finish k, the conductance K of the finish layer is then given by the equation [14]:

$$K = k\frac{2\pi\Delta r}{l} \qquad (7\text{–}8)$$

Generally, a concentration of antistatic agent of about 0.1% on weight of fiber is needed for the protection of textile fabrics. The exact amount depends on the efficiency of the antistat, relative humidity, temperature, diameter of the filament, fabric structure, and the degree of antistatic protection required.

7.2 MEASUREMENT OF STATIC ELECTRICITY

In determining the amount and type of antistat to be used in a spin finish, it is necessary to have some general technique for accurately measuring the electrical charge. The ideal procedure for the measurement of this charge should, as nearly as possible, simulate the actual process conditions that generate the charge on the yarn or fabric. Unfortunately, some of these devices do not give reproducible results. A partial list of the available techniques is described in the subsequent sections.

7.2.1 Resistivity

Electrical resistivity is a measure of the restraint of current flow passing either across the surface of a body or through the volume of the material under a potential difference. Surface, volume, and bulk resistivity measurements of several materials have found use in materials science and attempts have been made to correlate these measurements with the static properties of these materials. Measurements of bulk and volume resistivity are of little significance since static electricity is essentially a surface phenomena. Surface resistivity is numerically equal to the surface resistance between two electrodes forming the opposite sides of a square. The size of the square is immaterial and thus the unit in this measurement is simply the ohm, although frequently the term ohm per square is used.

Surface resistivity measurements are used primarily because of their familiarity and not because they are the best way to determine antistatic properties [26]. A standard test method for insulating materials (ASTM D257–90) [27], however, describes a surface resistance technique suitable for plastics. The method states "*surface resistance or conductance cannot be measured accurately, only approximated, because some degree of volume resistance or conductance is always involved in the measurement. The measured value is also affected by the surface contamination.*" The AATCC [28][29] has published two standard test methods for fabrics and yarns, but in this case the distinction between surface and volume resistivity is much less clear. Separate measurements of these two resistance values are almost impossible because of the geometry of the samples. A table of antistatic properties and surface resistivity has been published [2] which approximates this correlation.

Table 7–2

Static Propensity and Surface Resistivity, R

Log R, Ω	Antistatic Protection
>13	nil
12–13	poor
11–12	moderate
10–11	good
<10	excellent

(Ref. [2]. Reprinted by permission of John Wiley & Sons, Inc.)

Another resistivity measurement, MSR, or Mass Specific Resistivity, has been described [30] which measures the DC resistance of a fiber. The electrical properties of fabric and carpet samples, however, cannot be determined by this technique.

7.2.2 Fabric to Metal Cling Test

The fabric to metal cling test is a procedure developed by AATCC [31] to be used primarily in investigating the static properties of lightweight tricot fabric. This test is based on the principal of frictional charge generation on the sample by rubbing with a fabric of different composition. For example, if the test fabric is nylon, the rubbing fabric will be a specific type of polyester taffeta. The charged fabric is then allowed to cling to a grounded metal plate and the time until the fabric releases is measured. Several problems may arise when using this method:

1. If a heavy fabric is used, the weight of the fabric will cause the material to fall from the metal plate and thus yield low cling times that are obviously in error.
2. The surface of the fabric is affected by rubbing, and this probably changes the cling properties.
3. Variation in technique among people running the test is a serious problem causing large errors that affect accuracy and precision.

If one wishes to examine a tricot fabric to determine if it has any antistatic properties, this method can be used, but the actual numbers reported should be reported with caution.

7.2.3 Sail Test

The sail test has been proposed by one fiber producer [32] as an effective guide to the cling properties of slips or other lingerie. In this test a section of polyester fabric resembling a square sail, with dimensions of about 4 feet by 5 feet, is mounted on two vertical insulated pipes and a charge is generated by an evaluator wearing the test garment. After backing up to the sail, the person rubs the garment on the sail until static causes the garment to cling. The evaluator then walks until the garment de–clings or for a maximum of 10 minutes. The problems with this method are obvious since all measurements are subjective.

7.2.4 The Ash Test

An extremely rough estimate of static propensity can be obtained by rubbing a fabric against another surface to generate a charge and then holding the fabric over a shallow dish containing ashes. The presence of a static charge will be shown by the ashes vaulting onto the charged surface.

The ash test has been correlated to fabric surface resistivity in a study of antistatic agent application amounts [25]. Samples of nylon taffeta, plain weave Orlon®, and a ribbed Dacron® shirting fabric were rubbed by hand on a piece of wool flannel after treatment with increasing concentrations of an antistatic agent. The samples were passed slowly at a height of about 1 inch over a pile of cigarette ashes. The amount of ashes transferred in this test is described in the following table.

Table 7–3

Surface Resistivity and Ash Attraction

Fabric	Resistance log ohms/square	Ashes
Nylon	13.8+	Much
Nylon	12.3	Slight
Nylon	11.7	None
Nylon	10.5	None
Nylon	9.4	None
Orlon	13.8+	Much
Orlon	11.7	Slight
Orlon	10.4	Slight
Orlon	9.1	None
Orlon	8.6	None
Dacron	13.8+	Much
Dacron	12.1	Slight
Dacron	11.0	None
Dacron	10.0	None
Dacron	9.2	None

(Reference [25]. Reprinted with permission of the Textile Res. J.)

7.2.5 Hayek–Chromey Wheel

The Hayek–Chromey [33] static half–life tester overcame many difficulties of the procedures listed above since the measurements are objective, but the problems with surface changes with applied friction and reproducibility were still present. In the Hayek–Chromey technique the sample test fabric was placed on a rotating wheel, a section of a rubbing material (either fabric or a shoe sole for carpet testing) under tension was allowed to contact the test fabric, and the motor to drive the wheel was started. The charge generation was detected by an electrometer. After a constant charge was attained, the rubbing material was removed and the charge decay was measured. The value reported is the time for the charge half–life.

An experiment in which the Hayek–Chromey wheel was used to measure the static decay on nylon and polyester fabric before and after treatment with antistatic agent demonstrated the value of this test. The fabric under test was threaded through a series of slots on the wheel that was rotated at about 270 R.P.M. As the wheel rotated it rubbed against a surface of Botany wool serge for two minutes, by which time the charge built up to the maximum and the voltage was recorded. The rubbing surface was removed and the time for the charge to decay to half its maximum value was noted. The results from this test are listed in Table 7–4.

Table 7–4

Antistatic Effects on Nylon and Polyester Fabric
(65% R.H. and 70°F)

Fabric	Nonionic Antistatic Agent Wt. %	Charge (arbitrary units)	Half–life sec.
Nylon	0	1240	37
Nylon	0.2	120	1
Polyester	0	2200	130
Polyester	0.2	80	1

(Reference [34]. Reprinted with permission of The Society of Dyers and Colourists)

This author also studied the effect of the loss of antistatic properties of nylon staple fabric after heat aging and the results are given in Table 7–5.

Table 7–5

Heat Aging of Antistat on Nylon Fabric

Experiment No.	Treatment	Charge, Arbitrary Units	Half–life, sec.
1	Untreated	1340	48
2	Heated	1460	103
3	Treated with 0.3% alcohol phosphate	178	7
4	As 3, and then heated	1120	63
5	As 3, heated and extracted with water	1120	38
6	As 3, heated, extracted with water, then extract reapplied	200	7
7	As 6, then re–heated	680	35

(Reference [34]. Reprinted with permission of The Society of Dyers and Colourists)

In this experiment the antistatic agent (the diethanolamine salt of the acid phosphate ester of a long–chain fatty alcohol) was applied to the fabric and it was heated in an oven at 150°C for 30 minutes. It was then extracted in a Soxhlet device for 7 hours with water to remove any antistatic agent and the aqueous extract then re–applied to the fabric. The antistatic effect was checked at each stage.

Data in Table 7–5 clearly show that heat aging has a significant effect in reducing the antistatic properties of textiles.

7.2.6 The Static Honestometer

In the early 1970s, several fiber producers developed an antistatic nylon textile yarn for undergarments and night wear. Some of the tests listed above were used to measure the static charges, but poor reproducibility was obtained. A prototype of a new instrument was first built in the United States [21], but it was later commercialized in Japan as the Static Honestometer. This device was so named by the inventor because of the reliability of the data [35]. It had been reported [36][37] that this instrument was superior to other methods of static determination.

The Static Honestometer can be used to determine the electrical properties of yarn, fabric, or carpet samples of any denier or construction. It operates with a corona discharge as the source of electrical energy with an oscilloscope as the detector. A sample is placed under the sample holder on a turntable that is then rotated at 1750 RPM. A 10–kV discharge is impinged on the sample from a needle electrode located about 20–25 mm from the sample surface. The charge developed on the sample is detected by a receiver connected to the input of an oscilloscope and placed 180° across the turntable from the driver electrode. It has been found that the maximum charge is developed after about 6 seconds, so the charging voltage is ended after that time. The charge buildup and decay on the fabrics are displayed as sine wave pulses on the oscilloscope and the sweep control adjusted to show a narrow band. The best technique for observing these data is to photograph the oscilloscope face with a time setting so the voltage is seen as an envelope of voltage versus time.

The data from the Honestometer can be presented in several ways, including (1) the photographs of the oscilloscope face, (2) the initial peak–to–peak charge, (3) the half–life, or time for half the initial charge to decay, and (4) percent of the initial charge retained after 44 seconds. The latter system is included to describe the static existing on normal, unmodified nylon or polyester since the half–lives on these substances are too long to measure on one oscilloscope sweep of 50 seconds. Four typical samples were measured at 30% RH [36] and the results are described below.

1. Aluminum. Since aluminum is an excellent electrical conductor, no charge build–up was observed with all of the electrical energy transmitted to ground.
2. Unmodified nylon 6,6. A charge of about 360 mV peak–to–peak voltage was developed after the 6–results second discharge and this decays to only 220 mV after 44 seconds. This product would obviously have poor antistatic properties
3. Cotton. Since cotton is very hygroscopic, it has a very low initial charge, 110 mV, and a rapid decay pattern showing a half–life of only 2 seconds and essentially zero charge after 44 seconds.
4. Antistatic nylon. The initial charge was quite low, only 60 mV, with a half–life of about 20 seconds.

7.2.7 Rothschild Static Voltmeter

The instrument that is currently used in many fiber testing laboratories is the Rothschild Static Voltmeter [38]. In this instrument a vibrator–condenser converts the electrostatic charge into a proportional voltage. An electromagnetic driving circuit with a constant amplitude excites one condenser plate. Because the capacity varies, an alternating voltage proportional to the applied DC voltage is generated on the condenser electrode. The alternating voltage obtained by this technique is amplified and displayed on a galvanometer. There are several electrodes that can be used with the Rothschild instrument. These are:

1. The field measuring electrode. This electrode has a design that consists of a central measuring rod surrounded by a semi–cylindrical metal shield that determines the direction of the electrostatic field. After this electrode has been placed on the input terminal, the direction of the field in the room can be determined by rotating the Static–Voltmeter.
2. Roller electrodes. The electrostatic charges on moving yarn samples can be measured with this electrode. The rollers are fitted with ball–bearings to reduce the friction. This electrode can be fastened directly to the contact electrodes so that measurements on processing equipment can be made.
3. Contact electrode. This electrode consists of handle with an insulated contact tube and is connected to the input terminal by a 1–meter cable. It can also measure the charge on a moving threadline.
4. Resistance electrode. Two spring clips are mounted on the triangular base plate. The yarn or fabric (about 10 cm long by 1

cm wide) to be measured is clamped between the two clips and the electrode placed on the input terminal. The mode switch on the body of the instrument is set on *R* and the measuring range switch at *150 V*. The grounding switch is then depressed, charging the input terminal to full deflection. The discharge time for the galvanometer needle to return to half–scale is noted.

5. Cup electrode. The cup electrode is used to measure the conductivity of staple fibers. After placing the electrode on the input terminal, it is filled to about two–thirds with the sample and the lid placed in the cup. The measurements are carried out as described in Section 4.

6. Carpet electrode. This electrode makes it possible to test both sides, independently, of a carpet sample. After the electrode is placed on the input terminal, the sample is placed on the two plates with the surface to be measured facing downward. It is then covered with the weight that is part of the electrode. Measurements are made by the procedure of Section 4.

There are two separate techniques that may be used to determine static electricity with the Rothschild Voltmeter [39]. In the first system the measuring electrode does not contact the sample and only the electrostatic field generated by the charge is measured. This is accomplished by either the screened field measuring electrode or, at higher voltages, the roller or the contact electrode. The sample must not touch the electrode. One must take care that stray filaments do not protrude from the body of the sample and make contact with the electrode.

When making comparison measurements, care must be taken that the distance between the sample and the electrode remains the same. While carrying out field measurements, the input terminal must never be grounded.

In the second procedure the measuring electrode does contact the sample. Yarn is pulled from a bobbin and allowed to run over a guide that is insulated from ground. It then passes through the roller electrodes and to a yarn transport system. A series of experiments devised to examine the relationship between the static electrical charge developed on the fiber and the level of spin finish were performed [22]. Polyester yarn was made with different levels of finish and the static measured with a Rothschild Static Voltmeter [39] over a four week period. The data is shown in Table 7–6 and displayed graphically on Figure 7–5.

The plot, and the data in the table, clearly show that the static charge developed on the fiber is inversely related to the finish on yarn; that is, lower finish induces a higher charge on the yarn. If the finish on yarn is at a level of about 0.5% or greater, the static charge is changed very little.

Table 7–6
Static Charge on Yarn With Time

Finish on Yarn	Static in Volts				
	Day 1	Day 4	Day 7	Day 14	Day 28
0.2	124.5	142.5	172.5	255.0	250.0
0.3	84.8	100.5	142.5	185.0	180.0
0.4	—	45.0	64.5	110.0	117.0
0.5	33.2	33.0	43.5	65.0	81.0
0.65	—	10.5	36.0	65.0	60.0
0.7	19.5	25.5	30.0	60.0	57.0

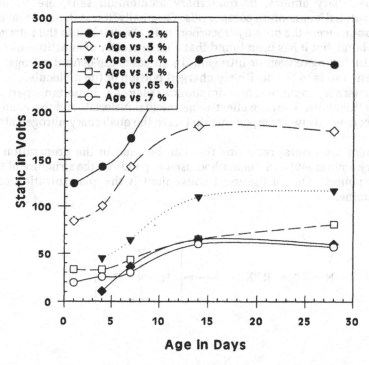

Figure 7–5. Effect of Static Charge versus Time for Polyester Yarn Made with Different Finish Levels.

Oxé and Keller [39] used the Rothschild static voltmeter to study the numerous phenomena related to the electrostatic charges developed while walking across carpets. They found that the charge was influenced by the electrical resistance of the shoe and the carpet, by the capacity of the condenser system person/carpet/mass, by the relative humidity, and by the step frequency of the walker.

7.3 CHEMICALS USED AS ANTISTATIC FINISH COMPONENTS

There are two general classes of chemicals that are effective as antistats, quaternary amines and inorganic–organic reaction products, principally phosphated alcohols. Some sulfates or sulfonated can be used also, but they are not especially effective.

7.3.1 Quaternary Amines

Quaternary amines, or quaternary ammonium salts, are usually tetra–substituted ammonium compounds. Originally, it was thought that the R–alkyl groups were the only hydrocarbon radicals attached to the nitrogen by a C–N bond, but it has been found that a large variety of substituents can be used, including oxygen or nitrogen. In all cases the central nitrogen is pentavalent and is in the positively charged portion of the molecule.

Quaternary ammonium compounds play an important part in biological functions. They are effective germicidal materials and the vitamin B complex contains two components that have the quaternary nitrogen atom [40].

There are several reactions that can be used in the preparation of quaternary amines [40] and the method used depends on the structure of the final compound. One of the most convenient is the quaternization of a tertiary amine.

$$
\begin{array}{c} R' \\ | \\ R-N-R'' \end{array} + R'''X \longrightarrow \left[\begin{array}{c} R' \\ | \\ R-N-R'' \\ | \\ R''' \end{array} \right]^{+} X^{-}
$$

The tertiary amines can be produced by several reactions with the those below accounting for many products.

Dimethyl alkylamines

$$RNH_2 + 2\ CH_2O + H_2 \longrightarrow RN(CH_3)_2 + 2\ H_2O$$

Or:

$$RCl + NH(CH_3)_2 \longrightarrow RN(CH_3)_2 + NaCl + H_2O$$

Ethoxylated dialkylamines

$$R_2NH + H_2C\!\!-\!\!CH_2 \longrightarrow R_2NCH_2CH_2OH$$
$$\overset{\diagdown\!\diagup}{O}$$

Primary or secondary amines can also be used to prepare quaternary ammonium compounds. In these cases a basic compound must be used to neutralize the acidic by–product of the reaction.

$$RNH_2 + 3\ RCl + 2\ NaOH \longrightarrow R_4N^+Cl^- + 2\ NaCl + 2\ H_2O$$

and

$$R_2NH + 2\ RCl + NaOH \longrightarrow R_4N^+Cl^- + NaCl + 2\ H_2O$$

7.3.2 Alcohol Phosphates

Most alcohol or ethoxylated alcohol phosphate esters are made by reacting the base alcohol with either P_2O_5, polyphosphoric acid, or phosphorus oxychloride, $POCl_3$ [41]. The reactions generally give a mixture of mono–, di–, and lesser amounts of triesters. When polyphosphoric acid is the phosphorylating agent, the yield of monoester is at a higher ratio than with phosphorus pentoxide. Phosphorus oxychloride is the phosphorylating agent chosen when high yields of the triester are desired, but this may be only about half of the product. The structures of these esters, in which the amount of ethylene oxide may range from zero for the alcohol ester to higher levels, are:

$$\underset{\overset{\displaystyle R(OC_2H_4)_nO\overset{\displaystyle \overset{O}{\|}}{P}-OH}{\underset{\displaystyle OH}{|}}}{}\qquad\qquad\text{Monoester}$$

$$\underset{\overset{\displaystyle R(OC_2H_4)_nO\overset{\overset{O}{\|}}{P}\,O(C_2H_4O)_nR}{\underset{\displaystyle OH}{|}}}{}\qquad\qquad\text{Diester}$$

$$\underset{\overset{\displaystyle R(OC_2H_4)_nO\overset{\overset{O}{\|}}{P}\,O(C_2H_4O)_nR}{\underset{\displaystyle O(C_2H_4O)_nR}{|}}}{}\qquad\qquad\text{Triester}$$

These phosphates can then be neutralized with KOH to form water–soluble salts.

7.4 COMMERCIAL ANTISTATIC AGENTS

7.4.1 Quaternary Amines

Some of the components added to early finish formulations to control static electricity were quaternary amines. These chemicals were, and still are, some of the most effective antistats to be incorporated in a finish, but some toxicity problems have limited their use. They also tend to remain on the fiber surface and can be removed in a customer's dyeing equipment. When extracted, they are toxic to bacteria and will cause problems in a waste disposal system.

One of the most widely used quaternary amine antistats is made by PPG as Larostat® 264A [42], CAS# 68308–67–8. The chemical name is quaternary ammonium compounds, ethyl dimethyl soya, ethyl sulfate salt. The fatty acids from soy bean oil contain about 90% unsaturated alkyl chains, so this antistat is very prone to the formation of varnishes on hot surfaces. The structure of this compound is probably:

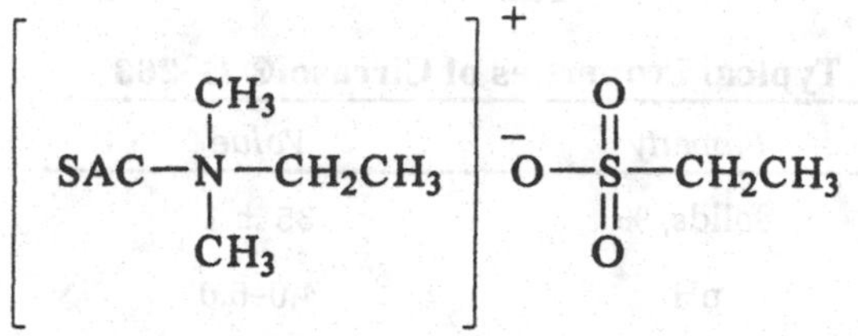

$$\left[\begin{array}{c} CH_3 \\ | \\ SAC-\overset{-}{N}-CH_2CH_3 \\ | \\ CH_3 \end{array} \right]^{+} \quad \overset{-}{O}-\overset{\displaystyle O}{\underset{\displaystyle O}{\overset{\|}{\underset{\|}{S}}}}-CH_2CH_3$$

In the above structure, SAC represents a soya alkyl chain. This antistat is normally supplied as a 35% amber aqueous solution, but it is also available as an anhydrous waxy solid. The properties of the solution are:

Table 7–7

Typical Properties of Larostat® 264A

Property	Value
Solids, %	35 ± 1
pH	5.0 – 7.0
Viscosity, cPs, 50°C	230 – 300
Color, Gardner	12 max

Several other quaternary amines are available from ICI Speciality Chemicals. These antistats vary in molecular structure, but are all effective in their reduction of static electricity.

N–cetyl–N–ethylmorpholinium ethosulfate, CAS# 78–21–7

Another quaternary amine supplied either in the anhydrous form or in a 35% aqueous solution is a material manufactured by ICI as Cirrasol® G–263 [43][44]. The structure is:

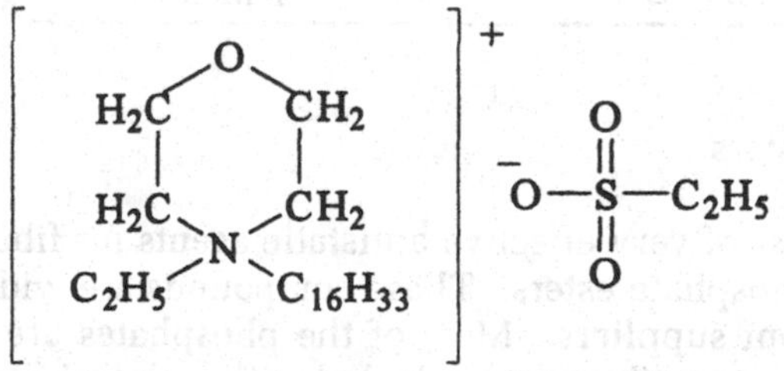

$$\left[\begin{array}{c} \overset{\displaystyle O}{H_2C \diagup \quad \diagdown CH_2} \\ | \qquad | \\ H_2C \diagdown \quad \diagup CH_2 \\ \underset{\displaystyle N}{} \\ C_2H_5 \diagup \quad \diagdown C_{16}H_{33} \end{array} \right]^{+} \quad \overset{-}{O}-\overset{\displaystyle O}{\underset{\displaystyle O}{\overset{\|}{\underset{\|}{S}}}}-C_2H_5$$

Some of the properties of the solution of this antistat are shown in Table 7–8.

Table 7–8

Typical Properties of Cirrasol® G–263

Property	Value
Solids, %	35 ± 1
pH	4.0–6.0
Viscosity, cPs, 25°C	Approx. 20

Polyalkoxylated Quaternary Amine, CAS# Proprietary

Both the composition and the CAS number of this quaternary amine are considered to be proprietary to the suppliers. It is a highly hydrophilic antistat that is not as hazardous as some products in this class of compounds since it is not irritating to the skin; however, it is corrosive to the eye. This compound, commonly used as an external antistatic agent on cellulosic and synthetic fibers, is made by ICI as Cirrasol® G–250 [43][45] and by Witco as Adogen® 66 [46]. Some of the attributes of the material are presented by Table 7–9.

Table 7–9

**Typical Properties of Polyalkoxylated
Quaternary Amine**

Property	Value
Solids, %	99
pH, 5% in 1:1 IPA:Water	7
Viscosity, cPs, 50°C	300–500
Color, Gardner	7 max

7.4.2 Phosphate Esters

Another class of very effective antistatic agents for fiber and textile applications are phosphate esters. These compounds are widely available from many different suppliers. Most of the phosphates are prepared by reacting an alcohol or an ethoxylated alcohol with a phosphorylating agent, as shown on page 295. In almost all instances the alcohol is aliphatic, but a few alkyl phenol phosphates are also available. The phosphate esters are sold in the free acid form or neutralized, usually with potassium hydroxide.

7.4.2.1 Aliphatic Alcohol Phosphates, Free Acids

Hexyl Alcohol Phosphate, CAS# 68511–03–5

The shortest aliphatic alcohol chain available in this class of antistats is the seven carbon hexyl alcohol. This liquid component can easily by neutralized by the user with a variety of bases to form water–soluble salts. The structure of the monoester is:

$$CH_3(CH_2)_6-O-\overset{\displaystyle \overset{-}{O}H^+}{\underset{\displaystyle \overset{-}{O}H^+}{P}}=O$$

Produced by Ethox as Ethfac® 106 [47][48], the properties are shown in Table 7–10.

Table 7–10

Typical Properties of Hexyl Alcohol Phosphate, Free Acid

Property	Value
Color, Gardner	2 max
Acid Value	350–390
Moisture, %	1.0 max
Activity, %	99

2–Ethylhexyl Alcohol Phosphate, CAS # 12645–31–7

2–Ethylhexanol is a branched chain alcohol that can be esterified with stearic acid to form an excellent low–viscosity lubricant. In this compound, the alcohol is reacted with a phosphorylating agent to yield a phosphate ester that is a good antistat base and a low foaming anionic emulsifier. It is a white to light yellow liquid that is oil soluble and can easily be neutralized. This chemical is made by Ethox with the name Ethfac® 104 [43][49]. The properties are listed in Table 7–11 and the structural formula of the monoester is:

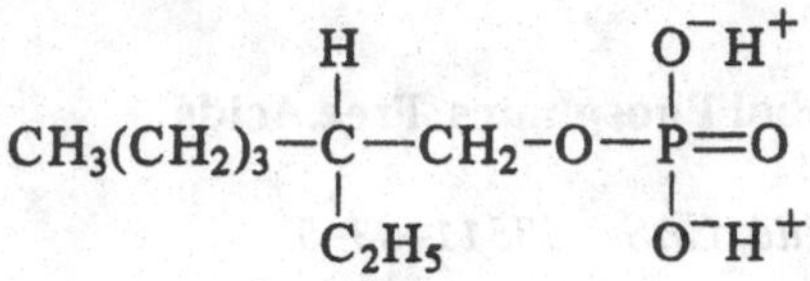

$$CH_3(CH_2)_3-\underset{\underset{C_2H_5}{|}}{\overset{\overset{H}{|}}{C}}-CH_2-O-\underset{\underset{O^-H^+}{|}}{\overset{\overset{O^-H^+}{|}}{P}}=O$$

Table 7–11

Typical Properties of 2–Ethylhexyl Alcohol Phosphate, Free Acid

Property	Value
Color, Gardner	2 max
Acid Value	290–320
Moisture, %	1.0 max
Activity, %	99

Linear Aliphatic Alcohol Phosphate, CAS# Proprietary

A free acid alcohol phosphate ester antistat that is similar to 2–ethyl–hexyl alcohol shown above is made by Henkel as Dacospin® PE–146 [50][51]. The composition and the CAS number are proprietary to the manufacturer. This product is a light yellow liquid with some of the attributes found in Table 7–12.

Table 7–12

Typical Properties of Dacospin® PE–146

Property	Value
Color, Gardner	3 max
Acid Value	300
Monoester, meq/g	1.85
Diester, meq/g	1.60
Viscosity, cSt, 100°F	115

Mixed Lauryl and Myristyl Alcohol Phosphate, CAS# 68412–59–9

Lauryl (C_{12}) and myristyl (C_{14}) alcohols can be obtained by the reduction of coconut fatty acids, in which they are major constituents. When they are reacted to produce the phosphate esters, a solid, oil–soluble component is formed that is an excellent antistatic agent and lubricant. The structure of the monoester, showing only the C_{12} and C_{14} moieties, is:

$$CH_3(CH_2)_{11-13}\ O-\underset{\underset{O^-\ H^+}{|}}{\overset{\overset{O^-\ H^+}{|}}{P}}=O$$

It is manufactured by Ethox as Ethfac® 102 [47][52] with several properties listed in Table 7–13.

Table 7–13

Typical Properties of Lauryl–Myristyl Alcohol Phosphate, Free Acid

Property	Value
Color, Gardner	2 max
Acid Value	200–230
Moisture, %	1.0 max
Activity, %	99

Stearyl Alcohol Phosphate, CAS# 68412–59–9

As with many other components based on stearyl alcohol or stearic acid, this chemical is useful as a lubricant and softener, after neutralization, in addition to being an antistat. It is an off–white to yellow waxy solid that is 99% active. It is made by Ethox with the trade name Ethfac® 1018 [47] [53]. The properties are presented by Table 7–14 and the structural formula is:

$$CH_3(CH_2)_{17}\text{-}O-\overset{\overset{\displaystyle O^-\,H^+}{|}}{\underset{\underset{\displaystyle O^-\,H^+}{|}}{P}}=O$$

Table 7–14

Typical Properties of Stearyl Alcohol Phosphate, Free Acid

Property	Value
Color, Gardner	5 max
Acid Value	160–200
Moisture, %	1.0 max
% Free Phosphoric Acid	1.0 max
% Monoester	47 min
% Diester	46 min

Several other aliphatic alcohol phosphates are available as the free acid from other suppliers, but their compositions and CAS numbers are considered to be proprietary. Further information about these products may be obtained by contacting the suppliers indicated. Hoechst–Celanese makes phosphates only with P_2O_5 as the phosphorylating agent. These components are Afilan® HIP and Afilan® ODP [54]. The Specialty Chemicals Group of ICI Americas produces two alkyl phosphate esters, Cirrasol® G–2199 and Cirrasol® G–2206 [43], and Witco makes Emphos® PS–900 [55].

7.4.2.2 Aliphatic Alcohol Phosphates, Potassium Salts

Although the acidic forms of alcohol phosphate antistats are readily neutralized by the purchaser, it may be more convenient to buy the salt from the manufacturer. These salts, generally the potassium salt, are supplied as their aqueous solutions. Hoechst–Celanese makes Afilan® PNL [54] that is a solution of a potassium alkyl phosphate at a 65% concentration. The exact composition is proprietary. ICI produces two materials [56], Cirrasol® G–2207, a 45% concentration of a proprietary potassium salt, and Cirrasol® G–2200, which has been identified as the potassium salt of hexyl alcohol phosphate, CAS# 68511–03–5. The structure of the monoester is:

$$\left[CH_3(CH_2)_5\text{-}O\text{--}\overset{\displaystyle O^-}{\underset{\displaystyle O^-}{\overset{|}{\underset{|}{P}}}}\text{=}O \right] 2\ K^+$$

Typical properties of this ester are displayed in Table 7–15.

Table 7–15

**Typical Properties of Hexyl Alcohol Phosphate,
Potassium Salt**

Property	Value
Color, APHA	150 max
Acid Value	15–20
Activity, %	45.5–50.0
Viscosity, cSt, 25°C	28

7.4.2.3 Polyoxyethylene Aliphatic Alcohol Phosphates, Free Acids

The acid forms of aliphatic alcohol phosphate esters as discussed in the previous section are usually insoluble in water. It is necessary to react the acids with base to produce water–soluble chemicals. Another class of alcohol phosphates with increased hydrophilic character can be synthesized by reacting ethoxylated alcohols with phosphorylating agents. While several suppliers consider the exact composition of these esters to be proprietary, many of the properties are well known.

Alkyl Alcohol (POE)$_n$ Phosphate, CAS# Proprietary

These proprietary phosphates are useful as anionic surfactant intermediates, lubricants, and antistats. They are manufactured by Ethox as Ethfac® 142W [47][57], by Henkel as Tryfac® 5552 [50][58], and by Hoechst–Celanese as Afilan® 2188 [54]. The properties of the Ethox component are shown in Table 7–16 and the Henkel material in Table 7–17.

Table 7–16

Typical Properties of Ethox Ethfac® 142W

Property	Value
Color, Gardner	2 max
Acid Value	155–175
Moisture, %	2.0 max
Viscosity, cSt, 25°C	650–900

Table 7–17

Typical Properties of Henkel Tryfac® 5552

Property	Value
Color, Gardner	1 max
Density, lbs/gal	8.8
Cloud Point, °C	76
Viscosity, cSt, 100°F	170

Butyl Alcohol (POE)$_n$ Phosphate, CAS# 14260–98–1

Ethoxylated butanol phosphate is a hydrophilic surfactant that is quite stable in caustic solutions and therefore is useful in alkaline cleaners. With the short aliphatic chain, the material is probably a better surfactant than an antistat, but it will have antistatic properties. Ethox makes two versions of the butyl phosphate, Ethfac® PB–1 [47][59] and Ethfac® PB–2 [47][60], the latter containing more ethylene oxide than the former. The structure is:

$$CH_3(CH_2)_3-O-(CH_2CH_2O)_n-\overset{\displaystyle O^-H^+}{\underset{\displaystyle O^-H^+}{P}}=O$$

The attributes of Ethfac® PB–1 are listed in Table 7–18 and those of PB–2 are in Table 7–19.

Table 7–18

Typical Properties of Ethox Ethfac® PB–1

Property	Value
Color, Gardner	4 max
Acid Value	580–640
Moisture, %	1.0 max
Activity, %	99 min

Table 7–19

Typical Properties of Ethox Ethfac® PB–2

Property	Value
Color, Gardner	4 max
Acid Value	490–540
Moisture, %	1.0 max
Activity, %	99 min

Hexyl Alcohol (POE)$_n$ Phosphate, CAS# 67989–06–4

Though this ester has the same base aliphatic chain as the hexyl alcohols previously described, the addition of polyoxyethylene results in the chemical being more water soluble than the non–ethoxylated version. It is a low foaming ester that is an excellent antistat and can be a co–emulsifier for many oils. The producer is Ethox who has assigned the name Ethfac® 136 [47][61] to that component. The structure is:

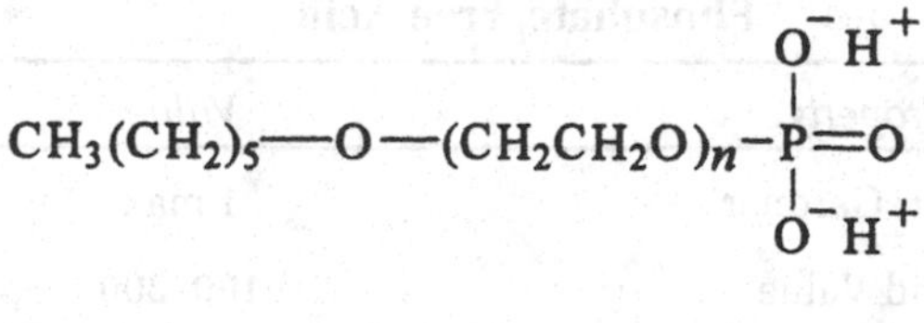

$$CH_3(CH_2)_5 - O - (CH_2CH_2O)_n - \overset{\overset{\displaystyle O^- H^+}{|}}{\underset{\underset{\displaystyle O^- H^+}{|}}{P}} = O$$

The typical properties of hexyl alcohol (POE) phosphate are presented by Table 7–20.

Table 7–20

Typical Properties of Hexyl Alcohol (POE)$_n$ Phosphate, Free Acid

Property	Value
Color, Gardner	2 max
Acid Value	170–210
Moisture, %	1.0 max
Activity, %	99 min

Octyl–Decyl Alcohol (POE)$_n$ Phosphate, CAS# 68130–47–2

Caprylic (C_8) and capric (C_{10}) fatty acids are the low molecular weight fractions removed by stripping coconut oil and isolated through saponification. These acids are then reduced to obtain the octyl–decyl alcohol mixture. After ethoxylation, they react with a phosphorylating agent to make the phosphate ester. Although this chemical is an effective antistat, a principal application is as a coupling agent for nonionic surfactants in alkaline systems. It is made by Ethox as Ethfac® 2684 [47][62]. Properties are shown in Table 7–21 and the structure is:

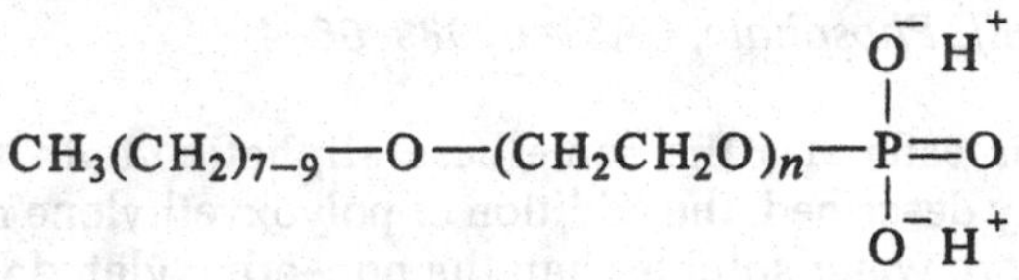

$$CH_3(CH_2)_{7-9}-O-(CH_2CH_2O)_n-\overset{\overset{\displaystyle O^-H^+}{\displaystyle |}}{\underset{\underset{\displaystyle O^-H^+}{\displaystyle |}}{P}}=O$$

Table 7–21

Typical Properties of Octyl–Decyl Alcohol (POE)$_n$ Phosphate, Free Acid

Property	Value
Color, Gardner	1 max
Acid Value	160–200
Moisture, %	0.5 max
Activity, %	99 min

Decyl Alcohol (POE)$_n$ Phosphate, CAS# 108818–88–8

Alkali metal or amine salts of the acid form of this chemical are excellent emulsifiers and antistats. The structure is similar to the octyl–decyl alcohol phosphate without the octyl fraction. It is manufactured by Ethox as Ethfac® 161 [47][63] with the properties listed in Table 7–22.

Table 7–22

Typical Properties of Decyl Alcohol (POE)$_n$ Phosphate, Free Acid

Property	Value
Color, Gardner	1 max
Acid Value	115–135
Moisture, %	1.0 max
Activity, %	99 min

Lauryl Alcohol (POE)$_n$ Phosphate, CAS# 68130–47–2

Applications for this product include antistatic agent, emulsifier, solubilizer, wet processing surfactant, and it is stable in caustic solutions at concentrations up to 130 g/l. This ester is made by Stepan with the name of Cedephos® FA–600 [64]. Miscellaneous characteristics are shown in Table 7–23 and the structure is:

$$CH_3(CH_2)_{11}-O-(CH_2CH_2O)_n-\overset{\displaystyle .\;\overset{-}{O}\,H^+}{\underset{\displaystyle \overset{-}{O}\,H^+}{\overset{|}{\underset{|}{P}}}}\!=\!O$$

Table 7–23

Typical Properties of Lauryl Alcohol (POE)$_n$ Phosphate, Free Acid

Property	Value
Color, Gardner	1 max
pH, 5% Aqueous	1
Moisture, %	0.3 max
Activity, %	99.7

Tridecyl Alcohol (POE)$_n$ Phosphate, CAS# 9046–01–9

Tridecyl alcohol is a widely used aliphatic base for lubricants, emulsifiers, and antistats. The acid form of this oil–soluble phosphate ester is an excellent lubricant and the salts are superior antistats. The structural formula is:

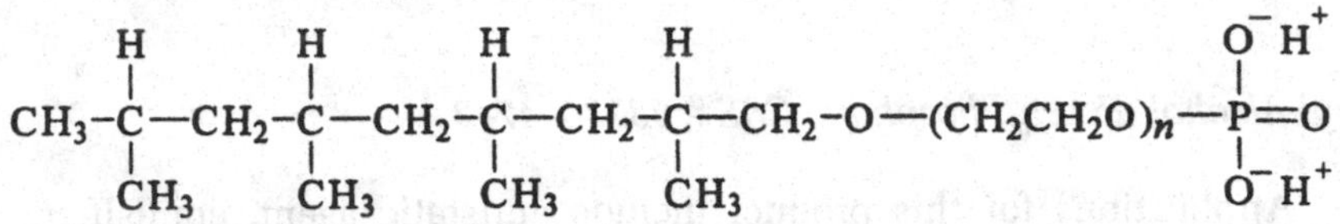

$$CH_3-\underset{\underset{CH_3}{|}}{\overset{\overset{H}{|}}{C}}-CH_2-\underset{\underset{CH_3}{|}}{\overset{\overset{H}{|}}{C}}-CH_2-\underset{\underset{CH_3}{|}}{\overset{\overset{H}{|}}{C}}-CH_2-\underset{\underset{CH_3}{|}}{\overset{\overset{H}{|}}{C}}-CH_2-O-(CH_2CH_2O)_n-\underset{\underset{O^-H^+}{|}}{\overset{\overset{O^-H^+}{|}}{P}}=O$$

Suppliers of this ester are Ethox with Ethfac® 133 [47][65] and Stepan with Cedephos® RE–613 [64]. The typical properties are presented in Table 7–24.

Table 7–24

Typical Properties of Tridecyl Alcohol (POE)$_n$ Phosphate, Free Acid

Property	Value
Color, Gardner	1 max
Acid Value	140–160
Moisture, %	1.0 max
Activity, %	99

Oleyl Alcohol (POE)ₙ Phosphate, CAS# 39464–69–2

The long–chain fatty alcohol in this phosphate ester makes the material an excellent lubricant. Unsaturation in the alkyl chain causes varnish and charred organic deposits on hot surfaces so its use should be limited to situations in which heat is not a factor. This phosphate has a unique structure:

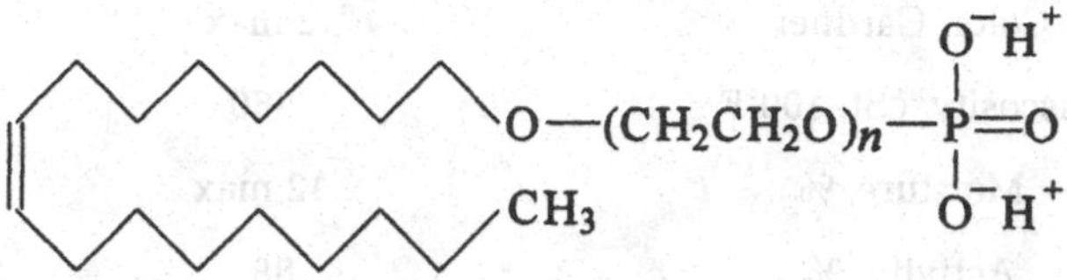

It is produced by Ethox using the trade name Ethfac® 140 [47][66] and by Witco with the name of Emphos® PS–900 [55]. Some properties of this ester are shown in Table 7–25.

Table 7–25

Typical Properties of Oleyl Alcohol (POE)ₙ Phosphate, Free Acid

Property	Value
Color, Gardner	8 max
Acid Value	135–150
Moisture, %	1.0 max
Activity, %	99

7.4.2.4 Polyoxyethylene Aliphatic Alcohol Phosphates, Potassium Salts

The potassium salts of ethoxylated alcohol phosphates are very effective as fiber and fabric antistats, wetting agents, and emulsifiers. Many of these have compositions that are proprietary to the suppliers, but the specifications of some of these are available. These salts have the advantage in not requiring neutralization, but the buyer must pay the shipping costs for the water that is present in the aqueous solutions. Some of these potassium salts of phosphated alkyl ethoxylates are Tryfac® 5559 [50][67], Tryfac® 5553 [50][68], Tryfac® 5554 [50][68], and Dacospin® PE–47 [50][69], all manufactured by Henkel. The typical properties are listed in the following tables.

Table 7–26

Typical Properties of Tryfac® 5559

Property	Value
Color, Gardner	2 max
Viscosity, cSt, 100°F	560
Moisture, %	12 max
Activity, %	88

Table 7–27

Typical Properties of Tryfac® 5553

Property	Value
Color, Gardner	1 max
Viscosity, cSt, 100°F	345
Acid Value	20 max
Moisture, %	10–15 max
pH, 5%	6.5–7.5

Table 7–28

Typical Properties of Tryfac® 5554

Property	Value
Color, Gardner	1 max
Viscosity, cSt, 100°F	340
Acid Value	10–14
pH, 5%	7
Moisture, %	9.0–12.0 max
Activity, %	90

Table 7–29

Typical Properties of Dacospin® PE–47

Property	Value
Color, Gardner	1 max
Viscosity, cSt, 100°F	102
pH, 5%	6.5
Moisture, %	20 max
Activity, %	80

Other ethoxylated alkyl ether phosphates, K^+ salts, are also summarized below.

POE(2) 2–Ethylhexyl Ether Phosphate, K^+ Salt, CAS# 68238–84–6

This phosphate ester is effective in wetting and scouring formulations and it has excellent antistatic properties for processing staple fibers. It is also an excellent antistat in spin finish formulations. The ester is made by Ethox as Ethfac® 324 [47][70] with characteristics listed in Table 7–30. The structure is:

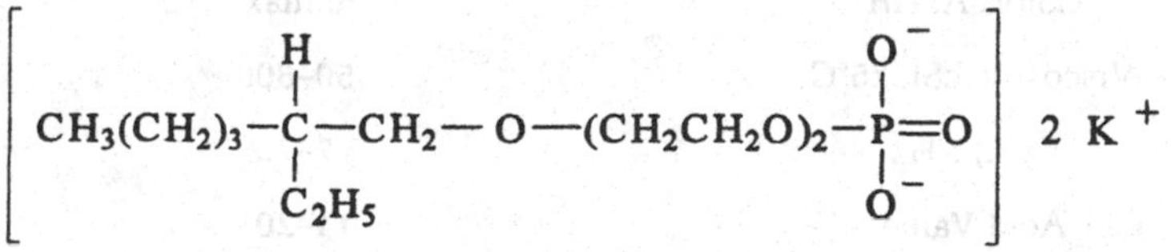

$$\left[CH_3(CH_2)_3 - \overset{\overset{\displaystyle H}{|}}{\underset{\underset{\displaystyle C_2H_5}{|}}{C}} - CH_2 - O - (CH_2CH_2O)_2 - \overset{\overset{\displaystyle O^-}{|}}{\underset{\underset{\displaystyle O^-}{|}}{P}} = O \right] \ 2\ K^+$$

Table 7–30

Typical Properties of POE (2) 2–Ethylhexyl Ether Phosphate, K^+ Salt

Property	Value
Color, APHA	100 max
Acid Value	20–30
Moisture, %	20 max
pH, 5%	6.5–7.2

(POE)ₙ Decyl Ether Phosphate, K⁺ Salt, CAS# 68071–17–0

There are two chemical compounds commercially available with this name and CAS number, one contains six moles of ethylene oxide and the other nine. Both are good emulsifiers and detergents in scouring textile fabric and can be used to inhibit corrosion in addition to being effective antistats. The base alcohol is *i*–decyl and the structure is:

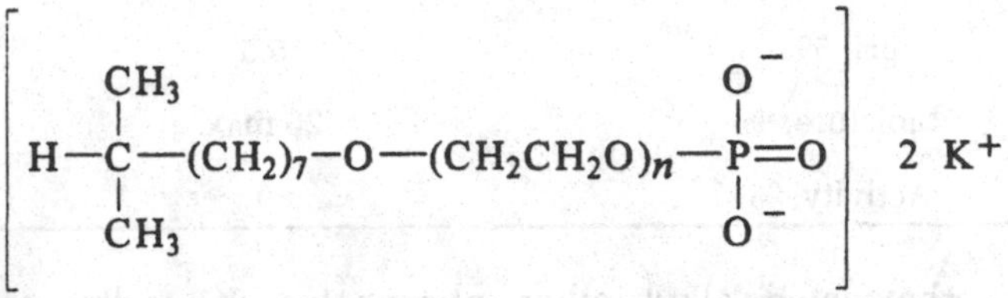

$$
\left[
\begin{array}{c}
\text{CH}_3 \qquad\qquad\qquad\quad \text{O}^- \\
| \qquad\qquad\qquad\qquad\quad | \\
\text{H}-\text{C}-(\text{CH}_2)_7-\text{O}-(\text{CH}_2\text{CH}_2\text{O})_n-\text{P}=\text{O} \\
| \qquad\qquad\qquad\qquad\quad | \\
\text{CH}_3 \qquad\qquad\qquad\quad \text{O}^-
\end{array}
\right] \quad 2\ \text{K}^+
$$

They are supplied by Ethox and named Ethfac® 361 [46][71] (six moles of EO) and Ethfac® 391 [47][72] (nine moles of EO). The properties are shown in the next two tables.

Table 7–31

Typical Properties of POE (6) Decyl Ether Phosphate, K⁺ Salt

Property	Value
Color, APHA	50 max
Viscosity, cSt, 25°C	550–800
pH, 5%	6.7–7.2
Acid Value	14–20
Moisture, %	11.0–14.0

Table 7–32

Typical Properties of POE (9) Decyl Ether Phosphate, K⁺ Salt

Property	Value
Color, Gardner	2 max
pH, 5%	6.8–7.2
Acid Value	10–14
Moisture, %	12 max

(POE)$_n$ Tridecyl Ether Phosphate, K^+ Salt, CAS# 68186–36–7

Another series of two similar compounds are the tridecyl ether phosphates, with five and six moles of ethylene oxide. These are some of the best antistatic agents for fibers and textiles and are very good co–emulsifiers. The monoester has a structural formula of:

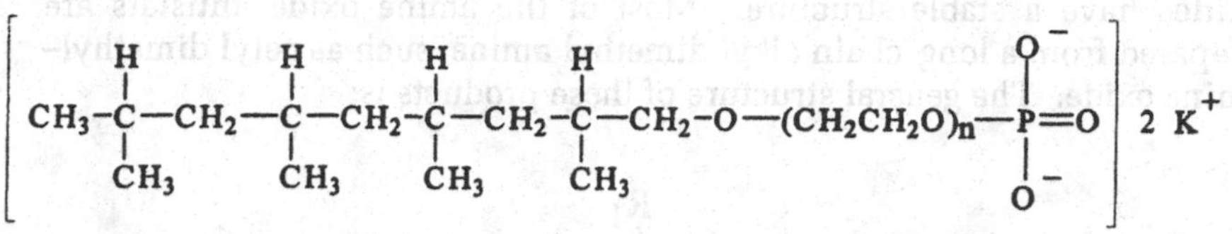

$$\left[CH_3-\overset{\overset{\displaystyle H}{|}}{\underset{\underset{\displaystyle CH_3}{|}}{C}}-CH_2-\overset{\overset{\displaystyle H}{|}}{\underset{\underset{\displaystyle CH_3}{|}}{C}}-CH_2-\overset{\overset{\displaystyle H}{|}}{\underset{\underset{\displaystyle CH_3}{|}}{C}}-CH_2-\overset{\overset{\displaystyle H}{|}}{\underset{\underset{\displaystyle CH_3}{|}}{C}}-CH_2-O-(CH_2CH_2O)_n-\overset{\overset{\displaystyle O^-}{\|}}{\underset{\underset{\displaystyle O^-}{|}}{P}}=O \right] 2\ K^+$$

These compounds are also made by Ethox with trade names of Ethfac® 353 [47][73] and Ethfac® 363 [47][74]. Typical properties are listed in Table 7–33 and Table 7–34.

Table 7–33

Typical Properties of POE (5) Tridecyl Ether Phosphate, K^+ Salt

Property	Value
Color, APHA	70 max
Viscosity, cSt, 25°C	1150–1500
pH, 5%	6.5–7.5
Acid Value	16–22
Moisture, %	12.5 max

Table 7–34

Typical Properties of POE (6) Decyl Ether Phosphate, K^+ Salt

Property	Value
Color, Gardner	2 max
Viscosity, cSt, 25°C	1000–1400
Acid Value	17–23
Moisture, %	9.5–11.5 max

7.5 OTHER ANTISTATIC AGENTS

Two other types of organic materials have been tried as antistats, but neither are as effective as quaternary amines or phosphate esters. These are amine oxides and betaines.

Amine oxides are synthesized through the oxidation of amines. Primary and secondary amines yield unstable oxides, but tertiary amine oxides have a stable structure. Most of the amine oxide antistats are prepared from a long–chain alkyl dimethyl amine, such as cetyl dimethyl–amine oxide. The general structure of these products is:

$$
\begin{array}{c}
R_1 \\
| \\
R_2-N:O \\
| \\
R_3
\end{array}
$$

Betaines, so named because of the simplest representative, betaine, are internal quaternary amines, one of the examples of zwitterions. Betaine occurs naturally in sugar beets and some other plants. The betaine antistats may be made by the reaction of a trialkylamine with chloroacetic acid. The structure of betaine antistats is:

$$
\begin{array}{c}
R_1 \quad\quad O \\
| \;+ \quad\quad || \\
R_2-N-CH_2-C-O^- \\
| \\
R_3
\end{array}
$$

REFERENCES

1. Edelstein, S. M. (1953). From Philosopher's Toy to Industry's Headache–The Story of Static Electricity, *Am. Dyestuff Rep.*, **42**: P70

2. Gur–Arieh, Z. and Reuben, B. J. (1978). Antistatic Agents in *Kirk–Othmer Encyclopedia of Chemical Technology, Third Edition*, Volume 3, John Wiley & Sons, New York, pp. 149–183.

3. Hersh, S. P. (1975). Resistivity and Static Behavior of Textile Surfaces, in *Surface Characteristics of Fibers and Textiles*, (M. J. Schick, ed.), Marcel Dekker, Inc., New York, pp. 225–288

4. Wilche, J. C. (1757). *Disputatio Physica Experimentalis de Electricitatibus Contrariis*, Academy of Rostok, Germany

5. Ballou, T. W. (1954). Static Electricity in Textiles, *Textile Res. J.* **24**: 146

6. Lehmicke, D. J. (1949). Static in Textile Processing, *Am. Dyestuff Rep.*, **38**: (24), 853

7. Hersh, S. P. and Montgomery, D. J. (1955). Static Electrification of Filaments. Experimental Techniques and Results, *Textile Res. J.*, **25**: 279

8. Henry, P. S. H. (1953). Survey of Generation and Dissipation of Static Electricity, *British J. App. Phys.*, *Suppl. No. 2, [Static Electrification]*, **4**: S31

9. Valko, E. I. and Tesoro, G. C. (1965). Antistatic Agents, in *Encyclopedia of Polymer Science and Technology, First Edition*, Volume 2, (J. I. Kroschwitz, ed.) John Wiley & Sons, New York, pp. 204–229

10. Harper, W. R. (1951). The Volta Effect as a Cause of Static Electrification, *Proc. Royal Soc.(London)*, **A205**: 83

11. Hersh, S. P. and Montgomery, D. J. (1956). Static Electrification of Filaments. Theoretical Aspects, *Textile Res. J.*, **26**: 903

12. Medley, J. A. (1954). The Discharge of Electrified Textiles, *J. Textile Inst., Transactions*, **45**: T123

13. Crovatt, L. W. and Garrett, U. C. (1972), assignors to Monsanto Company, St. Louis, MO, Polyamide Filaments Containing High Viscosity Antistatic Agents, *U.S. Patent 3,666,731*

14. Reck, R. A. (1985) Antistatic Agents in *Encyclopedia of Polymer Science and Engineering, Second Edition*, Volume 2 (J. I. Kroschwitz, ed.), John Wiley & Sons, New York, pp. 99–115

15. Boe, N. W. (1975). Man–made Textile Antistatic Strand, assignor to Monsanto Company, St. Louis, MO, *U.S. Patent 3,969,559*

16. Monsanto Company Brochure (1987). *No–Shock® Conductive Nylon*, Monsanto Chemical Company, Technical Center, Pensacola, FL

17. Hearle, J. W. S. (1957). The Relation Between Structure, Dielectric Constant, and Electrical Resistance of Fibres, *J. Textile Inst., Proceedings*, **48**: P40

18. Sereda, P. J. and Feldman, R. F. (1964). Electrostatic Charging on Fabrics at Various Humidities, *J. Textile Inst., Transactions*, **55**: T288

19. Hearle, J. W. S. (1960). Moisture and Electrical Properties in *Moisture in Textiles*, (J. W. S. Hearle and R. H. Peters, ed.), Butterworths, London, England, pp. 123–140

20. Keggin, J. F., Morris, G., and Yuill, A. M. (1949). Static Electrification in the Processing of Fibres: Variation with Moisture Regain During Carding, *J. Textile Inst. Transactions*, **40**: T702

21. Shashoua, V. E. (1958). Static Electricity in Polymers. I. Theory and Measurement, *J. Poly. Sci.*, **33**: 65

22. Slade, P. E. , unpublished data

23. Cunningham, R. G. and Montgomery, D. J. (1958). Studies in the Static Electrification of Filaments, *Textile Res. J.*, **28**: 971

24. Sprokel, G. J. (1975). Electrostatic Properties of Finished Cellulose Acetate Yarn, *Textile Res. J.*, **27**: 50

25. Steiger, F. H. (1958). Evaluating Antistatic Finishes, *Textile Res. J.*, **28**: 721

26. Rogers, J. L. (1973). Antistatic Tests and Agents for Plastics, *Soc. Plastic Eng. J.*, **29**: 28

27. ASTM Method D 257–90 (1991). Standard Test Method for D–C Resistance or Conductance of Insulating Materials, *Annual Book of Standards, Section 8 (Plastics), Vol. 08.01*, ASTM, Philadelphia, pp. 75–89

28. AATCC Test Method 76–1995 (1996), Electrical Resistivity of Fabrics, *AATCC Technical Manual*, American Association of Textile Chemists and Colorists, Research Triangle Park, NC, pp. 100–101

29. AATCC Test Method 84–1995 (1996). Electrical Resistivity of Yarns, *AATCC Technical Manual*, American Association of Textile Chemists and Colorists, Research Triangle Park, NC, pp. 108–109

30. Morton, W. D. and Hearle, J. W. S. (1962). *Physical Properties of Textile Fibres*, Butterworths, Manchester, England, p. 457

31. AATCC Test Method 115–1995 (1996). Electrostatic Clinging of Fabrics: Fabric–to–Metal Test, *AATCC Technical Manual*, American Association of Textile Chemists and Colorists, Research Triangle Park, NC, pp. 186–189

32. Wilkinson, P. R. (1970). Static Electricity: Domestic and Commercial Implications in Textile Materials, *Modern Textiles*, **51**: 35

33. Hayek, M., and Chromey, F. C. (1951). The Measurement of Static Electricity on Fabrics, *Am. Dyestuff Rep.*, **40**: 164

34. Henshall, A. E. (1960). Antistatic Agents in the Textile Industry, *J. Soc. Dyers and Colourists*, **76**: 525

35. Slade, P. E. (1973). The Static Honestometer: A New Approach to Static Measurements, *Modern Textiles*, **54**: (3) 68

36. Marumo, H. (1967). Antistatic Treatment and Antistatic Mechanism for Polymers, *Kobunshi*, Tokyo, Japan, **16**: 357

37. Marumo, H. (1970). *The Internal Antistatic Treatment of Thermoplastic Polymers*, Paper presented at the First International Congress of Static Electricity, Vienna, Austria, May 3–7

38. Rothschild Electronische Mess– und Steuergeräte, Zürich (1995), *Static–Voltmeter R–4021*, Lawson–Hemphill Sales, Inc., Spartanburg, SC

39. Oxé, J. and Keller, R. (1972). Die Prüfung und Beurteilung des Electrosta–tischen Verhaltens Textiler Bodenbeläge, *Textilveredlung*, Basel, Switzerland, **7**: 417

40. Reck, R. A. (1978). Quaternary Ammonium Compounds, *Kirk–Othmer Encyclopedia of Chemical Technology, Third Edition*, Volume 19, pp. 521–531

41. Burnett, L. W. (1966). Nonionics as Ionic Surfactant Intermediates, in *Nonionic Surfactants*, (M. J. Schick, ed.), Marcel Dekker, Inc., New York, pp. 372–394

42. PPG Industries (1992). *Larostat® Antistats from PPG Industries, Specialty Chemicals*, PPG Industries, Specialty Chemicals, Gurnee, IL

43. Specialties from ICI for Fiber Producers (1987). *Antistats*, ICI Americas, Inc., Wilmington, DE

44. ICI Product Information Bulletin (1995). *Cirrasol® G–263*, ICI Americas, Inc., Wilmington, DE

45. ICI Product Information Bulletin (1995). *Cirrasol® G–250*, ICI Americas, Inc., Wilmington, DE

46. Witco Technical Data (1993). *Adogen® 66 Quaternary Ammonium Compound*, Witco Corporation, Dublin, OH

47. Ethox Product Brochure (1994). *Phosphate Esters*, Ethox Chemicals, Inc., Greenville, SC

48. Ethox Specifications (1986). *Ethfac® 106*, Ethox Chemicals, Inc., Greenville, SC

49. Ethox Specifications (1986). *Ethfac® 104*, Ethox Chemicals, Inc., Greenville, SC

50. Henkel Corporation (1993). *Textile Chemicals, Technical Bulletin 103A, Phosphate Esters*, Henkel Corporation, Textile Chemicals, Charlotte, NC

51. Henkel Data Sheet (1993). *Dacospin® PE–146*, Henkel Corporation, Textile Chemicals, Charlotte, NC

52. Ethox Specifications (1986). *Ethfac® 102*, Ethox Chemicals, Inc., Greenville, SC

53. Ethox Specifications (1992). *Ethfac® 1018*, Ethox Chemicals, Inc., Greenville, SC

54. Hoechst–Celanese Product List (1995). *P_2O_5 Phosphate Esters*, Hoechst–Celanese Corporation, Specialty Chemicals Group, Charlotte, NC

55. Witco Fiber Production Auxiliaries (1994). *Antistats*, Witco Corporation, Greenwich, CT

56. ICI Specialty Chemicals (1987). Antistats, *Specialties from ICI for Fiber Producers*, ICI Americas, Wilmington, DE

57. Ethox Specifications (1993). *Ethfac® 142W*, Ethox Chemicals, Inc., Greenville, SC

58. Henkel Data Sheet (1993). *Tryfac® 5552*, Henkel Corporation, Textile Chemicals, Charlotte, NC

59. Ethox Specifications (1986). *Ethfac® PB–1*, Ethox Chemicals, Inc., Greenville, SC

60. Ethox Specifications (1986). *Ethfac® PB–2*, Ethox Chemicals, Inc., Greenville, SC

61. Ethox Specifications (1993). *Ethfac® 136*, Ethox Chemicals, Inc., Greenville, SC

62. Ethox Specifications (1986). *Ethfac® 2684*, Ethox Chemicals, Inc., Greenville, SC

63. Ethox Specifications (1989). *Ethfac® 161*, Ethox Chemicals, Inc., Greenville, SC

64. Stepan Textile Products, Technical Information (1995). *Phosphate Esters*, Stepan Company, Northfield, IL

65. Ethox Specifications (1986). *Ethfac® 133*, Ethox Chemicals, Inc., Greenville, SC

66. Ethox Specifications (1994). *Ethfac® 140*, Ethox Chemicals, Inc., Greenville, SC

67. Henkel Data Sheets (1993). *Tryfac® 5559*, Henkel Corporation, Textile Chemicals, Charlotte, NC

68. Henkel Data Sheet (1984). *Tryfac® 5553 and Tryfac® 5554*, Henkel Corporation, Textile Chemicals, Charlotte, NC

69. Henkel Data Sheet (1994). *Dacospin® PE–47*, Henkel Corporation, Textile Chemicals, Charlotte, NC

70. Ethox Specifications (1991). *Ethfac® 324*, Ethox Chemicals, Inc., Greenville, SC

71. Ethox Specifications (1993). *Ethfac® 361*, Ethox Chemicals, Inc., Greenville, SC

72. Ethox Specifications (1993). *Ethfac® 391*, Ethox Chemicals, Inc., Greenville, SC

73. Ethox Specifications (1989). *Ethfac® 353*, Ethox Chemicals, Inc., Greenville, SC

74. Ethox Specifications (1986). *Ethfac® 363*, Ethox Chemicals, Inc., Greenville, SC

CHAPTER 8

OTHER FINISH ADDITIVES: ANTIOXIDANTS, DEFOAMERS, ANTIMICROBIALS, AND WETTING AGENTS

Several other types of chemical compounds can be included in finish formulations to improve the effectiveness of those compositions. Generally, these materials are added in low concentrations, but they have very important effects on the finish and the yarn to which the finish is applied. This chapter consists of a discussion of these types of chemicals.

8.1 ANTIOXIDANTS

All components in spin finishes are susceptible to oxidation under certain conditions. Finish oils and yarn with finish oils applied to it are frequently exposed to heat or other situations where oxygen is also present. The reaction that occurs is known as *autoxidation* and will proceed through four steps; (1) initiation, (2) propagation, (3) autocatalysis, and (4) termination. In lubricants that might contain unsaturated double bonds, the reaction starts by addition to the double bond and formation of free radicals. These radicals then cross–link and form hard varnishes. If carboxyl,

319

carbonyl, or ether oxygens are a part of the molecule the reaction proceeds by the formation of a hydroperoxide that is a kinetically important initiator leading to a chain reaction. The hydroperoxide scheme for autoxidation can be described by the following equations [1]:

Initiation

$$RH \longrightarrow \text{Free Radicals, e.g, } R\cdot, ROO\cdot, RO\cdot, HO\cdot$$

Propagation

$$R\cdot + O_2 \longrightarrow ROO\cdot$$

$$ROO\cdot + RH \longrightarrow ROOH + R\cdot$$
$$\text{Hydroperoxide}$$

Autocatalysis

$$ROOH \longrightarrow RO\cdot + \cdot OH$$

$$2\,ROOH \longrightarrow RO\cdot + ROO\cdot + H_2O$$

$$ROOR \longrightarrow 2\,RO\cdot$$

Termination

$$2\,R\cdot \longrightarrow R{-}R$$

$$ROO\cdot + R\cdot \longrightarrow ROOR$$

$$2\,ROO\cdot \longrightarrow \text{Nonradical Products}$$

It had been known for several years that chemicals containing ethylene oxide decompose through this autoxidation mechanism [2][3]. Degradation of these molecules goes through the free radical mechanism as outlined below.

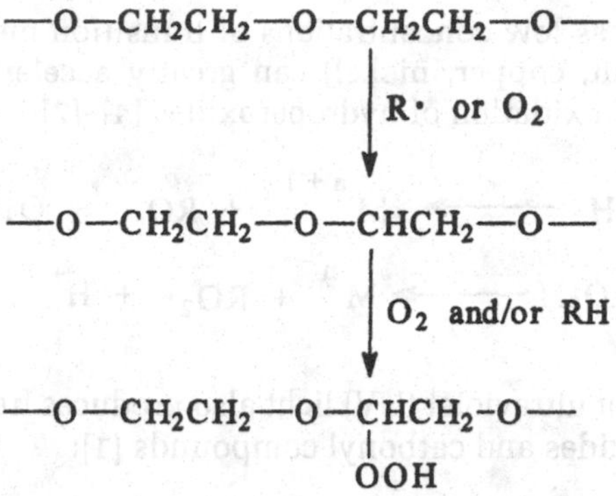

$$—O—CH_2CH_2—O—CH_2CH_2—O—$$

$$\Big\downarrow R\cdot \text{ or } O_2$$

$$—O—CH_2CH_2—O—CHCH_2—O—$$

$$\Big\downarrow O_2 \text{ and/or RH}$$

$$—O—CH_2CH_2—O—CHCH_2-O—$$
$$| $$
$$OOH$$

The hydroperoxide moiety then decomposes and the adjacent C–C bond is broken as:

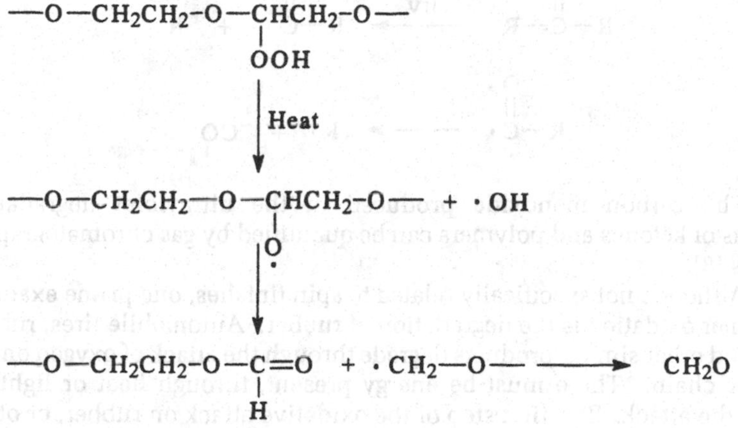

$$—O—CH_2CH_2-O—CHCH_2-O—$$
$$|$$
$$OOH$$

$$\Big\downarrow \text{Heat}$$

$$—O—CH_2CH_2—O—CHCH_2-O— \; + \; \cdot OH$$
$$|$$
$$O$$
$$\cdot$$

$$\Big\downarrow$$

$$—O—CH_2CH_2—O—C{=}O \; + \; \cdot CH_2—O— \longrightarrow CH_2O$$
$$|$$
$$H$$

Unzipping of the poly(ethylene oxide) chain continues with the evolution of formaldehyde and the formation of other free radicals. If the chain is poly(propylene oxide), the material evolved is acetaldehyde.

8.1.1 Initiation and Initiation Inhibitors

Free radical initiation can be produced by many different processes during manufacturing, including heat, shear stress, bimolecular reactions of compounds with oxygen, and photochemical reactions. In commercial practice it is almost impossible to make a chemical without the presence of traces of hydroperoxides. When one is concerned with the oxidation of polymers, the compounding steps alone contribute greatly to the autoxidation process.

Impurities such as low concentrations of transition metal ions (i.e., manganese, iron, cobalt, copper, nickel) can greatly accelerate initiation through reduction and oxidation of hydroperoxides [4]–[7].

$$M^n + RO_2H \longrightarrow M^{n+1} + RO\cdot + OH^-$$

$$M^{n+1} + RO_2H \longrightarrow M^n + RO_2\cdot + H^+$$

The absorption of ultraviolet (UV) light also produces free radicals by cleavage of hydroperoxides and carbonyl compounds [1]:

$$ROOH \xrightarrow{UV} RO\cdot + \cdot OH$$

$$R-\overset{\overset{\displaystyle O}{\|}}{C}-R \xrightarrow{UV} R-\overset{\overset{\displaystyle O}{\|}}{C}\cdot + R$$

$$R-\overset{\overset{\displaystyle O}{\|}}{C}\cdot \longrightarrow R\cdot + CO$$

The carbon monoxide produced in the ultraviolet degradation reactions of ketones and polymers can be quantified by gas chromatographic analysis [8].

Although not specifically related to spin finishes, one prime example of polymer oxidation is the degradation of rubber. Automobile tires, rubber bands and other similar products degrade through the attack of oxygen on the polymer chain. There must be energy present, through heat or light, to initiate the attack. The first step of the oxidative attack on rubber, or other polymeric materials, occurs when hydrogen is expelled from the polymer chain by this energy source. This forms a free radical on the body of the chain. At this time oxygen can combine with polymer free radicals to form hydroperoxides. The reaction then continues as shown in the equations above.

Several accelerated tests are used to determine the resistance of polymers to oxidation. Probably the most basic of these is the measurement of the absorption rate. Using a mercury manometer connected to a sample tube held at a constant elevated temperature, the amount of oxygen absorbed by the sample is measured volumetrically. This technique is valuable in studying the effects of vulcanization and the effectiveness of antioxidants. When as little as 1–2% oxygen has combined with most elastomers, products are usually no longer useful [9].

Oxidation of elastomeric polymers can result in two changes to the polymer form. A few products, such as rubber bands, soften and become sticky as they age. The softening is the result of chain scission. Natural rubber and a few synthetic rubber products follow this route. Most synthetic rubbers and other polymers follow the opposite route and harden as they oxidize. This change is due to oxidative cross–linking and the formation of very long polymer chains. An example that is familiar to all of us is the glazed surface that forms on an old pencil eraser. As oxidation begins, the surface glazes and with continued reaction it becomes totally glass–like and brittle. The effects of oxygen on various polymers, reported by Miller and Dean [9], are shown in Table 8–1.

Table 8–1

Effects of Oxygen Upon Polymers

Common Name or Trade Name	Chemical Designation	Effect of Oxidation	Max Operating Temp., °F
Natural Rubber	Natural Polyisoprene	Softens	212
Natsyn	Synthetic Polyiosprene	Softens	212
Neoprene	Chloroprene	Hardens	225
Plioflex	Styrene–Butadiene	Hardens	250
Chemigum	Acrylonitrile–Butadiene	Hardens	275
Butyl	Isobutylene–Isoprene	Softens	212
Thiokol	Polysulfide	Hardens	180
EPR	Ethylene–Propylene	Hardens	300
EPT	Ethylene–Propylene Terpolymer	Hardens	300
Hypalon	Chlorosulfonated Polyethylene	Hardens	250
Silicone	Methyl–Vinyl Siloxane		
Silicone	Phenyl–Methyl–Vinyl Siloxane		
Fluoro–silicone	Trifluoropropyl Siloxane	Hardens	450
Urethane	Polyurethane Diisocyanate	Hardens	212
Viton	Fluorinated Hydrocarbon	Hardens	450
Acrylic	Polyacrylate	Hardens	325

Since the autoxidation process is a free radical chain reaction, it can be inhibited at either the initiation or the propagation steps. Antioxidants can be classified based on their ability to accomplish either or both inhibition steps.

Two main classes of chemicals inhibit the initiation step in thermal oxidation. The peroxide decomposers, generally known as secondary antioxidants [10], function by breaking down hydroperoxides through polar reactions, thereby inhibiting the initiation step. Metal deactivators are strong metal ion complexing agents, such as ethylenediaminetetraacetic acid, which can stop the reactions promoted my transition metals.

Antioxidants that decompose hydroperoxides include the following [11].

Dialkyl Thiodipropionates

$$(RO_2CCH_2CH_2)_2S$$

Typical examples of these sulfur–based solid antioxidants are Lowinox® DLTDP, di–lauryl–3,3'–thio–dipropionate (CAS# 123–28–4), and Lowinox® DSTDP, di–stearyl–3,3'–thio–dipropionate (CAS# 693–36–7), manufactured by Great Lakes Chemical Corporation [12]. The structure of one of these, Lowinox® DLTDP, is:

$$S\begin{cases} CH_2-CH_2-COO-C_{12}H_{25} \\ CH_2-CH_2-COO-C_{12}H_{25} \end{cases}$$

A semi–solid thiopropionate is produced by Goodyear, identified as Wingstay® SN–1 [9][13], with the chemical composition of 1:11(3,6,9–tri-oxaudecyl)bis–(dodecylthio) propionate. Cytec makes several thioesters with trade names of Cyanox® LTDP, MTDP, STDP, 1212, and 711 [14]. A phenolic thiopropionate is manufactured by Ciba–Geigy as Irganox® 1035 [15] with a chemical name of 2,2' thiodiethyl–bis(3,5–di–tert–butyl–4–hydroxyphenyl)–propionate (CAS# 41484–35–9). The structural formula of Irganox® 1035 is:

Aryl and Alkyl Phosphites

$$(RO)_3P$$

A diverse series of alkyl and aryl phosphites is supplied by Great Lakes Chemicals [12]. Two are solids, Alkanox® 240 (tris(2,4–di–t–butyl–phenyl) phosphite, CAS# 31570–04–4) and Alkanox® P–24 (bis(2,4–di–t–butylphenyl) pentaerythritol diphosphite, CAS# 26741–53–7). The liquids are Lowinox® TNPP (tris(p–nonylphenyl) phosphite, CAS# 26523–78–4), Lowinox® OS–150 (diphenyl–isodecyl) phosphite, CAS# 26544–23–0), Lowinox® OS 300 (diisodecylphenyl) phosphite, CAS# 25550–98–5), Lowinox® OS 330 (triiso–decylphosphite, CAS# 25448–25–3), and Lowinox® OS 360 (trilaurylphosphite, CAS# 3076–63–9). As examples of the structural possibilities, the formulas for an aryl phosphite and an alkyl phosphite are shown:

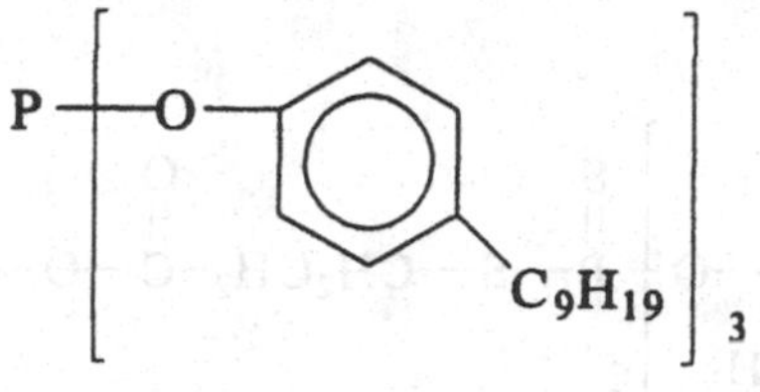

Lowinox® TNPP

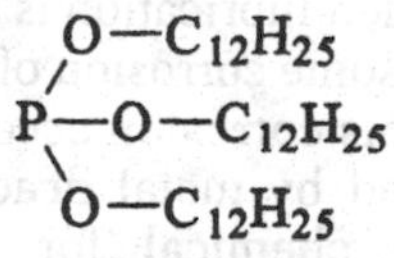

Lowinox® OS 360

Metal Salts of Dithioacids, such as Zinc Dithiocarbamates

$$(R_2NCS_2)_2Zn$$

Several solid dithiocarbamates are manufactured by Flexsys [16], primarily for use as accelerators in the curing of synthetic elastomers, but are also recommended as antioxidants. Examples are Perkacit® NDBC, nickel dibutyldithiocarbamate (CAS# 13927–77–0), Perkacit® ZEBC, zinc dibenzyldithio–dicarbamate (CAS# 14726–36–4), Perkacit® ZDBC, zinc dibutyldithiodicarbamate (CAS# 136–23–2), and Perkacit® ZDEC, zinc diethyldithiodicarbamate (CAS# 14324–55–1). The structural formula for Perkacit® ZDBC is:

$$\left[\begin{array}{c} \overset{S}{\underset{\parallel}{C_4H_9-N-C-S}} \\ \underset{C_4H_9}{\vert} \end{array} \right]^{-}_2 \; Zn^{+2}$$

Dithiophosphates

$(RO)_2PS_2R'$

A low–viscosity liquid, 5.2 cSt @ 40°C, with this composition is produced by Ciba–Geigy [15] and sold as an extreme pressure/antiwear agent as well as an antioxidant. The structure is:

$$\left[\begin{array}{c} \overset{H}{\underset{\vert}{CH_3-C-O}} \\ \underset{CH_3}{\vert} \end{array} \right]_2 \overset{S}{\underset{\parallel}{P-S-CH_2CH_2-C-O-C_2H_5}}$$

During any process in which lubrication is an important aspect, either fiber to metal or metal to metal, some corrosion of the metal surfaces occurs. The metal ions resulting from this corrosion accelerate the oxidation of the oils and these must be removed by metal deactivators [17] or chelating agents. The most common chemical for chelating metal ions is ethylenediaminetetraacetic acid (EDTA) (CAS# 60–00–4):

$$\begin{array}{ccc} \overset{O}{\underset{\parallel}{HO-C-CH_2}} & & \overset{O}{\underset{\parallel}{CH_2-C-OH}} \\ & N-CH_2CH_2-N & \\ \underset{\underset{O}{\parallel}}{HO-C-CH_2} & & \underset{\underset{O}{\parallel}}{CH_2-C-OH} \end{array}$$

Liquid aromatic triazoles are especially useful for deactivating copper surfaces even at concentrations as low as 50 ppm by forming a solid film on the surface [17]. These compounds are products of Ciba–Geigy and are sold as Irgamet® 39, which is oil soluble, and the water–soluble Irgamet® 42 [15]. The structures are:

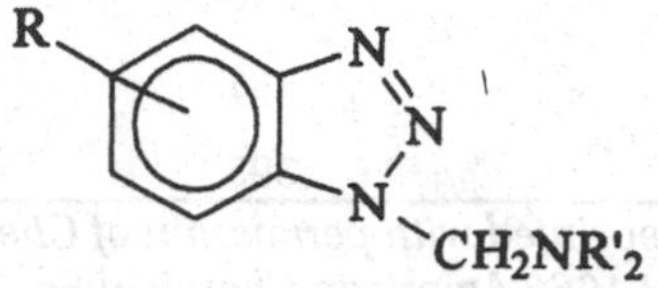

R—[benzotriazole ring]—N
                         ‖
                         N
                         |
                         N—CH₂NR'₂

Irgamet® 39

R—[benzotriazole ring]—N
                         ‖
                         N
                         |
                         N—CH₂N(CH₂CH₂OH)₂

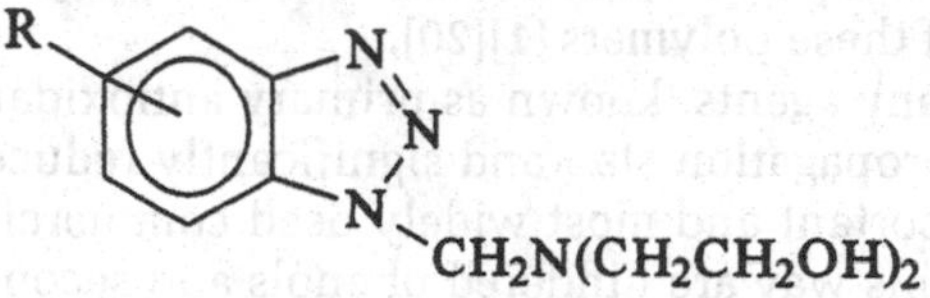

Irgamet® 42

Although many studies have been carried out to determine the exact mechanism of initiation inhibition, the results of these studies are not clear. A dialkyl ester of thiodipropionic acid is capable of decomposing at least 20 moles of hydroperoxide [18] by reducing the hydroperoxide moiety to an alcohol, but when used alone these thio–esters are weak antioxidants. Synergistic mixtures of thioesters with hindered phenols, however, are very effective. The phosphites also reduce the hydroperoxides to alcohols, but again they are not effective when used alone [1].

8.1.2 Propagation and Propagation Inhibitors

Propagation reactions can be repeated many times before being terminated by conversion of an alkyl or peroxy radical to a nonradical species. The rate of reaction of molecular oxygen to form peroxy radicals is much higher than the rate of reaction of peroxy radicals with a hydrogen atom of the polymer. The rate of the latter depends upon the dissociation energies (Table 8–2) and the stearic accessibility of the various carbon–hydrogen bonds and is an important factor in determining oxidative stability [19].

Table 8-2

Dissociation Energies of Carbon–Hydrogen Bonds

R–H	D_{R-H}, kJ/mol	Bond Type
$CH_2=CHCH_2-H$	356	allylic
$(CH_3)_3C-H$	381	tertiary
$(CH_3)_2CH-H$	395	secondary

(Reference [19]. Reprinted with permission of Chem. Reviews.
Copyright 1966 American Chemical Society)

Polybutadiene or polyisoprene containing allylic hydrogen atoms, oxidize much more readily than polypropylene, which contains tertiary hydrogen atoms. Linear polyethylene, which has secondary hydrogen atoms, is the most stable of these polymers [1][20].

There are many agents, known as primary antioxidants [10], which can interrupt this propagation step and significantly reduce the oxidation rate. The most important and most widely used commercial antioxidants which function in this way are hindered phenols and secondary alkylaryl– and diarylamines.

Hindered phenols can be mononuclear, as represented by the most widely used (approximately 1,400 metric tons per year [11]) antioxidant for food, BHT. The chemical name for BHT is 2,6–di–t–butyl–4–methyl–phenol (CAS# 128–37–0). It is made by Great Lakes using the name Lowinox® BHT. The structure is:

$$(CH_3)_3C-\underset{\underset{CH_3}{|}}{\overset{\overset{OH}{|}}{\bigcirc}}-C(CH_3)_3$$

Two other types of mononuclear phenol antioxidants are commonly used. One has a structure similar to BHT and is sold by Goodyear as Wingstay® T [9][21]. The structural formula is:

$$R_1-\overset{\overset{OH}{|}}{\bigcirc}\begin{matrix}-R_2\\-R_3\end{matrix}\qquad \begin{matrix}R_1, R_2, R_3 =\\ \\ H, C_4H_9, C_8H_{17}\end{matrix}$$

Another material is based on hydroquinone and is supplied by Flexsys as Santovar® A [16][22] and by Great Lakes with the trade name Lowinox® AH25 [12]. The chemical name is 2,5–di–(t–pentyl) hydroquinone (CAS# 79–74–3) with a structure of:

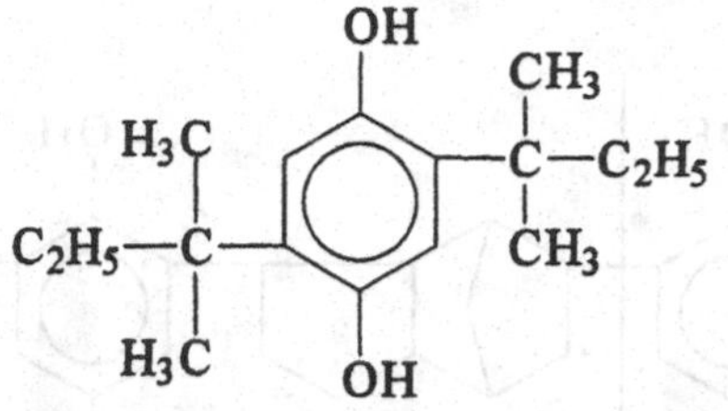

A large number of polynuclear phenols, such as bisphenols, diphenols and polyphenols, are also effective. Their low volatility is partially responsible for their high activity. One typical polynuclear phenol antioxidant is 4,4'–butylidene–bis–(6–t–butyl–3–methyl phenol) (CAS# 85–60–9). It is sold by Flexsys [22] as Santowhite® Powder and by Great Lakes [12] as Lowinox® 44B25. The structural formula is:

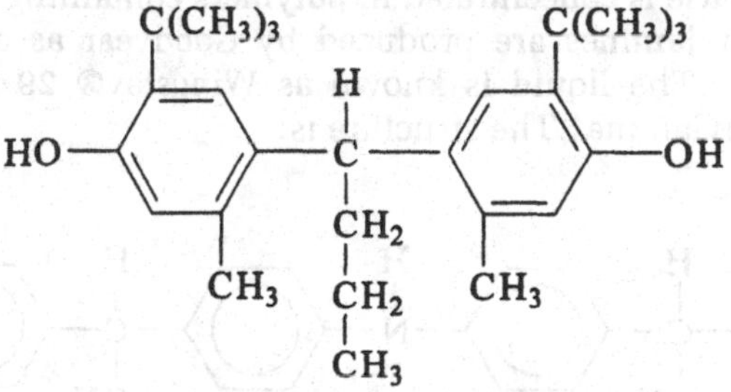

Another hindered phenol based on pentaerythritol is produced by Ciba–Geigy [15] as Irganox® 1010 and by Great Lakes [12] with a trade name of Anox® 20. The chemical name is tetrakismethylene (3,5–di–t–butyl–4–hydroxyhydrocinnamate) methane (CAS# 6683–19–8) and the structure is:

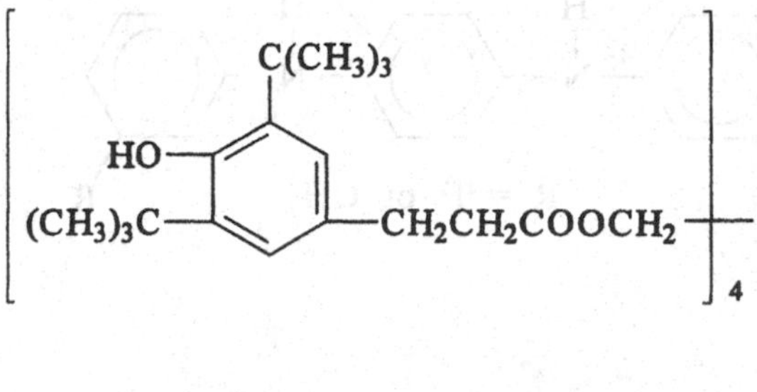

A polymeric hindered phenol prepared by building a chain onto the basic BHT structure is very effective because it retains much of the high antioxidant activity of BHT and the higher molecular weight decreases the volatility. It is made by Great Lakes [12] as Lowinox® 22CP46 and by Goodyear [9][23] as Wingstay® L HLS. The structure is:

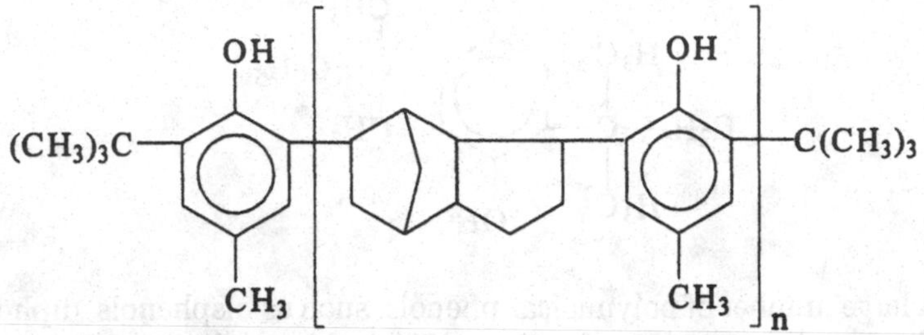

Other classes of primary antioxidants that inhibit the propagation step of autoxidation are alkylaryl- and diarylamines. These amines are generally more effective than the phenolic antioxidants in stabilizing easily oxidized unsaturated elastomers [1]. They do, however, discolor and will stain the polymers, so their use is concentrated in polymers containing carbon black. Some typical diarylamines are produced by Goodyear as either viscous liquids or solids. The liquid is known as Wingstay® 29 [24] and is a styrenated diphenylamine. The structure is:

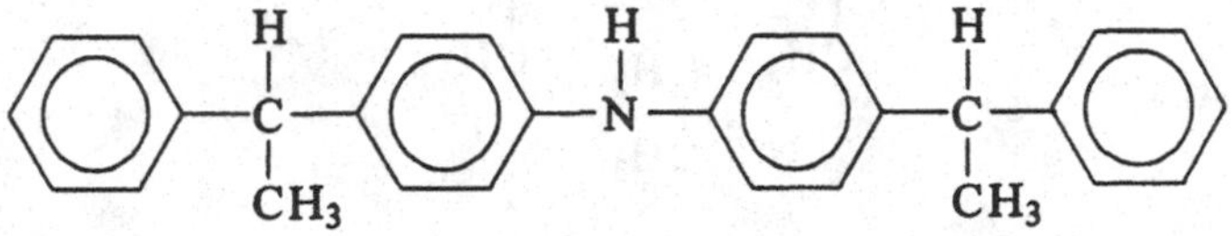

The solid amine is Wingstay® 100 [25], which is a mixed di-aryl-p-phenyl-enediamine with a structure of:

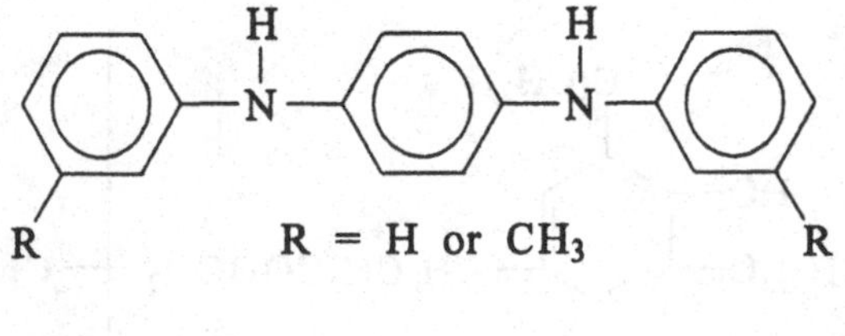

$R = H$ or CH_3

An alkylaryl diamine, N,N'–di–sec–butyl–p–phenylenediamine (CAS# 69796–47–0), is made by Flexsys [16][22] with the designation Santoflex® 44 and is used as a fuel additive. The structure is:

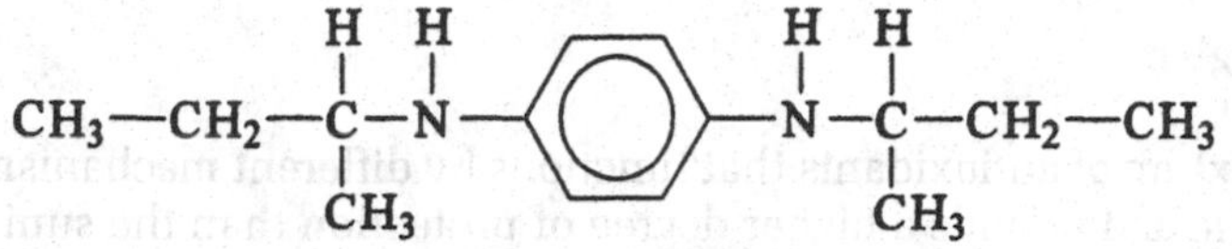

Polymerized 2,2,4–trimethyl–1,2–dihydroquinoline (CAS# 26780–96–1) is also made by Flexsys as Flectol® TMQ [16][22] and is an excellent antioxidant for polyethylene. A probable structure is:

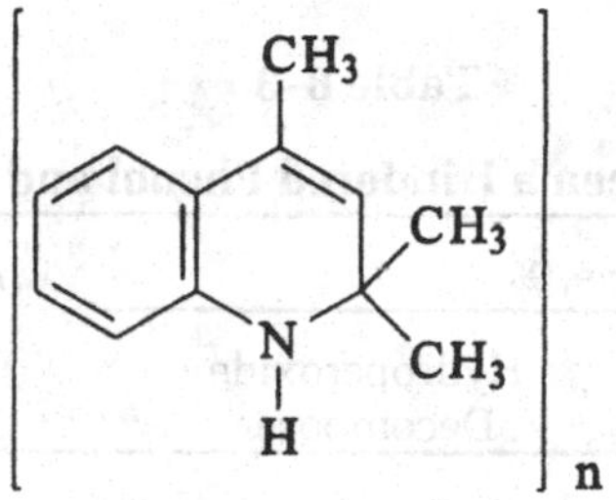

The mechanism of antioxidant inhibition during the propagation step involves the transfer of a hydrogen atom to the peroxy radical in a rate controlled reaction. A second peroxy radical is then often added to the antioxidant molecule. Hydrogen transfer to R• is kinetically unimportant[11]. The strongest support for this mechanism comes from deuterium isotope studies and subsequent analysis of the products as described by Shelton [18].

$$AH + RO_2\bullet \longrightarrow A\bullet + RO_2H$$

$$A\bullet + RO_2\bullet \longrightarrow AO_2R$$

$$2\,A\bullet \longrightarrow A{-}A$$

In this sequence AH is a hindered phenol or *sec*–amine antioxidant.

Other antioxidants, such as the zinc salts, inhibit the propagation step by electron transfer rather than hydrogen transfer, thus reducing the peroxy radical to a peroxide anion, RO_2^-. These materials, therefore, inhibit both the initiation and propagation steps.

8.1.3 Synergism

A mixture of antioxidants that functions by different mechanisms can be synergistic and provide a higher degree of protection than the sum of the stabilizing activities of each individual material. The combinations most frequently used are mixtures of radical scavengers and peroxide decomposers. The synergistic effect of a hydroperoxide decomposer, e.g., di–lauryl–3,3'–thio–dipropionate (CAS# 123–28–4), and a radical scavenger, e.g., tetrakis–methylene (3,5–di–t–butyl–4–hydroxyhydrocinnamate) methane (CAS# 6683–19–8) in protecting polypropylene in an oxygen uptake test at 140°C is shown in Table 8–3. The sum of the individual activities of these antioxidants was 20 days, whereas a mixture of the two stabilizers protected the polymer for 45 days [26].

Table 8–3

Synergism Between a Hindered Phenol and a Thioester

Additive, %		Induction Period, days
Radical Scavenger	Hydroperoxide Decomposer	
0.0	0.3	4
0.1	0.0	16
0.1	0.3	45

(Reference [26]. First published in Kunststoffe/plast Europe)

8.1.4 Photooxidation and Stabilization

As with oxidative degradation, hydroperoxides are important initiators in photooxidation. These hydroperoxides decompose both photochemically and thermally. Many photochemical reactions can produce hydroperoxides early in the photooxidation process. These reactions include radical formation from hydrocarbon–oxygen charge transfer complexes [27], Norrish scission of photoexcited carbonyls [28], scission through photoexcited conjugated and isolated double bonds [29], and other photo–mechanisms. The agents that protect substrates from photooxidation include those that reduce the amount of damaging radiation entering the substrate, UV absorbers, and those that deactivate photoexcited chromophores by energy transfer, quenching agents. Despite the importance of radical chain autoxidation in the photodegradation process, most hindered phenols alone provide little protection, probably because they rapidly decompose under UV irradiation. Nevertheless, combinations of UV absorbers and phenolic antioxidants can be synergistic and are commercially important [11].

8.1.5 Lubricant and Fuel Stabilization

Unstabilized, stored gasoline discolors slowly and gums are deposited through oxidation and polymerization. Gasoline containing such gums causes valve sticking, plugged fuel injection jets, and in general, lower motor efficiency. Metal ions, especially copper, catalyze the oxidation of gasoline since only trace amounts promote oxidation through peroxide decomposition. As little as 0.01 ppm can be harmful and such quantities are easily incorporated at the refinery.

The combination of an antioxidant capable of decomposing peroxides and a metal deactivator are required for optimum stabilization of gasoline. Based on cost–performance, the two most popular are butylated hydroxytoluene (2,6–di–t–butyl–4–methyl–phenol, CAS# 128–37–0), or BHT, and N,N'–di–sec–butyl–p–phenylenediamine (CAS# 101–92–2). They are usually added at a rate of 2.83 to 5.66 mg/L. The copper chelating agent is usually N,N'–disalicylidene–1,2–propanediamine (CAS# 94–91–7) used at about 5.7 to 14 mg/L of gasoline.

Although fuels make up the vast majority of products derived from crude petroleum, lubricants and lubricating oils are the second most important class. Engine crankcase oils are generally protected against oxidation by 1–2% of a zinc dithiophosphate or a terpene or other olefin–phosphorus pentasulfide reaction product mixture. Automatic transmission fluids operate at high temperatures for long periods and can best be protected by both phenolic and aromatic amine antioxidants in the range of 0.5–1.0%. Other lubricating oils, such as gas turbine engine lubricants, hydraulic fluids, gear oils, and greases, are also protected by antioxidant packages [11].

8.1.6 Vitamins and Antioxidants in Food

Oxygen causes a rapid deterioration in the flavor and odor in food. This change is normally referred to as the food becoming rancid. The most frequently used antioxidant is butylated hydroxytoluene, or BHT. Others are butylated hydroxyanisole, BHA, and vitamin C (CAS# 50–81–7) [30] or one of its derivatives. Foods also contain natural antioxidants that are fat soluble vitamins. Two of these are β–carotene (CAS# 7235–40–7) [31], and vitamin E (CAS# 1406–18–4) [32], which are free radical scavengers that function as inhibitors in both the initiation and promotion stage of carcinogenesis [33].

One of the antioxidants contained naturally in foods is vitamin C or ascorbic acid. The structure is:

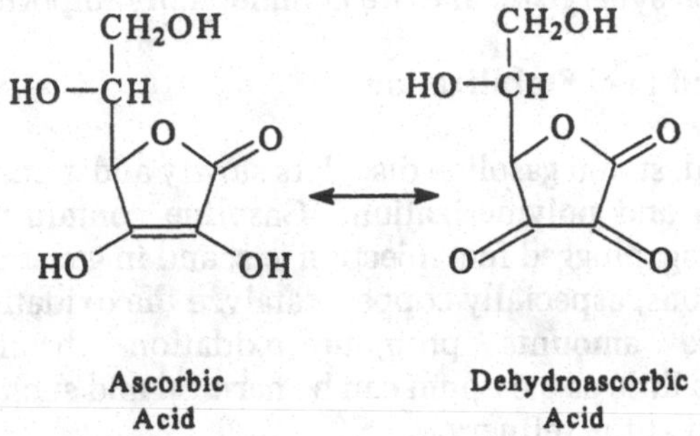

Ascorbic
Acid

Dehydroascorbic
Acid

This carbohydrate derivative of sorbase scavenges aqueous superoxide and hydroxyl radicals and acts as a chain–breaking antioxidant in lipid peroxidations. It may also act indirectly in protecting cell membranes by regenerating the active form of membrane–bound vitamin E. Ascorbic acid has been related to cholesterol and triglyceride metabolism by altering the hepatic conversion to other end products [31].

Beta–carotene is reported not only to be essential for good visual acuity but also may reduce the risk of some types of cancer [34], especially lung cancer. It has also been implicated in preventing the oxidation of low–density lipoproteins and consequently reducing the formation of atherosclerotic lesions [35]. The structure of β–carotene is:

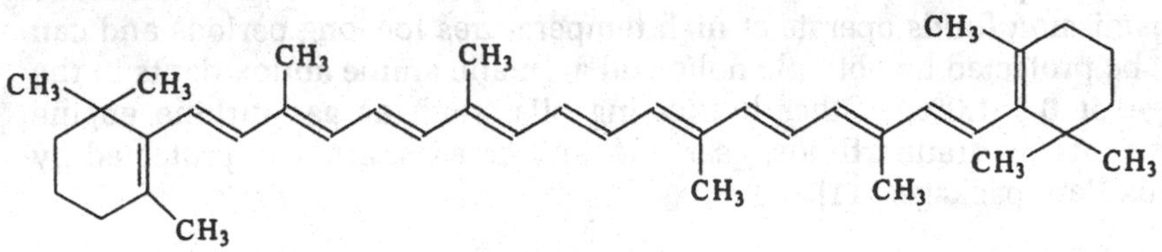

Vitamin E, a mixture of tocopherols, is another naturally occurring antioxidant that has been evaluated as a stabilizer for finish oils [3]. The principal application, however, is as an antioxidant in human and animal metabolism. There is widespread agreement that tocopherols can interrupt free radical chain decomposition and thus are capable of reducing some carcinogenicity [34], and can act to protect cellular membranes from oxidative destruction [36]. The structural formula for α–tocopherol (CAS# 59–02–9) is:

The structures of the β–, γ–, and δ–tocopherols are similar except for some rearrangement of the methyl groups.

8.2 DEFOAMERS

In the previous discussions, both in this chapter and in the preceding chapters, the emphasis has been on emulsions that are dispersions of a liquid within another liquid. A foam, in contrast, is a dispersion of a gas in a liquid. With foams, the liquid is the continuous phase and the gas the discontinuous phase. Bikerman [37] suggests that foams be classified into two morphological groups: (1) *kugelschaum* or spherical foams, which consist of widely separated spherical bubbles; and (2) *polyederschaum* or polyhedral foams, which consist of bubbles that are nearly polyhedral in shape having thin films of very low curvature separating the dispersed phase. When the concentration of the gas is low, the bubbles are spherical in shape, but at higher concentrations the bubbles form polyhedrons. These two-sided films are called the *lamellae* of the foam. Where three or more gas bubbles meet, the lamellae are curved, concave to the gas cells, forming what is called the *Plateau border* or *the Gibbs triangle*. The pressure difference across a curved interface due to the surface tension of the solution is given by the Laplace equation, Eq. 8–1:

$$\Delta P = \gamma \left(\frac{1}{R_1} + \frac{1}{R_2} \right)$$

(8–1)

where R_1 and R_2 are the radius of curvature of the surface. Since the curvature in the lamellae is greatest in the Plateau borders, there is a greater pressure across the interface in these regions than elsewhere in the foam. As the gas pressure inside an individual cell of a foam is equal throughout the cell, the liquid pressure inside the lamella at the highly curved Plateau border (the junction point of the three bubbles) must be lower than in the adjacent, less curved regions at each bubble surface. This causes drainage of the liquid from the lamellae into the Plateau border. In a column of foam, liquid also drains because of hydrostatic pressure, with the result that the lamellae are thinnest in the upper regions of the column and thickest in the lower region. Foams are destroyed when the liquid drains out from between the two parallel surfaces of the lamellae, causing it to become progressively thinner. When it reaches a critical thickness of 50–100 Å, the film collapses [38].

Absolutely pure liquids do not foam. Foam is also not important in mixtures of materials of similar chemical structure. For true foaming to occur, a solute capable of being adsorbed at the liquid–gas interface must be present. The presence of this surface–active solute produces lamellae between the gas cells of the foam that have adsorbed monomolecular films of surfactant molecules on each side of the liquid–gas interface. These absorbed films provide the system with the property that distinguishes foaming from non–foaming systems—the ability of the former system to resist excessive *localized* thinning of the lamellae surrounding the bubbles, while general thinning proceeds. This property is known as *film elasticity* and it is necessary for the production of foam. Theories of film elasticity are described completely in Rosen [38] or Myers [39] and therefore will not be pursued in this book.

8.2.1 Foaming and Chemical Structure

The relationship between the generation of foams with surfactants and the structure of those surfactants can be quite complex. A distinction must also be made between foam production, which is measured as the height of the foam when initially formed, and foam stability, which is the height after a period of time. Most of the foaming data have been produced by use of the Ross–Mills method [40]. In this procedure 200 ml of a solution of surfactant is placed in a special pipette with certain dimensions, including an exit orifice of 2.9 mm inside diameter and allowed to fall 90 cm onto 50 ml of the same solution contained in a cylindrical vessel at constant temperature. The height of the foam produced is read immediately after all the solution has run out of the pipette (initial foam height) and then again after a given length of time, generally after 5 minutes.

Foam height customarily increases with an increase in surfactant concentration below the critical micelle concentration (CMC) until the neighborhood of the CMC is reached, at which the foam height reaches a maximum. Thus, the CMC of a surfactant is a good measure of the efficiency as a foaming agent; the lower the CMC, the more efficient the surfactant is as a foaming agent.

The effectiveness of a surfactant as a foaming agent appears to depend on both its abilities to reduce the surface tension of the foaming solution and on the magnitude of its intermolecular cohesive forces. The volume of foam produced when a given amount of work is done on an aqueous solution of surfactant to create foam depends on the surface tension of the solution, since the minimum amount of work required to produce foam is $\gamma(\Delta A)$, the product of the surface tension and the change in area of the liquid–gas interface as a result of foaming [37]. Ionic surfactants are normally much more effective as foaming agents than nonionic surfactants.

Nonionic surfactants must, by their chemical structure, have large surface areas per molecule, and it becomes difficult for these absorbed molecules to interact to a significant degree, resulting in lower interfacial elasticity. Polyoxyethylene (POE) nonionic surfactants exhibit a strong sensitivity of foaming ability to the length of the POE chain. A short chain length may not have sufficient water solubility to lower the surface tension and produce foam. A chain that is too long, on the other hand, will greatly expand the surface area required to accommodate the absorbed molecules and will also reduce the interfacial elasticity. This characteristic has made it possible to design highly effective yet low foaming surfactant formulations. Even more dramatic effects can be obtained in which both ends of the POE chain are substituted. Usually a single methyl or propylene oxide group on the end of the chain can significantly reduce foaming [41].

If the solubility of a surfactant is highly temperature dependant, as often happens with most nonionics, it will be found that foaming ability will increase as in the same direction as its solubility. Nonionic POE surfactants, for example, will exhibit a decrease in foam production as the cloud point is approached [38].

8.2.2 Chemical Structure of Antifoaming Agents

Antifoaming agents probably operate by replacing the foam–producing surface film by a completely different type of film. To accomplish this change, antifoaming materials must displace any foam stabilizer, such as a surfactant, which is a part of the film. These agents then must have a surface tension low enough in the pure state so that they can spread spontaneously over the existing film. This means that their spreading coefficient, $S_{L/S}$, must be positive (see the section on wetting agents). Equation 8–2 presents the formula to calculate the spreading coefficient:

$$S_{L/S} = \gamma_{SA} - \gamma_{SL} - \gamma_{LA} \qquad (8-2)$$

In this equation, γ_{SA} = interfacial tension at the solid–air boundary, γ_{SL} = tension at the solid–liquid interface, and γ_{LA} = tension at the liquid–air interface. In addition, these antifoaming agents must also maintain a high concentration at the surface while being used at low concentrations. These materials then must be quite insoluble in the foaming solution, but still not so insoluble in it that they do not become a component of the surface film.

Two classes of antifoaming agents are used: foam breakers and foam inhibitors. Foam breakers are materials that destroy existing foams. They work by (1) reducing the surface tension in local areas to very low values, thus causing these local areas to be thinned rapidly to the breaking point by the pull of the surrounding higher surface tension regions; and (2) by promoting drainage of the liquid from the foam film and by that shortening its life. Examples of chemicals that work by the first mechanism are ethyl ether (γ = 17) and t-amyl alcohol (γ = 24) [42]. Tributyl phosphate is typical of the materials that operate by mechanism (2). Its large cross–sectional area in the interfacial film may cause it to reduce cohesive forces between surfactant molecules when it fits between them in the surface interfacial film.

Foam inhibitors are materials that prevent a foam from being formed. They act by inhibiting surface elasticity. These materials produce a surface that has a substantially constant surface tension when expanded or contracted. Some inhibitors do this by swamping the surface with nonfoaming, rapidly diffusing, non–cohesive, and only moderately surface–active molecules, so that any transient rises in surface tension caused by expansion is rapidly annulled. Some wetting agents and ethylene oxide–propylene oxide block copolymers appear to work in this manner. Other materials act by replacing the elastic surface film with a brittle, close–packed surface film. Metal salts of long chain fatty acids, or soaps, work by this technique, especially with foams of anionic surfactants, by forming the metal salts that tend to have a "solid", brittle film having no elasticity. These metal soaps thus produce an unstable foam [37].

At least occasionally, the foam–breaking and foam–inhibiting properties may be additive, and mixtures of a foam breaker and a foam inhibitor show remarkably effective foam–destroying properties [41].

Materials that are effective as defoaming agents can be classified into nine general chemical classifications, with the best choice of the material depending on such factors as cost, the nature of the liquid phase, the nature of the foaming agent present, and the nature of the environment to which it may be subjected. These chemical types are: (1) polar highly branched aliphatic alcohols; (2) fatty acids with limited water solubility; (3) fatty esters with limited water solubility; (4) metallic salts of carboxylic acids; (5) water insoluble amides; (6) alkyl phosphate esters; (7) EO–PO block copolymers; (8) organic silicones; and (9) fluorinated alcohols and acids. To elaborate:

1.   One of the most common classes of antifoaming agents are polar, highly branched aliphatic alcohols. Linear alcohols in intimate mixtures with surfactants can result in increased foam production

and stability due to the mixed monolayer formation and enhanced film strength. Branched materials, in contrast, reduce the lateral cohesive strength of the interfacial film, which increases the rate of bubble collapse. The higher alcohols also have limited water solubility and are strongly adsorbed at the air–water interface, displacing surfactant molecules in the process.

2 and 3. Fatty acids and fatty esters with limited water solubility can also be used as foam inhibitors. Their actions in foam inhibition are similar to those of the branched alcohols. Additionally, their low toxicity makes them attractive for use in food application. Organic compounds with multiple polar groups are also found to be effective foam inhibitors. The presence of several polar groups generally acts to increase the surface area per molecule of the adsorbed antifoamer and results in a loss of stabilization.

4. Metal salts of carboxylic acids, especially the water–insoluble polyvalent salts such as calcium, magnesium, and aluminum, can be effective as defoamers in both aqueous and non–aqueous systems. In water, they are usually employed as solutions in organic solvents, or as a fine dispersion in the aqueous phase.

5. Water–insoluble organic compounds containing amide groups are effective antifoaming agents in several systems, especially for use in boiler systems. It is generally found that greater effectiveness is obtained with materials containing at least 36 carbon atoms when compared to simple fatty acid amides.

6. Alkyl phosphate esters possess good antifoaming characteristics in many circumstances due to their low water solubility and large spreading coefficients. They also find wide application in non–aqueous systems as inks and adhesives.

7. The ethylene oxide–propylene oxide block copolymers act by swamping the surface with nonfoaming, rapidly diffusing, non–cohesive, only moderately surface–active molecules. Any transient rises in surface tension caused by expansion is rapidly cancelled.

8. Organic silicone compounds are also usually found to be outstanding antifoaming agents in both aqueous and organic systems. Because of their inherently low surface energy and limited solubility in many organic compounds, the silicone materials are one of the two types of compounds that are available to modify the surface properties of many organic liquids.

9. The fluorinated alcohols and acids are another class of defoaming agents. Due to their very low surface energies they are active in liquids where hydrocarbon materials have no effect. They are, in general, quite expensive, but their activity at very low levels and in harsh environments can overcome their initial cost barrier [39].

8.2.3 Commercial Antifoaming Agents

Because all spin finishes contain relatively large amounts of surfactants, foam generation is a significant problem. Although a large number of defoamers are manufactured and are available in commercial quantities, many are not appropriate for textile applications. Several types of chemical materials are used as antifoaming agents in varying degrees of effectiveness, both as foam inhibitors and foam breakers.

Ethylene Oxide–Propylene Oxide Block Copolymers, CAS# 9003–11–6

These copolymers are normally used at elevated temperatures since their cloud point causes them to lose water solubility and thus enhance their defoaming properties. An extensive discussion of these types of surfactants can be found in Chapter 4, Section 4.2. They are made in various molecular weight ranges by BASF [43], Ethox [44], and Hoechst–Celanese [45].

Oleic Acid Diethanolamide, CAS# 93–83–4

An example of the water insoluble amide defoamers, this material is supplied by Henkel as Emid® 6545 [46]. Primarily used as a foam suppressant in textile dyeing operations, this amide is also effective as an emulsifier for mineral oil.

Silicone–Based Defoaming Agents

A very large class of materials utilized to prevent and destroy foam in aqueous systems for textile applications are based on silicone products. Some of these are emulsions of reacted silica in a water–dispersible base and others are water–soluble silicone glycols. One of the primary suppliers for these products is Dow Corning, with the following materials available.

1. Dow Corning® Antifoam B [47][48]. The active ingredient in this product is a reacted silica dispersed in water with a nonionic emulsifier. Since it is a mixture, no CAS number has been assigned. This is a 10% active emulsion that is very effective at concentrations of 1–100 parts per million (ppm). It can be used in cool aqueous systems and should be diluted with cool water. This is a very effective antifoam emulsion for spin finish compositions.
2. Dow Corning® Antifoam H–10 [47][49]. A silicone emulsion with a pH of about 4, it is a low viscosity liquid that is easily diluted with cool water. This defoamer is primarily designed in hot aqueous applications, such as dye baths. Useful at concentrations of 1 to 10 ppm, it is supplied as a 10% emulsion.

3. Dow Corning® Antifoam Y–30 [47][50]. This silicone emulsion is a medium viscosity (2,500 centipoise at 25°C) product that must be diluted prior to use. The emulsion contains 30% active silicone material with a pH of about 3. Concentrations of 1–100 ppm are sufficient to eliminate most foams. Recommended uses include fermentation, ceramic manufacturing, antifreeze systems, textiles, and inks and dyes.

4. Dow Corning® Antifoam 1410 [47][51] and Antifoam 1430 [47][52]. Two silicone emulsion defoamers that are intended for use in hot procedures, one is a 10% emulsion (1410) and the other a 30% concentration (1430). The Antifoam 1410 has a viscosity of 3,500 centipoise @ 25°C and a pH of 7, but the 1430 is more viscous at 8,000 centipoise @ 25°C and has a pH of 4. These are useful in aqueous systems at wide ranges of extreme pH.

5. Dow Corning® Antifoam 2210 [47][53]. In contrast with the silicone emulsions described in the previous four sections, this component is based on a silicone–glycol product. The silicone–glycol compounds are modified dimethylsiloxane polymers in which one or more methyl groups are replaced with ethylene oxide or propylene oxide derivatives. Antifoam 2210 is an emulsion of 10% activity intended for use in hot applications, including concrete manufacturing and chemical gas processing. It is also effective in systems with widely varying pH.

6. Dow Corning® 544 Antifoam Compound [47][54]. This is a 100% active, water–soluble silicone glycol chemical that is a foam–preventing agent rather than a defoaming one. It is useful in Beck and jet dyeing, washing operations, and distillation processes. Recommended for hot aqueous systems, 1 to 100 ppm concentrations are very effective.

7. Dow Corning® Q2–3183A Antifoam [47][55]. This is another foam–preventing silicone–glycol component that can be used in either aqueous or non–aqueous systems. It is a 100% active liquid that is dispersible in water and organic solvents and can be used in concentrations as low as 1 ppm.

8. Witco Bubble Breaker® 3056A [56]. Although this material does not contain a silicone product, it is a dispersion of reacted silica in mineral oil. Widely used in many textile applications, typical properties are listed in Table 8–4.

Table 8–4

Typical Properties of Bubble Breaker® 3056A

Property	Value
Active Content, %	99.5
Viscosity, cPs @ 25°C	50
pH, 5% Aqueous	4–6

Nonsilica or Nonsilicone Defoamers

One serious problem with silicone or silica–based anti–foaming compounds is the persistence in the products in which they have been added. In some instances, a silicone defoamer will be retained on the substrate and affect further processing. When this effect is possible, defoaming agents that do not contain a silicone or silica are the products of choice. Some of these highly effective anti–foaming compounds are listed in the following sections.

1. Henkel Foamaster® SD [57]. Although this proprietary defoamer contains a small amount of silicone, less than 1.5%, it is effective when used in atmospheric and pressure dyeing of nylon, polyester, and wool fibers or blends of these three. It is stable in a pH range of 2–12 and to electrolytes up to a 2% concentration. The product is water dispersible and has a pH of 7.
2. Henkel Foamaster® 340 [58]. This proprietary composition does not contain any silicone and is very effective in the control of foam in jet dyeing machines. It is 100% active and has a pH of 7. It is also stable in dye baths in which the pH varies from 3 to 11.
3. Henkel Foamaster® DCD–2 [59]. A proprietary electrolyte stable defoamer composition that is 100% active, Foamaster® DCD–2 is designed for reactive and direct dye bath applications. The pH of a 5% solution is 3.0.
4. Witco Bubble Breaker® 748 [56]. One of the series of blends of metal carboxylates dispersed in mineral oil, this material is a dispersion of a semi–solid waxy metal salt, such as aluminum stearate. Other carboxylates that can be used are salts of polyvalent metals as calcium and magnesium. Some of the properties are shown in Table 8– 5

Table 8–5

Typical Properties of Bubble Breaker® 748

Property	Value
Active Content, %	99.5
Viscosity, cPs @ 25°C	100
pH, 5% Aqueous	9–10

5. Witco Bubble Breaker® 900 [56]. This defoamer is similar to Bubble Breaker® 748. It is a creamy liquid blend that is also recommended for textile processing. Attributes of the blend are shown in Table 8–6.

Table 8–6

Typical Properties of Bubble Breaker® 900

Property	Value
Active Content, %	99.5
Pour Point, °C	<−18
Viscosity, cPs @ 25°C	100
pH, 5% Aqueous	8.5–10

6. Witco Bubble Breaker® 913 [56]. An opaque oily liquid, the dispersion of a metal carboxylate in mineral oil is useful in foam reduction for adhesives, coated papers, floor wax, and textile processing. Typical properties are presented in Table 8–7.

Table 8–7

Typical Properties of Bubble Breaker® 913

Property	Value
Active Content, %	99.8
Pour Point, °C	<0
Viscosity, cPs @ 25°C	90
pH, 5% Aqueous	5.7

8.3 ANTIMICROBIALS

No organic compound has an infinite life in the environment. Many of these are subject to attack by microorganisms that use these organic materials as a source of food. A spin finish emulsion is an excellent example of this type of system. Since the first finish was formulated, the finish chemist was required to guard against microorganism growth by the addition of some agent to kill these organisms. Fiber producers and component manufacturers have used various techniques to prevent organism growth including maintaining a sterile system, but most use an antimicrobial agent and change on a regular schedule. Some of the more effective chemicals suggested for antimicrobial action are detailed in the next section.

Silver Nitrate, CAS# 7761–88–8

The first antimicrobial agent to be used in spin finishes was probably silver nitrate. The activity of silver in controlling bacteria and fungi is well known, but it has the problem of extreme toxicity of the silver ions, and also the very high cost of the product. At one time in the early days of carpet processing, the single largest source of pollution in the wastewater in the Dalton, Georgia area was silver and its salts [60]. This antimicrobial material is no longer in use and will likely never be used again.

6–Acetoxy–2,4–dimethyl–m–dioxane, CAS# 828–00–2

Unlike many other biocides, this product hydrolyzes in aqueous systems to form two compounds which act to significantly reduce the presence of microorganisms in those systems. It is produced by Givaudan–Roure and sold using the trade name Giv–Gard DXN® [61]. Although it is suggested for use as a bacteriostat and fungistat, it is especially effective against gram–negative bacteria, the organisms primarily involved in bacterial spoilage. The structure of Giv–Gard DXN® is:

$$\begin{array}{c} O \\ \parallel \\ O-C-CH_3 \end{array}$$

(structure of 6-acetoxy-2,4-dimethyl-m-dioxane: H_3C and O and CH_3 in the dioxane ring)

When this chemical is added to water, it forms acetic acid and a transient intermediate, dioxinol. Acetic acid formation proceeds in an easily measurable rate and is 99% complete in 14 hours. Dioxinol rapidly decomposes to acetaldehyde and aldol. The aldol dimerizes to paradol and these two products exist in equilibrium. Reactions describing this process are:

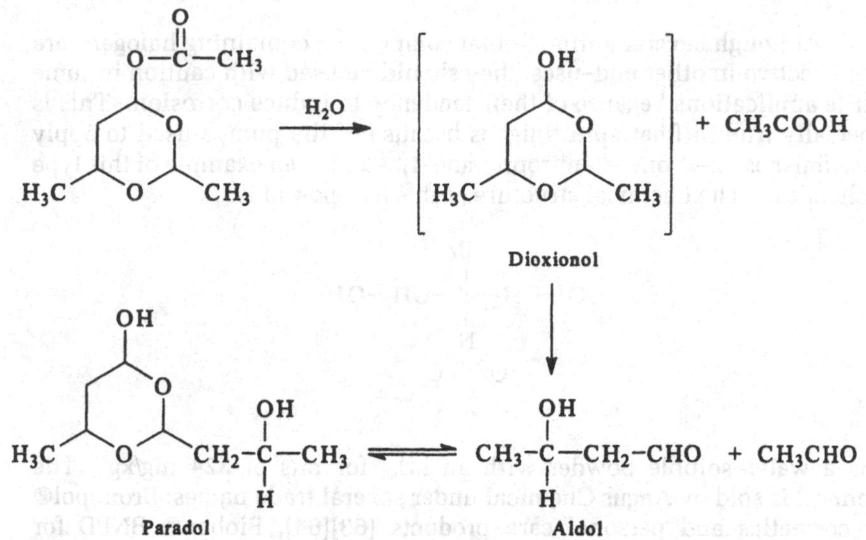

The reaction scheme shows the hydrolysis of the acetoxy dioxane compound:

$$\text{(6-Acetoxy-2,4-dimethyl-m-dioxane)} \xrightarrow{H_2O} \text{[Dioxionol]} + CH_3COOH$$

Dioxionol

Paradol ⇌ Aldol

$$CH_2-\underset{\underset{H}{|}}{\overset{\overset{OH}{|}}{C}}-CH_3 \;\rightleftharpoons\; CH_3-\underset{\underset{H}{|}}{\overset{\overset{OH}{|}}{C}}-CH_2-CHO \;+\; CH_3CHO$$

Paradol   **Aldol**

Since acetic acid is one of the byproducts of the initial hydrolysis, the pH of the system will decrease from near neutral to about 4 after 24 hours. Although GivGard® DXN is effective over a broad pH range, often this acidic aqueous liquid may corrode metals with which it is in contact. The manufacturer recommends neutralization with sodium carbonate instead of sodium or potassium hydroxide. Ammonium hydroxide should not be used because of the reaction of ammonia with aldehydes.

The acute oral toxicity (LD_{50}) in rats was found to be greater than 2.0 gm/kg, but no health hazards to humans are known. No irritation or sensitivity to human skin was observed in Repeated Insult Patch Tests. Some specifications [62] are given in Table 8–8.

Table 8–8

Typical Properties of 6–Acetoxy–2,4–dimethyl–m–dioxane

Property	Value
Activity, %	92 min
Acid Value	10.0 max
Specific Gravity, 25/25°C	1.060–1.075
Refractive Index, 20°C	1.430–1.437

2–Bromo–2–nitropropane–1,3–diol, CAS# 52–51–7

Although several antimicrobial compounds containing halogens are very effective in other end–uses, they should be used with caution in some textile applications because of their tendency to induce corrosion. This is especially true in fiber spin finishes because of the pumps used to apply these finishes. 2–Bromo–2–nitropropane–1,3–diol is an example of this type of chemical. The chemical structure of this compound is:

$$
\begin{array}{c}
\text{Br} \\
| \\
\text{HO}-\text{CH}_2-\text{C}-\text{CH}_2-\text{OH} \\
| \\
\text{N} \\
\diagup \diagdown \\
\text{O} \qquad \text{O}
\end{array}
$$

It is a water–soluble powder with an LD_{50} for rats of 324 mg/kg. The chemical is sold by Angus Chemical under several trade names: Bronopol® for cosmetics and personal care products [63][64], Bioban® BNPD for metalworking fluids [60][65], and Canguard® 409 [60][66] recommended for use in food packaging adhesives, printing inks, and paints. The antimicrobial action is due to the release of formaldehyde caused by the highly electronegative bromine and NO_2 groups adjacent to the diol moieties. This chemical is a broad spectrum antimicrobial agent and should be blended with another material for enhanced antifungal activity. General properties of this compound are shown in Table 8–9.

Table 8–9

Typical Properties of 2–Bromo–2–nitropropane–1,3–diol

Property	Value
Purity, % by weight	95 min
Water, %	1.0 max
pH, 20% aqueous solution	4.15
Melting Point, °C	118

Diiodomethyl–p–tolysulfone, CAS# 31350–46–6

One of the more valuable antifungal preservatives, this compound is useful against molds, yeast, and algae, and it is recommended for use as a mildewcide in latex paints. It is insoluble in water but soluble in aromatic organic solvents. The structure is:

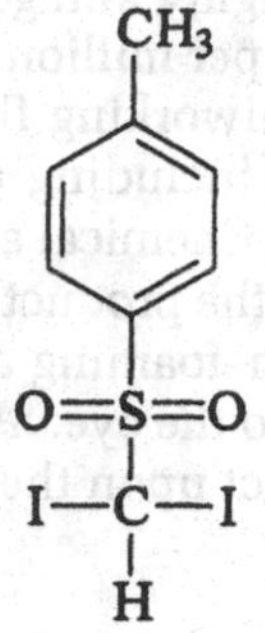

$$CH_3$$

$$O=S=O$$
$$I-C-I$$
$$H$$

It is sold by Angus Chemical using the trade name Amical® [60] either as a powder, a dispersion, or a wettable powder. The FDA has granted clearance for food packaging adhesives and can sealants.

4,4–Dimethyloxazolidine, CAS# 51200–87–4

Unprotected emulsions, including latex paints and finish emulsions, are subject to rapid degradation from microbial attack. Oxazolidines are effective in reducing both bacteria and fungi in these emulsions. One example of this water–soluble chemical is 4,4–dimethyloxazolidine that is produced by Angus Chemical and sold as Canguard® 327 [60][67]. Typical properties are listed in Table 8–10 and the structural formula is:

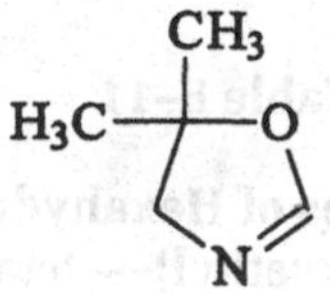

$$CH_3$$
$$H_3C \quad O$$
$$N$$

Table 8–10

Typical Properties of 4,4–Dimethyloxazolidine

Property	Value
Water, %	25.3
Viscosity, cp, 25°C	7.5
pH	10.5–11.5

Hexahydro–1,3,5–tris(2–hydroxyethyl)–s–triazine, CAS# 4719–04–4

One of the most effective antimicrobial agents for many applications is this triazine derivative. It is highly antagonistic to bacteria, fungi, and molds at levels as low as 200 parts per million. Although the manufacturers recommend this product for metalworking fluids and paints, it is also an excellent choice for textile uses, including spin finish emulsions. This antimicrobial is supplied by Angus Chemical as Canguard® 454 [60][64] and by Olin as Triadine® 3 [68]. Since the product is water soluble, it is supplied in an aqueous solution that is non–foaming and it has no odor, but in the concentrated solution is irritating to the eye. At concentrations of 5000 ppm or less, these is no discernible effect upon the eye. The structure is:

$$CH_2CH_2OH$$

$$H_2C \quad \overset{N}{\diagup} \quad CH_2$$

$$HOCH_2CH_2 - \overset{N}{\diagdown} \underset{H_2}{C} \diagup \overset{N}{\diagup} CH_2CH_2OH$$

Some of the attributes of this biocide are shown in Table 8–11.

Table 8–11

Typical Properties of Hexahydro–1,3,5–tris (2–hydroxyethyl)–s–triazine

Property	Value
Freezing Point, °F	–10
Water, %	21.5
Viscosity, cp, 25°C	300
pH	10–11
Specific Gravity, 25/25°C	1.16

Sodium Dimethyldithiocarbamate, CAS# 128–04–1, Sodium 2–Mercapto-benzothioazole, CAS# 2492–26–4, Zinc Dimethyldithiocarbamate, CAS# 137–30–4, and Zinc 2-Mercaptobenzothioazole, CAS# 155–04–4

The sodium salts are combined in a product sold by R. T. Vanderbilt intended for use as a preservative in starch adhesives for paper and paperboard and in textiles known as Vancide® 51 [69]. It is not proposed for use in paper that will be used for food–packaging purposes. To prepare mold–resistant cotton fabric, a 5% aqueous solution can be padded onto the fabric. Vancide® 51 is a solution of 27.6% dimethyldithiocarbamate, 2.4% mercaptobenzothioazole in 70% water. These components exhibit low toxicity with the acute oral toxicity LD_{50} for rats being 3,120 mg/kg and it showed no reactions during skin exposure. This pesticide is, however, toxic to fish, and should not be discharged into water containing this species. The structures are:

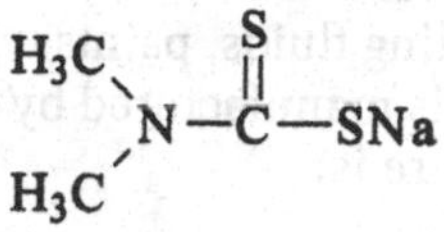

Sodium dimethyldithiocarbamate

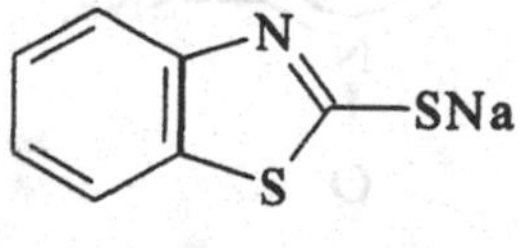

Sodium 2-mercaptobenzothioazole

Zinc salts are very valuable textile mildew and bacterial growth inhibitors for mattresses, pillow covers, canvas shoes, and shoe liners. Zinc dimethyldithiocarbamate, a white powder melting at 252–260°C, alone is sold as Vancide® MZ–96 [70] and a solution composed of 46.0% zinc dimethyldithiocarbamate, 4% zinc 2–mercaptobenzothioazole, and 50% water has the designation Vancide® 51Z Dispersion [71]. The structural formulas are:

$$H_3C \backslash \underset{H_3C}{N} - \overset{\overset{S}{\|}}{C} - S - Zn - S - \overset{\overset{S}{\|}}{C} - N \overset{CH_3}{\underset{CH_3}{\diagup}}$$

Zinc dimethyldithiocarbamate

Zinc 2-mercaptobenzothioazole

Sodium 2–Pyridinethiol–1–oxide, CAS# 3811–73–2 and Zinc 2–Pyridine-thiol–1–oxide, CAS# 13463–41–7

The sodium salt of this pyridinethiol is an industrial fungicide and bactericide that is water soluble and it is not a skin sensitizer. Applications for the chemical include cooling fluids, paints, vinyl acetate latex, and some synthetic fiber lubricants. It is manufactured by Olin with the name Sodium Omadine® [72]. The structure is:

Generally supplied as a 40% aqueous solution, some properties are listed in Table 8–12.

Table 8–12

Typical Properties of Sodium 2–Pyridinethiol–1–oxide

Property	Value
Color, Gardner	13 max
Water, %	58–60
pH, 10%	8.5–10.5

Zinc 2–pyridinethiol–1–oxide is one of the most effective antimicrobial agents available with concentrations as low as 4–8 ppm effective against gram–positive and gram–negative bacteria. It is equally useful in suppressing molds, yeasts, and algae. This zinc salt is one of the most effective anti–dandruff agents used in shampoos around the world, with the designation zinc pyrithione. Other applications include a preservative for cosmetics, plastics, paints, adhesives, and sealants. The structure is:

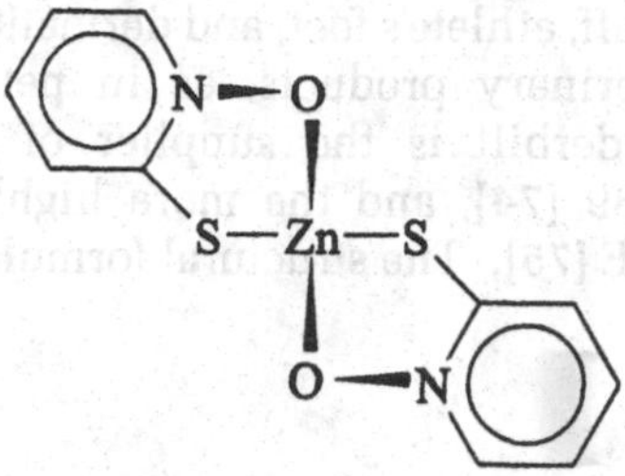

Pyrithione is an excellent chelating agent for heavy metal ions, especially iron, copper, mercury, and silver. Both the sodium and zinc salts form the chelates, some of which are highly colored. For example, only a few parts per million of ferric ion results in a complex that is deep blue. This transchelation can be prevented by adding a soluble zinc salt in a four– to tenfold excess over that amount of ferric ions present. Corrosion of finish pumps can produce enough iron to cause the complex formation.

This component is also made by Olin as Zinc Omadine® [73] and supplied either as a water insoluble powder or as an aqueous dispersion. Typical properties are presented by Table 8–13.

Table 8–13

Typical Properties of Zinc 2–Pyridinethiol–1–oxide

Property	Value	
	Powder	Dispersion
Assay, %	95–99	48–50
pH, 5%	6.5–8.5	6.5–8.5
Melting Point, °C (decomposes)	~240	—
Specific Gravity, 25/25° C	1.782	—

N–trichloromethylthio–4–cyclohexene–1,2–dicarboximide, CAS# 133–06–2

Sometimes known as Captan, this antimicrobial has a number of end–uses, including the restraint of the growth of mold in wallpaper paste, protecting vinyl polymer from mildew, and also in cosmetics. Since the molecule contains three chlorine groups, the use in fiber finishes may result in pump corrosion. This material is, however, very valuable in reducing the growth of fungi and molds. It is particularly suitable in preparations designed to combat dandruff, athlete's foot, and dermatitis caused by bacteria and fungi. Use in veterinary products, as in pet shampoos, is well established. R. T. Vanderbilt is the supplier of both the industrial preservative, Vancide® 89 [74], and the more highly purified cosmetic ingredient, Vancide® 89RE [75]. The structural formula for this component is:

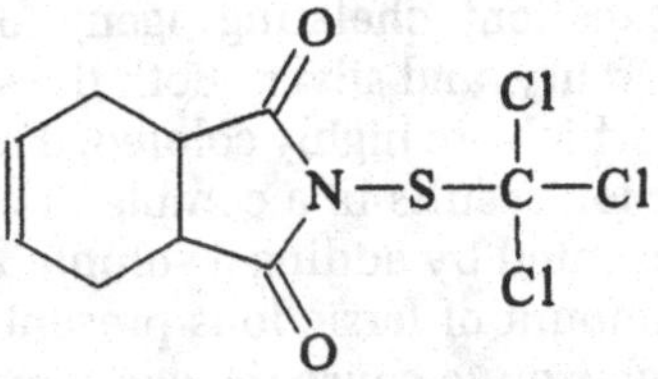

All of the antimicrobials described above are superior in textile applications to the esters of 4–hydroxybenzoic acid, the parabens, which are the most widely used preservatives in foods and cosmetics.

8.4 WETTING AND WETTING AGENTS

The forces existing between different objects in our natural world vary greatly in type and strength. From gravity to intra–atomic, attractive and repulsive forces maintain order and cohesion in all materials. These are especially true in chemistry since no molecules could form without various forces of attraction. Although the primary chemical forces are ionic, covalent, and coordinate bonding, when materials are attracted to polymer surfaces we must consider weaker and different attractive forces. In many applications, including the application of spin finish to a yarn bundle, it is very important that substances be attracted to or repelled from the polymer surface. The science of adsorption and adhesion is very complex and can be only partially explained through the mathematical perturbations of quantum mechanics. There are, however, much simpler approximations to adhesion theory that will be used in these discussions.

Attraction between a polymer surface and another substance cannot normally be attributed to primary chemical bonding. In almost every situation the forces are quasi–chemical (hydrogen bonding) or secondary forces. Secondary forces are divided into three categories: Keesom forces [76] arising from molecules containing permanent dipoles, Debye forces [77] that are caused by a molecule with a permanent dipole inducing a dipole in a neighboring molecule by the polarization of the electron cloud, and London dispersion forces [78] that result when instantaneous dipoles are generated by the motion of electrons within the molecule. Table 8–14 shows the relative bond energies of various types of molecular forces [79].

Table 8–14

Comparison of Bond Energies

Type of Force	Energy, kcal/mole
Primary Chemical Bonds	
Ionic	140–250
Covalent	15–170
Secondary Intermolecular Bonds	
Hydrogen Bonds	up to 12
Dipole–Dipole	up to 10
London Dispersion	up to 0.5
He–He (weakest measured)	about 0.02

(Reference [76])

The surface energy of the solid substrate has a very large influence on which types of attractive forces are involved in the mechanism of attraction and adsorption of materials onto polymers. Zisman [80]–[82] has classified solid surfaces into two categories: high energy and low energy. A high energy surface attracts the molecules of a solute or a liquid more strongly than the molecules attract each other, causing an adsorption pressure in the monolayer next to the surface as in Figure 8–1.

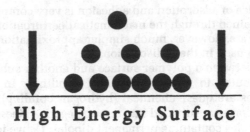

High Energy Surface

Figure 8–1. Representation of High Energy
Surface.

In contrast, a low–energy surface has less attraction to the liquid molecules
than they do for each other, as shown in Figure 8–2.

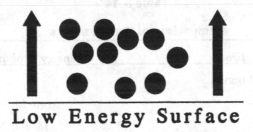

Low Energy Surface

Figure 8–2. Representation of Low Energy
Surface.

The molecules at the interface of a liquid phase with some other phase
are affected by an imbalance in the forces surrounding those molecules.
While the resultant forces on molecules are in equilibrium in the bulk of the
liquid (the net forces equal zero), the molecules at the interface are not in
equilibrium. There are forces attempting to pull these surface molecules
back into the bulk of the liquid and, in liquid–air systems, forces trying to
expel the molecules into the vapor phase above the liquid. This imbalance
in forces produces an apparent contractile effect that is called *surface tension*.
The forces involved are the same van der Waals interactions that account for
the liquid state and for most interactions between atoms and molecules.
Because the liquid state is of higher density than the vapor, surface
molecules are pulled away from the surface and into the bulk of the liquid,
causing the surface to contract spontaneously. For that reason, it is probably
correct to think of surface tension as the amount of work required to increase
the surface area at constant temperature and in a reversible system [39].
When we consider the liquid–air–solid system, if the surface tension

of the liquid is lower than that of the solid surface, the liquid will spread, as shown in Figure 8–3a. If the liquid surface tension is higher than that of a high energy surface, a drop is as shown in Figure 8–3b; and if the surface tension is higher than that of a low energy surface, the drop is as shown in Figure 8–3c. If a tangent is drawn to the drop at the drop–surface junction, the angle that is described is called the *contact angle*, and is denoted by the Greek letter Θ.

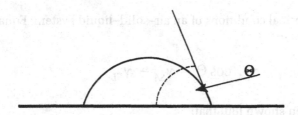

Figure 8–3a. Liquid Spreading.

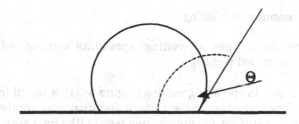

Figure 8–3b. Drop on High Energy Surface.

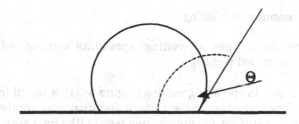

Figure 8–3c. Drop on Low Energy Surface.

8.4.1 Contact Angle

In his classic work on capillarity, first published in 1806, Laplace [83] described surface tension and suggested methods for measuring liquid surface tension. The Laplace treatment was abandoned because it could not explain many phenomena that were commonly observed. At about the same time Thomas Young [84] proposed an equation that has been the basis of much of the surface energy work today. Young's equation for the contact angle, Θ, of a liquid, L, on a solid surface, S, is:

$$\gamma_{LA} \cos \Theta = \gamma_S - \gamma_{SL} - \pi_e \qquad (8\text{-}3)$$

where:

γ_{LA}	=	surface tension of the liquid in air,
γ_S	=	surface energy of the solid,
γ_{SL}	=	interfacial energy at the solid–liquid junction, and
π_e	=	equilibrium film pressure of adsorbed vapor on the solid surface.

In the normal conditions of an air–solid–liquid system, Equation 8–1 can be rewritten as:

$$\gamma_{LA} \cos \Theta = \gamma_{SA} - \gamma_{SL} \qquad (8\text{-}4)$$

since it has been shown [80] that:

$$\gamma_S - \pi_e = \gamma_{SA} \qquad (8\text{-}5)$$

8.4.2 Thermodynamics of Wetting

There are three types of wetting: spreading wetting, adhesional wetting, and immersional wetting.

1. *Spreading Wetting.* In spreading wetting (Figure 8–3a), a liquid in contact with a solid surface spreads over the surface and displaces another phase, such as air, from that surface. Obviously, one would like for a spin finish to spread along the yarn bundle as far as possible while the water in the emulsion is still present. During the spreading process the interfacial area between the solid (S) and the initial phase (P1), which is frequently air, is decreased by the amount A, while the area between the solid and the wetting liquid phase (P2), or the finish emulsion, increases by an equal amount. The interfacial area between (P1) and (P2) also increases during the process. The change in interfacial area in each case will be the same, so that the total decrease in the energy of the system will be [80]:

$$-\Delta G = A\left(\gamma_{SP1} - \gamma_{SP2} - \gamma_{P1P2}\right) \qquad (8\text{--}6)$$

where:

γ_{SP1} = interfacial tension between the solid and air phases,

γ_{SP2} = interfacial tension between the solid and liquid phases, and

γ_{P1P2} = interfacial tension between the liquid and air phases.

The term in Equation 8–6 in parenthesis is called the spreading coefficient, S (see Equation 8–2), and can be rewritten as:

$$S = \gamma_{SA} - \gamma_{SL} - \gamma_{LA} \qquad (8\text{--}7)$$

If S is positive, then a fiber finish will spread over the surface to the greatest extent possible. If, for example, γ_{SA} is large, S will be greater and the liquid will spread more. If the solid surface is treated with a monomolecular layer of a material with a high surface tension, the finish will spread to the greatest degree possible.

If Young's equation (Eq. 8–3) is rewritten as:

$$\cos \Theta = \frac{\gamma_{SA} - \gamma_{SL}}{\gamma_{IA}} \qquad (8\text{--}8)$$

and combining it with the equation for the spreading coefficient (Eq. 8–7), it follows that:

$$S = \gamma_{LA}\left(\cos \Theta - 1\right) \qquad (8\text{--}9)$$

and if $\Theta > 0$, S cannot be positive or zero and spontaneous spreading will not occur.

2. *Adhesional Wetting.* In adhesional wetting, if a drop of liquid is added to a substrate and makes contact with that substrate, a change in free energy occurs. This change is:

$$\Delta G_w = A\left(\gamma_{SA} + \gamma_{LA} + \gamma_{SL}\right) \qquad (8\text{--}10)$$

where A is the surface area of the substrate in contact with the liquid. We can thus define the work of adhesion as:

$$W_a = \gamma_{SA} + \gamma_{LA} - \gamma_{LS} \qquad (8\text{--}11)$$

This equation was first proposed by Dupré [85], but according to Equation 8–4 $\gamma_{LA} \cos \Theta = \gamma_{SA} - \gamma_{SL}$, so the work of adhesion can be modified to read:

$$W_a = \gamma_{LA} \cos \Theta + \gamma_{LA} = \gamma_{LA} (\cos \Theta + 1) \qquad (8\text{--}12)$$

The work of self–adhesion of a liquid is known as the work of cohesion, and is:

$$W_c = 2 \gamma_{LA} \qquad (8\text{--}13)$$

The difference between the work of adhesion and the work of cohesion is the spreading coefficient, S.

$$W_a - W_c = \gamma_{SA} - \gamma_{LS} + \gamma_{LA} - 2\gamma_{LA} = \gamma_{SA} - \gamma_{SL} - \gamma_{LA} = S \qquad (8\text{--}14)$$

3. Immersional Wetting. This type of wetting is of less interest in finish application and spreading so that topic is not discussed. The reader is referred to Rosen [38] and Myers [39] for detailed information in immersional wetting.

8.4.3 The Effects of Surfactants on Wetting and Spreading

Since water has a high surface tension, about 72 milli–Newtons per meter (mN/m) (or formerly dynes/cm), it does not spontaneously spread over covalent solids that have a surface energy of less than 72 erg/cm². This is true because the molecules of water have a very strong intermolecular attraction for each other. The addition of a surface active agent to water, to modify the interfacial tensions of the system, is often necessary to enable water to wet a solid substrate. As seen in Equation 8–7, the spreading coefficient must be positive for spreading to occur. The addition of a surfactant to water acts by reducing the surface tension of the water, γ_{LA}, and perhaps the interfacial energy between the water and the solid surface, γ_{SL}, and may cause the spreading coefficient to be positive and spreading wetting will occur. The mechanism of surfactant reduction of water surface tension is due to the surfactant molecules at the surface interfering with the energy of attraction of the water molecules. The surfactant molecules orient themselves with the hydrophilic end toward the water and the hydrophobic end toward the air, or another non–polar surface.

If an anionic surfactant of the type Na^+R^- is added incrementally to water and the equivalent electrical conductivity plotted against the square root of the normality of the solution, the curve obtained has a sharp break at low concentrations instead of the smooth curve obtained with ionic electrolytes. This sharp break in the curve, with the reduction in conductivity, indicates that there has been a dramatic increase in the mass per unit charge of the material in solution. The larger mass particles in the solution are called *micelles*, or the association of individual surfactant molecules into a group of surfactant molecules neutralized with counter ions. The concentration at which this occurs is called the *critical micelle concentration* (CMC). Similar breaks are seen in almost every measurable property, such as interfacial tension, osmotic pressure, surface tension, density, and detergency. This pattern of behavior is observed with all types of surfactants: nonionic, anionic, cationic, and zwitterion. The determination of the CMC can be made by measuring almost any of these properties, but the most common is the surface tension.

In most aqueous systems the structure of the micelle is such that the hydrophobic groups are oriented toward the center of the structure and the hydrophilic groups toward the water phase. Changes in temperature, concentration of the surfactant, additives in the liquid phase, and structural groups in the surfactant all may cause change in size, shape, and aggregation number of the micelle. The structure can vary from spherical to rodlike or to lamellar.

In concentrated solutions, ten times the CMC or more, micelles are generally non–spherical. In some circumstances, the surfactant molecules are believed to form extended parallel sheets two molecules thick (lamellar micelles) with individual molecules oriented perpendicularly to the plane of the sheet. In aqueous solution, the hydrophilic heads of the surfactant molecules form the two parallel sheets and the hydrophobic tails comprise the interior region.

Micellization is an alternate mechanism to adsorption at an interface for removing hydrophilic groups from the bulk of the solution and thus reducing the total free energy of the system. Although the removal of the hydrophilic groups from the solution may reduce the free energy, the surfactant molecule may experience some loss of freedom in being confined to the micelle.

Two surfactants that are commonly identified as wetting agents, dioctyl sodium sulfosuccinate and POE (6) decyl alcohol, were used in experiments to determine typical CMC data [3]. The surface tension at various concentration was measured at 25°C and plots of this data were made. The initial slope and the final slope were extrapolated and the junction of these lines is defined as the CMC.

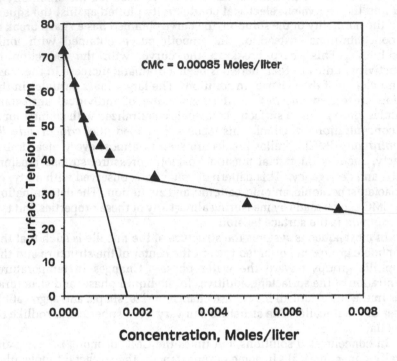

Figure 8–6. CMC Plot for Dioctyl Sodium Slufosuccinate.

The value calculated from Figure 8–7, 1100 μmoles per liter, is very similar to the ones reported by Becher [86] for this surfactant that varied from 640 to 1140 μmoles per liter, depending on the temperature of measurement and the method employed.

The amount of material adsorbed per unit area can be indirectly calculated from plots of surface (or interfacial) tension against the concentration of surfactant, as presented above, by use of the Gibbs adsorption equation [87]. In its most general form this equation is:

$$d\gamma = -\sum_{i} \Gamma_i d\mu_i \qquad (8\text{–}15)$$

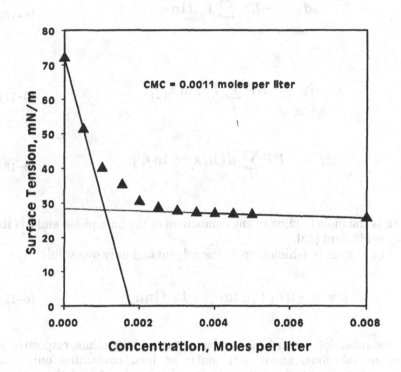

Figure 8–7. CMC Plot for POE (6) Decyl Alcohol.

where:

$d\gamma$	=	change in surface tension of the solvent,
Γ_i	=	surface excess concentration of any component of the system, and
$d\mu_i$	=	change in chemical potential of any component of the system.

and is fundamental to all adsorption processes. At equilibrium between the interfacial and bulk phase concentrations, $d\mu_i = RTd\ln a_i$, where a_i = activity of any component in the bulk (liquid) phase, R = gas constant, and T = absolute temperature.

Thus:

$$dy = -RT \sum_i \Gamma_i d\ln a_i \qquad (8\text{-}16)$$

and

$$dy = -RT \sum_i \Gamma_i d\ln x_1 f_1 \qquad (8\text{-}17)$$

$$dy = -RT \sum_i d(\ln x_i + \ln f_1) \qquad (8\text{-}18)$$

where x_i is the mole fraction of any component in the bulk phase and f_i is its activity coefficient [38].

In a system in which there is the solvent and only one solute:

$$dy = -RT(\Gamma_1 d\ln a_1 + \Gamma_2 d\ln a_2) \qquad (8\text{-}19)$$

where the subscripts 1 and 2 refer to the solvent and the solute, respectively. For dilute solutions, about .001 molar or less, containing only one non-dissociating surfactant, the activity of the solvent and the activity coefficient of the solute can be considered to be constant and the mole fraction of the solute x_2 may be replaced by the molar concentration C_2. The Gibbs equation can now be stated as:

$$dy = -RT \Gamma_2 d\ln C_2 \qquad (8\text{-}20)$$

or

$$dy = -2.303 RT \Gamma_2 d\log C_2 \qquad (8\text{-}21)$$

For surfactants dissolved in a normal solute, such as water, the surface excess concentration, Γ_2, can be considered equal to the actual surface concentration without significant error. The concentration of the surfactant at the interface is calculated from surface or interfacial tension data by the rearranged Gibbs equation. This form of the equation is:

$$\Gamma_2 = -\frac{1}{2.303\,RT}\left(\frac{\partial \gamma}{\partial \log C_2}\right) \qquad (8\text{--}22)$$

and the surface concentration can be obtained from a plot of γ versus $\log C_2$ at constant temperature. In this equation $R = 8.31 \times 10^7$ ergs/mole/°K if the surface tension is in mN/m and Γ_2 is in moles/cm^2.

The area per molecule at the surface gives information on the degree of packing of the adsorbed molecule. The area per molecule, a_2, in square angstroms (Å^2) is calculated from:

$$a_2 = \frac{10^{16}}{N\,\Gamma_2} \qquad (8\text{--}23)$$

where N = Avogadro's number and Γ_2 is in moles/cm^2. When these plots are made, the slope is determined below but near the CMC, since the surface concentration has reached a constant maximum value.

In the experiments for determining the CMC as shown in Figures 8–6 and 8–7, the surface tension was measured by the maximum bubble pressure method. If the bubble rate is varied, a change in the surface tension, indicating a change in the diffusion coefficient of the surfactant is observed [88]. An equation for this estimate, derived at the Textile Research Institute [89], is:

$$D = \frac{b}{2a}\left(\frac{\Delta x}{\Delta c}\right) \qquad (8\text{--}24)$$

where:

D	=	diffusion coefficient,
b	=	bubble rate, in these experiments 1,
a	=	area per molecule (Eq. 8–23),
Δx	=	distance traveled, about 100 Å, and
Δc	=	concentration change as the surfactant moves, estimated to be 0.2×10^{-4} M.

As an example of these calculations, the values determined for POE (6) decyl alcohol are shown below. A plot of surface tension versus the log of the molar concentration was made and the slope of the tangent line near but just below the CMC was calculated. This slope was –19.6. The Gibbs excess concentration, Γ_2, from Equation (8–22) is then:

$$\Gamma_2 = - \cfrac{1}{2.303 * 8.31\,x\,10^7 * 298}(-19.6)$$

$$= 3.4\,x\,10^{-10}\ moles/cm^2 \qquad\qquad (8\text{--}25)$$

The literature [90] gives a value of Γ_2 for this material as 3.0 x 10^{-10} moles/cm², so reasonable agreement was attained. The area per molecule, Equation (8–23), can then be calculated:

$$a_2 = \cfrac{10^{16}}{6.023\,x\,10^{23} * 3.0\,x\,10^{-10}} \qquad\qquad (8\text{--}26)$$

$$= 48.3\ \text{Å}^2$$

with good agreement with the literature value of 55 Å² [86]. The diffusion coefficient D, was calculated to be 8.6 x 10^{-9} cm/sec by Equation (8–24). This number is a slower rate than reported for POE (6) dodecyl alcohol [91] as measured by the Draves procedure [88], which was 7.8 x 10^{-7} cm/sec. The difference may reflect the concentration and dissociation of the micelles into the individual molecules, since that would be the rate controlling step in surfactant diffusion.

8.4.4 Fiber, Yarn, and Textile Wetting

There are two distinct processes involved in the wetting of fabric and yarn by finishes, or by other dispersions, emulsions, or solutions. The first of these is the initial wetting of the dry yarn bundle at the finish applicator by the emulsion, and the second is the wicking of the finish emulsion along the yarn bundle.

If we consider the first process, the initial wetting, we can assume that it is similar to textile wetting in the Draves test [92]. In this test a 5-gram skein of cotton yarn is attached to a 3-gram hook and totally immersed in a tall cylinder of a surfactant solution or another liquid. The surfactant solution displaces the air in the skein by a spreading wetting process and when sufficient air has been displaced, the skein will sink. The better the wetting agent, the shorter will be the time for sinking to occur. A similar process will take place during the spinning of a synthetic fiber. The finish emulsion will displace the air surrounding the yarn and attempt to penetrate the moving threadline. Since in fiber producer finishes that are emulsions, and the surfactants are present in concentrations well above the CMC, the Gibbs adsorption equation will not hold. As the finish adsorbs onto the yarn, the emulsion containing the surfactant is adsorbed onto the yarn. Fowkes reported [93] that the rate of wetting was determined not by the bulk phase concentration of the surfactant, but by the rate of diffusion of the surfactant to the wetting front. In this case the concentration of surfactant at the wetting front was depleted by adsorption on the substrate and consequentially the wetting time was dependant solely by the rate at which the surfactant arrives at the front. For nonionic surfactants used above their CMC and present as micelles, the rate controlling step is the dissociation of the micelles into their monomolecular units.

In the second process, the wicking of finish along the yarn bundle, a considerable amount of data is available. If we consider yarn to be a group of parallel rods with the rods touching at their perimeters, the spaces between the rods act as capillaries and the finish can proceed to wick as in a normal capillary rise experiment (see Chapter 3, Section 3.2.4). Of course, if the yarn has a modified cross-section, the capillaries can assume very irregular shapes. Figure 8–8 is a graphical example of capillary wicking of a surfactant solution/emulsion.

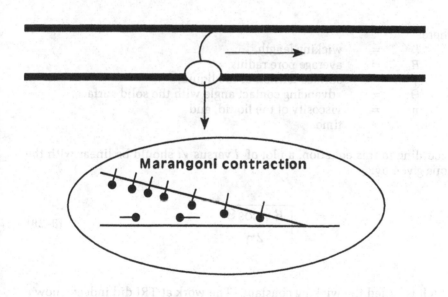

Figure 8–8. Diagram of the Marangoni Contraction of a Liquid Meniscus.
(Ref. [94]. Reprinted by permission of the Textile Research Journal)

When the meniscus moves within a pore under capillary pressure, surfactant molecules can be adsorbed on the solid surface, resulting in fewer surfactant molecules at the leading edge and thus a higher surface tension will result. This effect is called the Marangoni contraction. The diffusion rate of a surfactant is very important in the wetting of yarn and fabric; however, occasionally one might want greater adsorption and higher surface tension at the leading edge of the meniscus.

Experiments at the Textile Research Institute [94] showed that increasing the amount of fluorocarbon wetting agents in an aqueous solution reduced the surface tension but significantly retarded the wicking rate of that solution on a horizontal section of yarn. Although the data may appear to be unusual, that is, reduction of surface tension slows wicking rates, this agrees with the Washburn equation [95]. This equation states:

$$l = \sqrt{\frac{R}{2}} \sqrt{\frac{\gamma \cos \Theta}{\eta}} \sqrt{t} \qquad (8\text{--}27)$$

where:

l	=	wicking length,
R	=	average pore radius,
γ	=	surface tension of the liquid,
Θ	=	advancing contact angle with the solid surface,
η	=	viscosity of the liquid, and
t	=	time.

According to this equation, a plot of l versus $\sqrt{t}$ should be linear with the slope given by:

$$\sqrt{\frac{R \gamma \cos \Theta}{2 \eta}} \qquad (8\text{--}28)$$

which is called the wicking constant. The work at TRI did indeed show a straight line relationship.

Surfactants with a centrally located hydrophilic group are especially good textile wetting agents, presumably because of their rapid diffusion to and orientation at the wetting front. The excellent wetting properties of dioctyl sodium sulfosuccinate are probably due to this mechanism.

For nonionic polyoxyethylenenated alcohols, wetting times go through a minimum with an increase in the number of oxyethylene units in the polyoxyethylene chain and optimum wetting power is generally shown by those surfactants cloud points are just above the temperature at which they are used.   In distilled water at 25°C, materials having an effective hydrophobic chain length of 10–11 carbon atoms and a polyoxyethylene chain of 6–8 EO units appear to be the best wetting agents [96]. Ethoxylated alcohols also are considered better wetting agents than the corresponding ethoxylated fatty acids [97].

8.4.5 Commercially Available Wetting Agents

Several surfactants, including anionic, nonionic, and fluorochemical, are available from various suppliers and are commonly used as wetting agents. These are detailed in the following sections.

Dioctyl Sodium Sulfosuccinate, CAS# 577–11–7

This anionic surfactant, generally known by the acronym DOSS, is actually di(2–ethylhexyl) sulfosuccinic acid, sodium salt. It is one of the most frequently used wetting agents in the textile industry, and is very effective because of the centrally located hydrophilic group and the long hydrophobic tails.   This unique structure is also responsible for the formation if an inverse micelle [98], that is, one in which the hydrophilic end is in the interior of the micelle and the hydrophobic tail on the exterior of the agglomerate.   The structure of di(2–ethylhexyl) sulfosuccinic acid, sodium salt is:

$$
\begin{array}{ccc}
 & O & C_2H_5 \\
 & \| & | \\
H_2{-}C{-}C{-}O{-}CH_2CH(CH_2)_3CH_3 \\
 & | & \\
Na^+\ {}^-O_3S{-}C{-}C{-}O{-}CH_2CH(CH_2)_3CH_3 \\
 & \| & | \\
 & O & C_2H_5
\end{array}
$$

It is available from American Cyanamide with the trade name Aerosol® OT [99] and from Witco as Emcol® 4500 [100]. Typical properties are listed in Table 8–15.

Table 8–15

Typical Properties of Dioctyl Sodium Sulfosuccinate

Property	Value
Acid Number	2.5
Iodine Number	0.3
Color, APHA of a 50% Aqueous Solution	35
Solids, wt%	69–76

POE (4) and POE (6) Decyl Alcohol, CAS# 26183–52–8

These short EO chain components are dispersible in water and mineral oil, and are superior wetting agents at concentrations of 0.1% or less. The POE (4) ether is sold by Ethox under the designation Ethal® DA–4 [101][102] and by Henkel with a trade name Trycol® 5950 [103][104]. POE (6) decyl alcohol, widely used as a wetting agent in textile fabric dyeing systems, is sold by Ethox as Ethal® DA–6 [100][105], by Henkel as Trycol® 5952 [102][106], and Stepan as Stepantex® DA–6 [107]. Some properties of these compounds are presented in Table 8–16.

Table 8–16

Typical Properties of POE (4) and POE (6) Decyl Alcohol

Property	Value	
	POE (4)	POE (6)
Acid Value	1 max	1 max
Hydroxyl Value	160–175	120–135
pH, 5% in Distilled Water	6–8	5–8
Moisture, %	1 max	3 max
Draves Wetting, Seconds (0.1% @ 25°C)	3	6

The chemical structure of decyl alcohol ethoxylates is:

$$CH_3(CH_2)_8CH_2O(CH_2CH_2O)_xH$$

POE (2) and POE (5) 2–Ethylhexanol, CAS# 26468–86–0

2–Ethylhexanol is manufactured by a commercial oxo process in which propylene is reacted with carbon monoxide and hydrogen in contact with a cobalt salt at 150°C and 200 atmospheres pressure [108]. The initial step is the formation of *n*–butraldehyde and *i*–butraldehyde in a ratio of 3:2.

$$CH_3CH{=}CH_2 + CO + H_2 \xrightarrow[\text{200 atm}]{\text{150° C}} CH_3CH_2CH_2CHO$$

and

$$\underset{\underset{\displaystyle CHO}{|}}{CH_3CHCH_3}$$

2–Ethylhexanol is formed when there is probably an aldol addition of two moles of *n*–butraldehyde, dehydration, and hydrogenation.

$$2\ CH_3CH_2CH_2CHO \longrightarrow \underset{\underset{\displaystyle C_2H_5}{|}}{CH_3CH_2CH_2CHOHCHCHO} \longrightarrow$$

$$\underset{\underset{\displaystyle C_2H_5}{|}}{CH_3CH_2CH_2CH{=}C{-}CHO} \longrightarrow \underset{\underset{\displaystyle C_2H_5}{|}}{CH_3CH_2CH_2CH_2CHCH_2OH}$$

Addition of two or five moles of ethylene oxide to the alcohol form surfactants that are low foaming wetting agents and an intermediate in the production of dioctyl sodium sulfosuccinate. These surfactants are made by Ethox with the name Ethal® EH–2 [100][109] and Ethal® EH–5 [100][110]. Some typical properties of these wetting agents are presented in Table 8–17.

Table 8–17

Typical Properties of POE (2) and POE (5) 2–Ethylhexanol

Property	Value	
	POE (2)	POE (5)
Acid Value	1 max	1 max
Hydroxyl Value	240–285	145–170
pH, 5% in Distilled Water	5.5–7.5	5.0–6.5
Moisture, %	0.5 max	3 max
Draves Wetting, Seconds (0.1% @ 25°C)	—	6

POE (9) Nonylphenol, CAS# 9016–45–9

The nine mole ethylene oxide alkylphenol is an effective wetting agent, emulsifier, and general purpose detergent. It is produced by Ethox with the trade name Ethal® NP–9 [100][111], by Henkel where it is called Trycol® 6964 [112][113], and also by Witco as Witconol® NP–90 [114]. Classical properties are displayed in Table 8–18.

Table 8–18

Typical Properties of POE (9) Nonylphenol

Property	Value
pH, 5%	5.5–8
Color, APHA	100 max
Acid Value	1.0 max
Hydroxyl Number	85–95
Viscosity, cSt, 100°F	112
Cloud Point, 1%, °C	52–56
Draves Wetting, Seconds (0.1% @ 25°C)	9

The structure is:

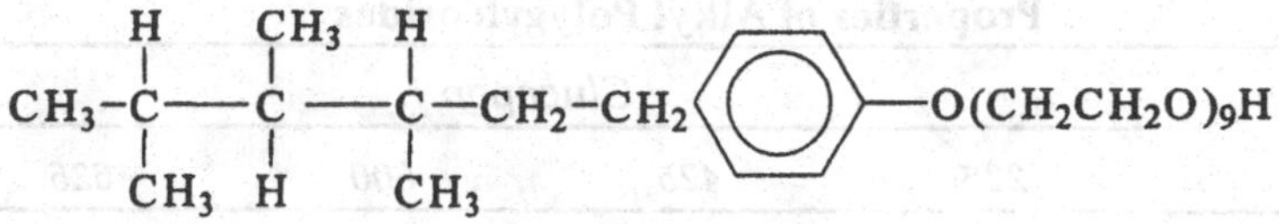

H   CH₃   H
│   │   │
CH₃-C——C ——C-CH₂-CH₂⟨○⟩-O(CH₂CH₂O)₉H
│   │   │
CH₃   H   CH₃

Alkyl Polyglycosides

Alkyl polyglycosides are new classes of nonionic surfactants based on glucose and fatty alcohols. The structure is:

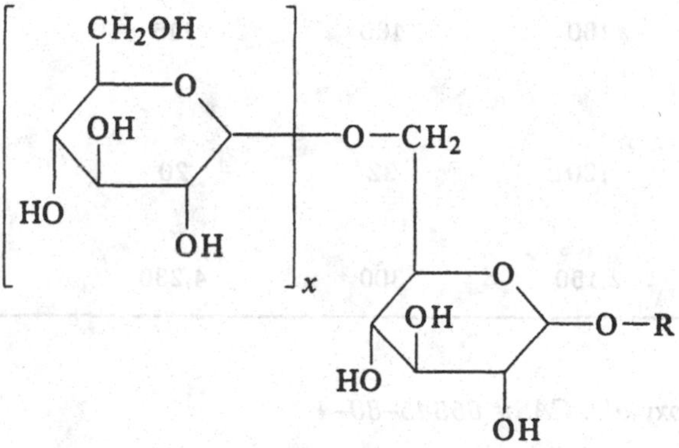

In this structure x is predominantly zero to three carbohydrate units and R is an alkyl chain between 8 and 16 carbon atoms. All of the raw materials are based on renewable sources, glucose from corn and the fatty alcohols from coconut and palm kernel oils. They are excellent detergents and wetting agents, exhibiting good surface tension/interfacial tension reduction, and also exhibit mildness to the skin and eyes. Although these compounds were initially suggested for formulation into cleaning products, they are also useful in textile processing. Information about these products is listed in Table 8–19. They are made by Henkel as Glucopon® surfactants [115][116].

Table 8–19

Properties of Alkyl Polygylcosides

	Glucopon			
	225	425	600	625
CAS#	68515–73–1	68515–73–1	110615–47–9	110615–47–9
Alkyl Chains Present	8,10	8,10,12,14,16	12,14,16	12,14,16
Average Alkyl Chain	9.1	10.3	12.8	12.8
% Active	70	50	50	50
pH, 10% in Water	6–9	7–9.5	11.5–12.5	11.5–12.5
Ross Miles Foam Height, mm, 0.1%, 49°C	160	160	115	150
Draves Wetting, sec, 0.1% @ 25°C	120	32	20	23
Viscosity, cPs@ 35°C	2,150	300	4,230	6,250

Fluorinated Ethoxylate, CAS# 65545–80–4

Fluorinated alkyl ethoxylates are one type of the class of highly effective wetting agents based on fluorochemical surfactant technology. This chemical, called Fluowet® OT [117], was developed in Germany by Hoechst and is valuable in reducing the surface tension in almost all water–based systems. It is recommended for use in fiber and textile manufacturing processes. The chemical structure of this surfactant is approximately:

$$C_2F_5(CF_2)_nCH_2CH_2\text{—}O\text{—}(CH_2CH_2O)_xH$$

Surface tensions of solutions of Fluowet® OT in water are recorded in Table 8–20 and other properties in Table 8–21.

Table 8–20

Surface Tension of Aqueous Solutions of Fluowet® OT

Surface Tension, 20°C mN/m	Concentration, weight percent					
	0	0.003	0.01	0.03	0.1	0.3
	72	24	23	21	21	20

Table 8–21

Typical Properties of Fluowet® OT

Property	Value
pH, 1% Aqueous	5–8
Acid Value	1.5 max
Hydroxyl Number	100–120
Activity, %	95 min

Ammonium Perfluoroalkyl Sulfonate, CAS# 67906–42–7

An anionic fluorochemical surfactant, this chemical is supplied as a 25% solution in an equal mixture of 2–butoxy ethanol and water. Although the C_{10} fluoroalkyl chain is the predominate alkyl moiety present, the surfactant does contain a small amount of C_9 ammonium perfluoroalkyl sulfonate, CAS# 17202–41–4. It is especially recommended for use in aqueous coating systems, particularly clear polishes based on alkali soluble polymers. This material is sparingly soluble in water, but its effectiveness in the reduction of surface tension allows it to be useful in textile applications. The chemical is produced by 3M as Fluorad® 120 [118]–[120]. The general structure is:

$$C_{10}F_{21}SO_3^- \ NH_4^+$$

Surface tension properties are shown in Table 8–22 and other properties are listed in Table 8–23.

Table 8–22

Surface Tension of Aqueous Solutions of Fluorad® FC–120

Surface Tension, 20°C mN/m	Concentration, weight percent			
	0	0.004	0.04	0.4
	72	50	26	20

Table 8–23

Typical Properties of Fluorad® FC–120

Property	Value
pH, as is	8.5–9.5
Activity, %	25
Viscosity, cPs, 25°C	10
Specific Gravity	1

Potassium Fluoroalkyl Carboxylate, CAS# 2991–51–7

This chemical is another anionic surfactant that contains about 40–45% of a C_8 fluoroalkyl chain (CAS# 2991–51–7), 1–3% of a C_7 chain (CAS# 67584–62–4), 1–5% C_6 (CAS# 67584–53–6), 1–5% C_4 (CAS# 67584–51–4) and lesser amounts of other chains. It is supplied by 3M as Fluorad® FC–129 [118][119][121] in a 50% active solution of water, 2–butoxyethanol, and ethanol. Normally suggested for floor polish and alkaline cleaners, it is also effective in other water–based systems. Typical properties are displayed in Table 8–24 and the effects on surface tension in Table 8–25.

Table 8–24

Typical Properties of Fluorad® FC–129

Property	Value
pH, 0.8% Aqueous	8–11
Activity, %	50
Viscosity, cPs, 25°C	30
Specific Gravity	1.3

Table 8–25

Surface Tension of Aqueous Solutions of Fluorad® FC–129

Surface Tension, 20°C mN/m	Concentration, weight percent			
	0	0.002	0.02	0.2
	72	49	23	17

The general structure is:

$$
\begin{array}{ccc}
 & C_2H_5 & O \\
 & | & || \\
C_8F_{17}SO_2\text{-N} & \text{-CH}_2\text{-C} & \text{-O}^-\ K^+
\end{array}
$$

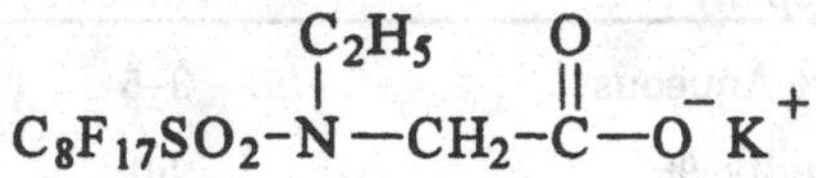

Fluoroalkyl Quaternary Ammonium Iodide, CAS# 1652–63–7

In contrast with the two preceding fluorochemical surfactants, this compound is a cationic material dissolved in a 2:1 mixture of isopropanol: water. The distribution of alkyl chains is similar to the other products. It is substantive to a variety of metal and silicone surfaces and is used in polishing and cleaning products. Since low concentrations are quite effective in reducing surface tensions, it can be used in many water–based processes. Another end use in which this product is very valuable is in addition to herbicide dispersions or solutions to enhance the wetting of weed leaves [3]. Also manufactured by 3M, it has a trade name of Fluorad® FC–135 [118] [119][122]. The general structure is:

$$
\begin{array}{c}
H \\
| \\
C_8F_{17}SO_2\text{-N}\text{-CH}_2CH_2CH_2N^+(CH_3)_3I^-
\end{array}
$$

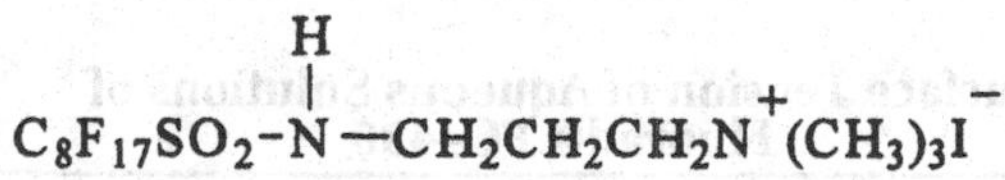

Surface tension effects are shown in Table 8–26 and the other typical properties in Table 8–27.

Table 8–26

**Surface Tension of Aqueous Solutions of
Fluorad® FC–135**

Surface	Concentration, weight percent			
Tension, 20°C	0	0.002	0.02	0.2
mN/m	72	28	18	17

Table 8–27

Typical Properties of Fluorad® FC–135

Property	Value
pH, 1 % Aqueous	3–5
Activity, %	50
Viscosity, cPs, 25°C	30
Specific Gravity	1.2

Fluoroaliphatic Polymeric Esters, CAS# Proprietary

This nonionic fluorosurfactant is suggested as a wetting, leveling, and flow control agent for both water– and solvent–based coating systems. It is reported to eliminate surface defects in coatings, such as cratering and crawling. Promotion of pigment wetting and dispersion often results in reduced grinding times and better color development. In cases in which systems have been contaminated with oils or silicones they can be reclaimed with the use of the product. It is made by 3M as Fluorad® 430 [118][119][123]. The surface tension reduction is shown in Table 8–28 and other properties in Table 8–29.

Table 8–28

**Surface Tension of Aqueous Solutions of
Fluorad® FC–430**

Surface	Concentration, weight percent			
Tension, 20°C	0	0.1	0.2	0.5
mN/m	72	30	28	28

Table 8–29

Typical Properties of Fluorad® FC–430

Property	Value
pH, 1 % Aqueous	6
Activity, %	100
Viscosity, cPs, 25°C	7000
Specific Gravity	1.1

REFERENCES

1. Dexter, M. (1985). Antioxidants in *Encyclopedia of Polymer Science and Engineering* (J. I. Kroschwitz, ed.), John Wiley & Sons, New York, pp. 73–91

2. Streitwieser, A. and Heathcock, C. H. (1985). *Introduction to Organic Chemistry*, Macmillian, New York, p. 510

3. Slade, P. E., unpublished data

4. Kamiya, Y., Beaton, S., Lafortune, A., and Ingold, K. W. (1963). The Metal-Catalyzed Autoxidation of Tetralin. I. The Cobalt–Catalyzed Autoxidation in Acetic Acid, *Can. J. Chem.*, **41**: 2020

5. Kamiya, Y., Beaton, S., Lafortune, A., and Ingold, K. U. (1964), The Metal-Catalyzed Autoxidation of Tetralin. II. The Cobalt–Catalyzed Autoxidation of Undiluted Tetralin and of Tetralin in Chlorobenzene, *Can. J. Chem.* **41**: 2034

6. Kamiya, Y. and Ingold, K. U. (1964). The Metal–Catalyzed Autoxidation of Tetralin. III. Catalysis by Manganese, Copper, Nickel, and Iron, *Can. J. Chem.*, **42**: 1027

7. Kamiya, Y., and Ingold, K. U. (1964). The Metal–Catalyzed Autoxidation of Tetralin. IV. The Effect of Solvent and Temperature, *Can. J. Chem.*, **42**: 2424

8. Guillet, J. E., Dhanra, J., Golemba, F. J., and Hartley, G. H. (1968). Fundamental Processes in the Photodegradation of Polymers, *Stabilization of Polymers and Polymer Processes, Advances in Chemistry Series No. 85* (R. F. Gould, ed), American Chemical Society, Washington, D. C., pp. 272–286

9. Miller, D. E. and Dean, P. R., II (1988). *Antioxidants and Antiozonants for Rubber and Rubber–Like Products*, Goodyear Chemicals, Akron, OH

10. Ainsworth, S. J. (1992). Plastic Additives, *Chem. and Eng. News*, **70**: August 31, p. 34

11. Nicholas, P. P., Luxeder, A. M., Brooks, L. A., and Hammes, P. A. (1978). Antioxidants and Antiozonants, in *Kirk–Othmer Encyclopedia of Chemical Technology, Third Edition, Volume 3*, John Wiley & Sons, New York, pp. 128–148

12. Great Lakes Chemical Corporation (1995). *The Great Lakes Polymer Stabilizers Collection*, Polymer Stabilizers Business Unit, Great Lakes Chemical Corporation, West Lafayette, IN

13. Goodyear Chemicals (1995). *Wingstay® SN-1*, Goodyear Chemicals, Akron, OH

14. Cytec (1995). *Polymer Additives*, Cytec Industries, Inc., Stamford, CT

15. Ciba–Geigy (1994). *Additives for Lubricants*, Additives Division, Ciba–Geigy Corporation, Hawthorne, NY

16. Flexsys (1995). *Product Guide*, Flexsys America L. P., Akron, OH

17. Hamblin, P. C., Kristen, U., and Ardsley, D. (1959). Ashless Antioxidants and Corrosion Inhibitors: Their Use in Lubricating Oils, *Lubrication Science*, 2: 287

18. Shelton, J. R. (1972). Stabilization Against Thermal Oxidation, in *Polymer Stabilization*, (W. L. Hawkins, ed.), Wiley–Interscience, New York, pp. 29–116

19. Kerr, J. A. (1966). Bond Dissociation Energies by Kinetic Methods, *Chem. Rev.* **66:** 465

20. Schnabel, W. (1981). *Polymer Degradation*, Carl Hanser, München, GDR, pp. 195–199

21. Goodyear Chemicals (1995). *Wingstay® T*, Goodyear Chemicals, Akron, OH

22. Flexsys (1995). *Antioxidants for Specialty Chemicals Applications*, Flexsys America L. P., Akron, OH

23. Goodyear Chemicals (1995). *Wingstay® L HLS*, Goodyear Chemicals, Akron, OH

24. Goodyear Chemicals (1995). *Wingstay® 29*, Goodyear Chemicals, Akron, OH

25. Goodyear Chemicals (1995). *Wingstay® 100*, Goodyear Chemicals, Akron, OH

26. Rysavý, D. (1970). Oxidative Thermal Stability of Isotactic Polypropylene as a Function of the Composition of the Stabilizer Mixture and the Temperature of Observation, *Kunststoffe*, Munich, Germany, 60: 118

27. Tsuji, K. and Seiki, T. (1970). Absorption Spectra Due to Charge Transfer Complexes of Polymers with Oxygen and their Possible Contribution to Radical Formation in Polymers by Ultraviolet Radiation, *J. Poly. Sci.*, *Part B*, **8:** 817

28. Guillet, J. E., Dhanraj, J., Golemba, F. J., and Hartley, G. H. (1968), Fundamental Processes in the Photodegradation of Polymers, in *Stabilization of Polymers and Stabilizer Processes, Advances in Chemistry Series*, No. 85, (R. F. Gould, ed.), American Chemical Society, Washington, pp. 272–286

29. Morand, J. L. (1972). Oxidation of Photoexcited Elastomers, *Rubber Chem. Technol.*, **45:** 481

30. Jacob, R. A. (1994). Vitamin C, in *Modern Nutrition in Health and Disease, Eighth Edition*, (M. E. Shils, J. A. Olson, and M. Shike, eds.), Lea and Febiger, Philadelphia, pp. 432–448

31. Pitt, G. A. J. (1985). Vitamin A, in *Fat–Soluble Vitamins: Their Biochemistry and Applications*, (A. T. Diplock, ed.), Technomic Publishing Co., Lancaster,

U. K., pp. 1–75

32. Diplock, A. T. (1985). Vitamin E, in *Fat–Soluble Vitamins: Their Biochemistry and Applications*, (A. T. Diplock, ed.), Technomic Publishing Co., Lancaster, U. K., pp. 154–224

33. Thomas, J. A. (1994). Oxidative Stress, Oxidant Defense and Dietary Constituents, in *Modern Nutrition in Health and Disease, Eighth Edition*, (M. E. Shils, J. A. Olson and M. Shike, eds.), Lea and Febiger, Philadelphia, pp. 501–512

34. Kane, J. P. (1988). The Judicious Diet, in *Cecil Textbook of Medicine*, (J. B. Wyngaarden and L. H. Smith, eds.), W. B. Saunders, Philadelphia, pp. 42–45

35. Olsen, J. A. (1994). Vitamin A, Retinoids and Carotenoids, in *Modern Nutrition in Health and Disease, Eighth Edition*, (M. E. Shils, J. A. Olson, and M. Shike, eds.), Lea and Febiger, Philadelphia, pp. 287–307

36. Farrell, P. M. and Roberts, R. J. (1994). Vitamin E, in *Modern Nutrition in Health and Disease, Eighth Edition*, (M. E. Shils, J. A. Olson, and M. Shike, eds.), Lea and Febiger, Philadelphia, pp. 326–341

37. Bikerman, J. J. (1973). *Foams*, Springer–Verlag, New York, pp. 1–2

38. Rosen, M. J. (1989). *Surfactants and Interfacial Phenomena, Second Edition*, John Wiley & Sons, New York, pp. 277, 278–282, 294, 250–252, 65

39. Myers, D. (1991). *Surfaces, Interfaces, and Colloids*, VCH, New York, pp. 251–270, 139–140, 358, 360–362

40. ASTM Method D1173–53 (1991). Standard Test Method Foaming Properties of Surface–Active Agents, *Annual Book of ASTM Standards, Volume 15.04*, Philadelphia, pp. 108–109

41. Gantz, G. M. (1966). Foaming in *Nonionic Surfactants* (M. J. Schick, ed.), Marcel Dekker, Inc., New York, pp. 733–752

42. Okazaki, S. and Sasaki, T. (1960). Two Types of Antifoamers and their Cooperating Action, *Bull. Chem. Soc. Japan*, Tokyo, Japan, **33**: 564

43. BASF (1989). *Pluronic® and Tetronic® Surfactants*, BASF Corporation, Mount Olive, NJ

44. Ethox Product Brochure (1994). *Block Copolymers*, Ethox Chemicals, Inc., Greenville, SC

45. Hoechst–Celanese Product Data Sheet (1993). *Afilan® 4PF*, Hoechst–Celanese Specialty Chemicals Group, Charlotte, NC

46. Henkel Corporation (1993). *Textile Chemicals, Technical Bulletin 103A, Miscellaneous Chemicals*, Henkel Corporation, Textile Chemicals, Charlotte, NC

47. Dow Corning Corporation (1993). *Dow Corning® Silicone Foam Control Agents*, Dow Corning Corporation, Midland, MI

48. Dow Corning Corporation (1990). *Information About Silicone Antifoams, Dow Corning® Antifoam B Emulsion*, Dow Corning Corporation, Midland, MI

49. Dow Corning Corporation (1990). *Information About Silicone Antifoams, Dow Corning® Antifoam H–10 Emulsion*, Dow Corning Corporation, Midland, MI

50. Dow Corning Corporation (1990). *Information About Silicone Antifoams, Dow Corning® Antifoam Y–30 Emulsion*, Dow Corning Corporation, Midland, MI

51. Dow Corning Corporation (1990). *Information About Silicone Antifoams, Dow Corning® Antifoam 1410*, Dow Corning Corporation, Midland, MI

52. Dow Corning Corporation (1990). *Information About Silicone Antifoams, Dow Corning® Antifoam 1430*, Dow Corning Corporation, Midland, MI

53. Dow Corning Corporation (1990). *Information About Silicone Antifoams, Dow Corning® Antifoam 2210*, Dow Corning Corporation, Midland, MI

54. Dow Corning Corporation (1990). *Information About Silicone Antifoams, Dow Corning® 544 Antifoam Compound*, Dow Corning Corporation, Midland. MI

55. Dow Corning Corporation (1992). *Information About Silicone Antifoams, Dow Corning® Q2–3183A Antifoam*, Dow Corning Corporation, Midland, MI

56. Witco Corporation (1996). *Bubble Breaker® Foam Control Agents*, Witco Corporation, Greenwich, CT

57. Henkel Textile Technology (1995). *Foamaster® SD*, Henkel Corporation, Textile Technology, Charlotte, NC

58. Henkel Textile Technology (1995). *Foamaster® 340*, Henkel Corporation, Textile Technology, Charlotte, NC

59. Henkel Textile Technology (1995). *Foamaster® DCD–2*, Henkel Corporation, Textile Technology, Charlotte, NC

60. Tincher, W., Personal Communication, School of Textile & Fiber Engineering, Georgia Institute of Technology, Atlanta, GA

61. Givaudan–Roure (1996). *Giv–Gard DXN® Technical Literature*, Givaudan–Roure Corporation, Clifton, NJ

62. Givaudan–Roure (1996). *Giv–Gard DXN® Specifications/Procedures*, Givaudan–Roure Corporation, Clifton, NJ

63. Angus Chemical (1995). *Guide to Our Products*, Angus Chemical Company, Buffalo Grove, IL

64. Angus Chemical Technical Data Sheet TDS 29 (1991). *Bronopol®–Cosmetic Grade*, Angus Chemical Company, Buffalo Grove, IL

65. Angus Chemical Technical Data Sheet TDS 29A (1990). *Bioban® BNPD Preservatives*, Angus Chemical Company, Buffalo Grove, IL

66. Angus Chemical Technical Data Sheet TDS 39 (1994). *Canguard® 409*, Angus Chemical Company, Buffalo Grove, IL

67. Angus Chemical Technical Data Sheet TDS 37 (1992). *Canguard® 327 and Canguard® 454*, Angus Chemical Company, Buffalo Grove, IL

68. Olin Chemicals Biocides Application Data (1996). *Triadine® 3 Bactericide*, Olin Corporation, Norwalk, CT

69. R. T. Vanderbilt Technical Data (1991). *Vancide® 51*, R. T. Vanderbilt Company, Inc. Norwalk, CT

70. R. T. Vanderbilt Technical Data (1995). *Vancide® MZ–96*, R. T. Vanderbilt Company, Inc., Norwalk, CT

71. R. T. Vanderbilt Technical Data (1993). *Vancide® 51Z Dispersion*, R. T. Vanderbilt Company, Inc., Norwalk, CT

72. Olin Chemicals Biocides Product Information (1996). *Sodium Omadine®*, Olin Corporation, Norwalk, CT

73. Olin Chemicals Biocides Product Information (1996). *Zinc Omadine®
    Bactericide–Fungicide*, Olin Corporation, Norwalk, CT

74. R. T. Vanderbilt Technical Data (1991). *Vancide® 89*, R. T. Vanderbilt
    Company, Inc., Norwalk, CT

75. R. T. Vanderbilt Technical Bulletin (1991). *Vancide® 89RE*, R. T. Vanderbilt
    Company, Inc, Norwalk, CT

76. Keesom, W. H. (1921). Van der Waals Attractive Forces, *Physik. Z.*, Leipzig,
    Germany, **22**: 129

77. Debye, P. J. W. (1920). Van der Waals Cohesive Forces, *Physik. Z.*, Leipzig,
    Germany, **21**: 178

78. London, F. (1937). The General Theory of Molecular Forces, *Trans. Faraday
    Soc.*, **33**: 8

79. Good, R. L. (1967). Intermolecular and Interatomic Forces in *Treatise on
    Adhesion and Adhesives*, (R. L. Patrick, ed.), Marcel Dekker, Inc., New York,
    pp. 28

80. Zisman, W. A. and Ellison, A. H. (1954). Wettability Studies of Nylon,
    Poly(ethylene terephthalate), and Polystyrene, *J. Phys. Chem.*, **58**: 503

81. Zisman, W. A. (1963). Influence of Constitution on Adhesion, *Ind. Eng.
    Chem.*, **55**: 19

82. Zisman, W. A. (1964). Relation of Equilibrium Contact Angle to Liquid and
    Solid Constitution, *Contact Angle, Wettability, and Adhesion*, ACS Advances
    in Chemistry Series, Vol. 43, American Chemical Society, Washington, pp.
    1–51

83. Laplace, P. S. (1966). On Capillary Attraction, *Celestial Mechanics*,(English
    Translation by N. Bowditch in 1821), Vol. 4, Supplement to the Tenth Book
    of the *Méchanique Céleste*, Chelsea Publishing Co., Bronx, NY, pp. 685–1018

84. Young, T. (1855). *Miscellaneous Works*, (G. Peacock, ed.), Murry, London,
    England, p. 418

85. Dupré, A. (1869) *Théorie Méchanique de la Chaleur*, Gauthier–Villars, Paris,
    France, p. 369

86. Becher, P. (1966). Micelle Formation in Aqueous and Nonaqueous Solutions,
    Nonionic Surfactants, (M. J. Schick, ed.), Marcel Dekker, Inc., New York, pp.
    478–515

87. Gibbs, J. W. (1928). *The Collected Works of J. Willard Gibbs, Vol 1*. Longmans,
    Green and Company, London, England, p. 119

88. Hirt, D. E., Prud'homme, R. K., Miller, B., and Rebenfeld, L. (1990). Dynamic
    Surface Tension of Hydrocarbon and Fluorocarbon Surfactants Using the
    Maximum Bubble Pressure Method, *Colloids Surfaces*, **44**: 101

89. Kamath, Y. K., Research Director, Textile Research Institute, personal
    communication

90. Carless, J. E. , Challis, R. A., and Mulley, B. A. (1964). Nonionic Surface
    Active Agents. V. The Effect of the Alkyl and the Polyglycol Chain Length on
    the Critical Micelle Concentration of Some Monoalkyl Polyethers, *J. Colloid
    Sci.*, **19**: 201

91. Cohen, A. W. and Rosen, M. J. (1981). Wetting Properties of Nonionic Surfactants of Homogeneous Structure $C_{12}H_{25}(OC_2H_4)_xOH$, *J. Amer. Oil Chemists' Soc.*, **58**: 1062

92. Draves, C. W. (1939). Evaluation of Wetting Agents– Official Method, *Am. Dyestuff Reptr.*, **28**: 425

93. Fowkes, F. M. (1953). Role of Surface Active Agents in Wetting, *J. Phys, Chem.*, **57**: 98

94. Kamath, Y. K., Hornby, S. B., Weigmann, H. –D., and Wilde, M. F. (1994). Wicking of Spin Finishes and Related Liquids into Continuous Filament Yarn, *Textile Res. J.*, **64**: 33

95. Washburn, E. W. (1921). The Dynamics of Capillary Flow, *Phys. Rev,*, **17**: 273

96. Crook, E. H., Fordyce, D. B., and Trebbi, G. F. (1964). Molecular Weight Distribution of Nonionic Surfactants. III. Foam, Wetting, Detergency, Emulsification, and Solubility Properties of Normal Distribution and Homogeneous p–tert–Octyl–phenoxyethanols, *J. Am. Oil Chem. Soc.*, **41**: 231

97. Wrigley, A. N., Smith, F. D., and Stirton, A. J. (1957). Synthetic Detergents from Animal Fats. VIII. The Ethoxylation of Fatty Acids and Alcohols, *J. Am. Oil Chem. Soc.*, **34**: 39

98. Heitz, M. P. and Bright, F. V. (1996). Rotational Reorientation Dynamics of Aerosol–OT Reverse Micelles Formed in Near Critical Propane, *Appl. Spec.*, **50**: 732

99. Cyanamid MSDS (1992). *Aerosol® OT 75% Surfactant*, American Cyanamid Company, Wayne, NJ

100. Witco Fiber Production Auxiliaries (1944). *Desizing Agents*, Witco Corporation, Greenwich, CT

101. Ethox Product Brochure (1994). *Ethoxylated Alcohols and Alkyl Phenols*, Ethox Chemicals, Inc., Greenville, SC

102. Ethox Specifications (1993). *Ethal® DA–4*, Ethox Chemicals, Inc., Greenville, SC

103. Textile Chemicals, Technical Bulletin 103A (1993). *Ethoxylated and Ethoxylated / Propoxylated Alcohols*, Henkel Corporation, Textile Chemicals, Charlotte, NC

104. Henkel Specifications (1989). *Trycol® 5950*, Henkel Corporation, Textile Chemicals, Charlotte, NC

105. Ethox Specifications (1993). *Ethal® DA–6*, Ethox Chemicals, Inc., Greenville, SC

106. Henkel Specifications (1989). *Trycol® 5952*, Henkel Corporation, Textile Chemicals, Charlotte, NC

107. Stepan Company (1995). Products for the Textile and Fiber Industries, Catalog on Computer Disk, *Stepantex® DA–6*, Stepan Company, Northfield, IL

108. Noller, C. R. (1958). *Textbook of Organic Chemistry*, W. B. Saunders Co., Philadelphia, pp 172–173

109. Ethox Specifications (1993). *Ethal® EH–2*, Ethox Chemicals, Inc., Greenville, SC

110. Ethox Specifications (1993). *Ethal® EH–5*, Ethox Chemicals, Inc., Greenville, SC

111. Ethox Specifications (1986). *Ethal® NP–9*, Ethox Chemicals, Inc., Greenville, SC

112. Textile Chemicals, Technical Bulletin 103A (1993). *Ethoxylated Alkylphenols*, Henkel Corporation, Textile Chemicals, Charlotte, NC

113. Henkel Specifications (1989). *Trycol® 6964*, Henkel Corporation, Textile Chemicals, Charlotte, NC

114. Witco Fiber Production Auxiliaries (1994). *Oil Emulsifiers*, Witco Corporation, Greenwich, CT

115. Textile Chemicals, Technical Bulletin 103A (1993). *Alkyl Polyglycosides*, Henkel Corporation, Textile Chemicals, Charlotte, NC

116. Henkel Technical Data Sheet (1993). *Glucopon® Alkyl Polyglycoside Surfactants*, Henkel Corporation, Textile Chemicals, Charlotte, NC

117. Hoechst Data Sheet E HOE 4611E (1983). *Fluowet® OT Fluorinated Surfactant*, Hoechst Aktiengesellschaft, Frankfurt am Main, Germany

118. 3M (1992). *Presenting the Solutions for your Problems*, 3M, Specialty Chemicals Division, St. Paul, MN

119. 3M (1994). *Fluorad® Fluorochemical Surfactants*, 3M, Specialty Chemicals Division, St. Paul, MN

120. 3M Technical Information (1994). *Fluorad® Fluorochemical Surfactants, FC–120*, 3M, Specialty Chemicals Division, St. Paul, MN

121. 3M Technical Information (1994). *Fluorad® Fluorochemical Surfactants, FC–129*, 3M, Specialty Chemicals Division, St. Paul, MN

122. 3M Technical Information (1994). *Fluorad® Fluorochemical Surfactants, FC–135*, 3M, Specialty Chemicals Division, St. Paul, MN

123. 3M Technical Information (1994). *Fluorad® Coating Additives, FC–430*, 3M, Specialty Chemicals Division, St. Paul, MN

CHAPTER 9

SOIL– AND STAIN–RESISTANT TEXTILE FINISHES

Two distinct processes dramatically affect the appearance of textiles, soiling and staining. In soiling, particulate matter approaches the fiber or yarn and is held there either by physical entrapment or by attraction to the surfaces by some chemical force. When the fabric is a carpet most of the soil is deposited by direct contact with shoe soles on the surface, while additional soiling is caused by atmospheric particulates. In contrast, staining is defined as the dyeing of a fabric, especially carpets, by the massive spill of common household materials, including red wine or solutions of powdered soft drinks. Staining is a more serious problem with fabric made from polyamide fibers, such as wool or nylon, than with polyester or olefin fiber because the dye causing the color is probably acidic and readily attracted to the basic dye sites.

9.1 SOILING AND SOIL–RESISTANT FINISHES

The soiling of textile materials is very familiar to residents of every continent on the earth and has been a nuisance for centuries. Washing of these textiles is a major source of domestic drudgery. Soiled clothing and other household furnishings are unattractive and some functional effectiveness may be decreased [1]. Since a large percentage of the fabrics intended for apparel and home uses were made from hydrophilic cotton fibers for hundreds of years [2], laundering was never considered a significant problem before the advent of wash and wear and durable press finishes. In about 1964, hydrophobic fibers, particularly polyesters, were blended with cotton and the resulting fabrics treated to impart durable press properties. At this time, consumers realized that fabrics with durable press finishes soiled more readily and were more difficult to clean than the products used previously [3]. Compton and Hart [4] suggest that there are three, and only three, ways that a fabric can retain soil. The three binding mechanisms for formation of a soil–fabric complex are:

1. Macro–occlusion, or entrapment in the intra–yarn and inter–yarn spaces.
2. Micro–occlusion, or entrapment of particles in the irregularities of the fiber surfaces.
3. Sorptive bonding of the soil to the fiber by van der Waals' or coulombic forces.

Carpets, in common with other fabrics, soil during the life of that material on floors. Often the effect of carpet soiling, when compared to apparel, is more pronounced since the amount of soil deposited is usually greater and cleaning is less frequent. Carpets are generally soiled by both direct contact and by atmospheric deposition, with the first method the most important [5]. About 80% of the total soil accumulated on carpets consists of sand and dry particulate matter that are usually of a large particle size, about 10μ, and can easily be removed by vacuuming [6].

A further division of carpet soiling can be described as being real or apparent. Real soil is bound to the fibers by the mechanisms proposed by Compton and Hart [4], but additionally it can cause abrasion and wear thus producing a dull appearance. Apparent soil can also affect the appearance of a carpet, making it look dirty and having an undesirable effect. Conditions that contribute to apparent soil include:

1. Dark Areas or Traffic Lane Grey. The carpet appears to be soiled, but it is actually clean. This is often due to a combination of factors such as fiber surface abrasion or distortion due to traffic and wear, which cause light to reflect at different angles rather than uniformly.
2. Shading, Water Marks, Pooling and Nap Reversal. A condition in cut pile only where the pile lays in a different direction than the rest of the carpet, causing light reflection that makes the surface appears soiled or wet.
3. Corn Rowing. This is also a condition that is a property of cut pile carpet in which alternate rows of tufts bend over to fill the density voids in the carpet, thus creating a row like effect at right angles to the traffic pattern.

4. Reflections and Shadows. These conditions are caused by reflections that make the carpet look light and dark due to non–obvious light sources. Examples might be pin holes in draperies, light fixtures, and window coverings that allow concentrated light to focus on a specific area of carpet.

5. Wear. Wear is a reduction in the amount of face fiber in traffic areas, especially as compared to non–traffic areas. A carpet fiber is said to have wear when there is an actual loss of face fiber [7].

A standard procedure for determining the change in appearance of pile carpet using photography and a visual comparison before and after traffic has been established by the American Society for Testing and Materials [8].

During the time when research personnel were attempting to define the mechanism of soiling and to formulate new finishes to resist soil, many papers were published in the technical literature and a large number of patents were issued. Several very good reviews of textile soiling literature are available, and two of these that are quite comprehensive are by Venkatesh *et al.* [9] and Ranney [10]. This research began in the late 1950s and extended through the 1970s. Since 1980, very little information has been reported. This probably results from the commercialization of antisoiling finishes and the manufacturers considered their compositions to be highly proprietary.

9.1.1 Types of Soil and the Character of Soiling

Although soiling of textiles by atmospheric particle deposition is essentially identical for both apparel and carpet fabric, contact soiling of these textile types arises from two difference types of soil. Wearing apparel constructed from various fiber types are naturally soiled by body oils, known as sebum, which may contain some particulate matter similar to carbon black or clays. Microscopic evaluation of test polyester fabric show little deposition of dry particulates alone [11]. In contrast, carpets are soiled by contact with material adhering to shoes and pets tracked in from outside our homes [6].

Several independent studies have been conducted that show the contribution of oily soil to the discoloration of various fabrics. Bowers and Chantry [12] prepared a synthetic sebum, containing both vegetable and petroleum substances, with the composition of:

20% Olive oil	10% Paraffin wax
15% Spermacetic wax	5% Stearic acid
15% Coconut oil	5% Squalene
10% Oleic acid	5% Cholesterol
10% Palmitic acid	5% Linoleic acid

When this combination was emulsified and mixed with airborne soil collected from air conditioner filters, it was applied to various fabrics. The results of their experiments showed that the parameters contributing to the soiling of polyester–cotton blends were the hydrophobicity of the fiber substrate, fabric construction, fabric finishes, and the type and concentration of the detergent used for washing the fabrics. They also established that body sebum is largely responsible for the pickup and retention of the particles that were suspended in this aqueous medium. Polyesters and cotton absorb similar amounts and types of oily body sebum, but since they are harder to wet, sebum is more difficult to remove from the polyesters. Nonionic detergents were more effective in removing sebum soil from polyester than anionic or cationic detergents because polyester fiber has a minimum of active, ionic dye sites to attract these ionic surfactants. Two other studies [13][14] also found that oil is very important in soiling fabric made from one fiber instead of blends. Other experiments [15] suggested that soil is more easily removed from a polyester filament than from polyester staple or polyester/cotton blends. This was attributed to the entrapment of fatty particles between individual staple filaments.

An important investigation into the desorption by detergent action of carbon–14 labeled fatty acid, fatty alcohol, hydrocarbon, and triglycerides from cellulose, nylon, polyethylene terephthalate (PET), and tetrafluoroethylenehexafluoropropylene (TFEP) surfaces [16] were carried out. The average rates of desorption of soils increased in the order: hydrocarbon <triglycerides <fatty alcohols <fatty acid. Relative rates of soil removal from the four polymeric substrates depended on both the temperature and the detergent. Thus, at 20°C removal became faster in the order TFEP~PET~nylon<< cellulose. At 60°C, with cationic and anionic detergents, soil removal rates increased in the order PET < TFEP < nylon < cellulose. With nonionic detergents at this same temperature, ease of cleanability increased in the order TFEP < cellulose < PET < nylon.

Many chemicals present in human body oils are unsaturated and some might change color after aging. Experiments by Park and Obendorf [17] showed that aging of oily soils on polyester/cotton fabric with unsaturated oleic acid, triolein, and squalene as model soils revealed changes by one of the materials. Yellowing was observed only when squalene soiled fabric swatches were aged for 2 to 46 weeks. The yellow material formed upon aging was not removed after laundering or organic solvent extraction.

The composition of the mixtures used for carpet soiling are completely different from that for apparel soiling. Getchell [2] reported that the collection of street dirt from several cities resulted in the formulation of a synthetic soil specifically designed for carpet treatment. The composition of this synthetic street dirt is:

35% Humus	0.25% Iron oxide
15% Cement	1.6% Stearic acid
15% Silica	1.6% Oleic acid
15% Clay	3.0% Palm oil fatty acids
5% Sodium chloride	1.0% Lanolin
3.5% Gelatin	1.0% n–Octadecane
1.5% Carbon black	1.0% i–Octadecane

As one might expect, this mixture is more complex than the soil used in apparel studies. Reports from this investigation also contain data on the composition of typical carpet dirt removed from vacuum cleaner bags. This material is:

3% Moisture	12% Cellulosic materials
45% Sand, clay	10% Resins, gums, etc.
5% Limestone	6% Fats, oils, tar, etc.
5% Gypsum	2% Undetermined
12% Animal Fibers	

Other carpet soiling studies by Florio and Mersereau [6] found that in addition to the retention of soil by individual fibers it was important to review the soiling of the entire carpet construction. Cross–sections of tufts of carpet fibers revealed that soiling occurs from the top down, with the section directly under the walking area, or that carry out the wiping action on shoe soles, becoming saturated first. After that each successive area below it becomes saturated until all of the available sites down the tuft are full.

9.1.2 Measurement of Textile Soiling

Despite the type and composition of soil or methods of application to textile fabric, improvement by an additive or other treatment cannot be assessed unless some technique is used to measure the fabric appearance before and after soiling. Experimental methods for determining the amount of soil may be grouped into four areas: (1) visual observation, (2) microscopical inspection, (3) determination of the amount of soil removed during detergent washing, and (4) instrumental evaluation of the light reflection from the fabric surface.

Trained observers can differentiate between different degrees of soil deposition and assign a number according to a gray scale. The human eye is an excellent tool for observing inconsistencies, but unfortunately different people may assess observed soiling differently. Benisek [5] related that there is a fairly good agreement between visual grading by a panel of observers and instrumental measurement of reflectance; however, the visual technique can lead to uncertain ratings and is not recommended as a viable technique.

Both light and electron microscopy have been used to evaluate the amount of soil and soil distribution on various fabric types. The location of various soiling agents on plain weave cotton fabric was studied with both types of microscopy equipment [18]. The authors found that gross and fine geometry of both fiber and fabric play a major role in the distribution of soil in most cases. Mechanical entrapment of soil particles and agglomerates between the fibers near the surface of the fabric is repeated in the accumulation of soil in the irregularities of natural fiber surfaces. Sorption of the soil particles on apparently smooth surfaces also occurs, especially when soft films were present on the fiber surface. Experiments by Fort, Billica, and Sloan [11] with an electron microscope found both organic and inorganic particulate dirt on soiled polyester and cotton fibers. Soil deposited during use is concentrated at the exposed surface of the fabric, while soil redeposited during laundering is distributed over each filament and concentrated at the fiber–fiber junction points. In all cases a sheath of organic material appears to spread over the fiber surfaces with particulate soil embedded in the sheath. Similar results were obtained by Hoffman [19] during an investigation of oil release from polyester and polyester/cotton fabric. Soil redeposition during laundering is a persistent problem, especially with naturally soiled fabric. A scanning electron microscope was used to evaluate this phenomenon [20] and the authors found that artificial soil did not redeposit to the degree found with natural soil that contains an organic compound.

The measurement of the soil removed during detergent action in washing by weighing the residue is quite difficult, but several studies have been made using carbon–14 labeled oily soil. In one experiment Wagg and Britt [21] found that a hydrocarbon, octadecane, was removed from various fibers more readily than glyceryl tristearate, cholesterol, and stearic acid. The relative removal of the different types of fatty compounds did not depend on the type of fiber, but it seemed to be more difficult to remove fatty compounds from cotton than nylon or polyester. Another investigation [22] found that the type of fabric was the major contributor to soil removal. In a major survey [23] Grindstaff, Patterson, and Billica used a carbon–14 labeled amorphous carbon suspended in a synthetic sebum composed of 30% glyceryl tristearate, 30% stearic acid, 20% octadecyl alcohol, and 20% octadecane. The particulate carbon as received is considered hydrophobic but was made to be hydrophilic by treating with wet air containing 1.5% ozone at 90°C for six hours. This research showed that hydrophilic carbon was easier to remove than hydrophobic carbon, but the presence of fatty soil did not significantly decrease the extent of carbon removal. The addition of hypochlorite bleach to anionic surfactant solutions increased their ability to remove hydrophobic carbon, presumably by converting it to hydrophilic carbon. Particulate carbon was much more difficult to remove from cotton fabric than from fabrics made from polyester or nylon yarn. The probable cause is entrapment of the carbon particles by the fissures in the cotton fibers and fabric.

Among all of the techniques that are available for measuring the amount of soil, instrumental determination of light reflected from a fabric surface yield the most accurate and precise results. As early as 1953 Hart and Compton [24] reported on using the Kubelka–Munk equation [25] for measuring soil as applied to fabric. This equation states:

$$\frac{K}{S} = \frac{(1 - R)^2}{2R} \tag{9-1}$$

where:
  R = reflectance,
  K = light absorption coefficient, and
  S = light scattering coefficient.

While employing reflectance methods in their experiments, Weatherburn and Bayley [26] found that the soil retention increased in the order cotton < acetate < viscose rayon < nylon < wool. The same order was maintained when the time of contact with the soil varied from 5 to 80 minutes. Soil retention of all the fibers increased with decreasing moisture content of the yarn and the presence of even small amounts of oily material substantially increased the amount of soil retained. Dry soiling has also been investigated [27] with similar results.

It has been pointed out, however, that the Kubelka–Munk equation is valid only for monochromatic light [5] and the evaluation of dyed carpets does not give reliable results. In his work on carpet soiling, Lamb [28] reported on the use of a Hunter ColorQuest® colorimeter that evaluates the color in terms of a red–green number a and blue–yellow number b, where the numbers may be diluted with white light. The Hunter Lab color difference formula [29], which calculates the reflectance from fabrics before and after soiling, is:

$$\Delta E_{Hunter} = (\Delta L^2 + \Delta a^2 + \Delta b^2)^{\frac{1}{2}} \tag{9-2}$$

The values for L (light–dark), a (red–green), and b (blue–yellow) are derived from Hunter Lab uniform color space [29] in rectangular coordinates and calculated as:

$$L_{lightness} = 100 \left(\frac{Y}{Y_0}\right)^{\frac{1}{2}}$$

$$a_{red-green} = 175 \left(\frac{0.0102X_0}{Y/Y_0}\right)^{\frac{1}{2}} \left(\frac{X}{X_0} - \frac{Y}{Y_0}\right) \qquad (9\text{--}3)$$

$$b_{yellow-blue} = 70 \left(\frac{0.00847Z_0}{Y/Y_0}\right)^{\frac{1}{2}} \left(\frac{Y}{Y_0} - \frac{Z}{Z_0}\right)$$

where X, Y, and Z and X_0, Y_0, and Z_0 are the tristimulus values for the sample and illuminate respectively. Most laboratories in which soiling studies are performed generally have developed their own equations for calculating ΔE for soiled and unsoiled fabric.

9.1.3 Effect of Fiber, Yarn, and Fabric Structure on Soiling

Since cotton fiber has a ribbon shaped cross–section that does not lie in one plane and has an irregular surface, macro– and micro–occlusion on the fiber surface would be expected. Tripp et al. [18] in their investigation into the distribution of dry soil did indeed find this to be true. Their observations make it clear that fiber and fabric geometry governs the overall distribution of soil to a considerable degree. The natural imperfections of the fiber surface, which persisted through modifications such as mercerization and partial acetylation, were preferred locations for the buildup of soil. Mechanical entrapment of soil between the fibers also occurred. Smith and Sherman [30] reported that irregularities in the fiber surface served as sinks for the original deposition of soil and removal of soil from these sinks is much slower during laundering. Untreated (greige) cotton was found to be more resistant to soiling than bleached cotton [31] and the total surface area of a fabric is directly related to soiling. Any treatment that reduces the total available surface will impart some degree of soil resistance to the fabric.

The soiling of cotton can also be improved by chemical treatment. Desized, scoured, and bleached 80 x 80 cotton fabric was treated with cross–linking agents in a series of soiling experiments [32]. The fabric was cross–linked with either dimethylol cyclic ethyleneurea (DMEU) or a triazine–formaldehyde condensate. Both of these, especially when carboxymethylcellulose (CMC) was included in the formulations, improved the wet soil resistance and soil removal from finished fabric. Weight add-ons of approximately 3–5% cross-linking agent and 2% CMC produced the best improvement without impairing wrinkle resistance. Other experiments by the same investigators [33] showed that cotton was partially swollen with 0.6% formaldehyde and highly swollen with 1.2% formaldehyde and then cross-linked with various agents. Both types of swollen and cross-linked fabric picked up less nonaqueous oily carbon black than the unmodified fabric.

Cotton is not the only fiber type for which macro–occlusion plays an important role. Hart and Compton [24] studied the soiling of cotton, linen, nylon, silk, and wool and found that inter–fiber and inter–yarn entrapment of soil particles in the fabric structure was a major factor in all cases, for both primary and secondary soil deposition. In general, as the weight and complexity of the fabric structure increased, the number of macro–occluded soil particles increased. During experiments by Weatherburn and Bayley [34] with acetate, nylon, polyester, and rayon it was found that soil retention increased with decreasing fiber denier. These authors also reported that circular cross–section fibers retained less soil than those of a similar denier having a serrated cross–section. Another interesting study was carried out on the effect of twist of acetate yarn containing 34 filaments of 3.5 denier/filament [35]. When the twist increased, soil retention also increased to a maximum and then decreased to a level below that of untwisted yarn. Microscopic examination of the soiled yarn revealed that at 20 turns/inch soil particles had penetrated to the center of the yarn, whereas at 60 turns/inch the soil was located on the exterior surface only.

9.1.4 Fiber Surface Properties Affecting Soiling

Although it has been established that macro– and micro–occlusion can account for much of the soiling on cotton and other fibers with rough surfaces, these mechanisms do not completely account for the soiling of other fibers with smooth surfaces. The third mechanism proposed by Compton and Hart [4] is sorptive bonding to the fiber by van der Waals or electrical forces. In the detergency experiments in which carbon–14 labeled particulate carbon was removed in washing [23], hydrophilic carbon was more easily removed than hydrophobic carbon. This indicates that hydrophobic fragments are attracted more strongly to the fiber surface than the hydrophilic particles, probably through coulombic attraction. Polypropylene fabric with an almost completely hydrophobic surface was grafted with an air–aqueous solution of acrylic acid after exposure to gamma radiation. These experiments [36], in which the grafted propylene changed continuously from hydrophobic to hydrophilic with the degree of grafting, the fabrics were soiled and the soil removed in a washing procedure. The degree of grafting was correlated with the work of adhesion of an oil drop to the fabrics in water and to the zeta potential on the fabric–water interface. The authors assumed that London/van der Waals attraction between the polypropylene fabric and soil decreases with grafting. This is true because the work of adhesion of oil to the fabric and that the work of adhesion is based on London/van der Waals attraction [37]. Thus London/van der Waals attraction in water between polypropylene fabric and oil decreases as grafting increases, that is, grafting and an increase in hydrophilic character of the fiber leads to good detergency since the soil becomes more easily removable in water.

A very simple experiment to determine the hydrophilic–hydrophobic nature of a yarn or fabric is to measure the capillary rise in a yarn sample. In standard physical chemistry laboratory tests it is well known that water will rise in a glass capillary because water will wet the hydrophilic glass surface. In contrast, water will not rise in a hydrophobic polyethylene capillary. Hoffman [19] studied the capillary rise with cotton and polyester–cotton blends, both treated and untreated with permanent–press resins. This author was able to correlate soil retention with low capillary rise in some instances. These conclusions were verified in several laboratories [38]. Most of these studies were made with yarn in which capillaries exist between individual filaments. A technique was devised to measure capillary rise with single filaments [39] and this data related to the free energy of wetting.

Capillarity and the simple occasion of having a hydrophobic or hydrophilic fiber surface do not completely explain the adsorption of soil onto textiles. A better clarification of this complex process is in terms of work of adhesion of soil to the surface and differences in surface energy. In studies of oily soil removal during washing, Saito, Otani, and Yabe [40][41] showed that there was a correlation between the work of adhesion of soil and removal during washing. Their data showed that as the work of adhesion increased, the percent of soil removed decreased significantly. Other experimental investigations on laundering confirm that the surface energy of the fabric substrate is one of the more important aspects of soil removal [30].

After cotton fabric was treated with various finishing agents that were more hydrophobic than cellulose, the critical surface tension of the finished fabric was measured [42] using the method described by Zisman [43]. Cellulose has a high critical surface tension for wetting in air (high surface energy), but it has a low critical surface tension for wetting in water. Due to this low surface energy in water, cellulose has a high resistance to wet soiling by hydrophobic soils. This also accounts for the ease of removal of hydrophobic soil during washing. Silicone and fluorocarbon finishes have low surface energies in air but form high energy surfaces in water. This leads to a strong tendency for the finishes to attract hydrophobic soil in water and, correspondingly, to be difficult to remove during laundering. If these finishes are soft [31][44] the soil particles adhere even more tenaciously to the fiber surface.

Perhaps even more important than the total surface energy are the dispersion and polar segments of that energy (see Chapter 3, Section 3.2.5 for a more thorough discussion of surface energy and its components). Using the techniques and equations suggested by Fowkes [45] and Wu [46], Saito and Yabe [47] measured the dispersion and polar forces of cellophane and cellulose acetate with increasing degrees of acetylation. Their studies showed that the dispersion forces become larger with increasing acetylation while the polar forces become smaller. These authors also determined [48] the dispersion and polar components of oily soil and showed that triglycerides have the lowest polar force component of all chemical materials in sebum and thus would result in low soiling. From these observations one can conclude that when the polar component of the surface energy is low, oily soil would have less attraction to the surface and less soiling would occur.

Another factor to consider in deposition and extraction of soil is the electrical charge that exists on both the fabric substrate and the soil particles. Most particles in an aqueous suspension have a negative charge [44] so positively charged substrate surfaces should be avoided. Thus it is apparent that electrokinetic forces are very important in soil deposition and removal. One electrical property that describes the difference in potential between the immovable layer attached to the surface of the solid phase (either soil or substrate) and the movable part attached to that surface is called the zeta potential [49]. In detergency systems it was shown that it is desirable for the zeta potential of the soil and substrate be nearly equal to reduce the attractive forces causing them to adhere. Jacobasch [50] determined that soiling tendency is proportional to the maximum zeta potential. A neutralization of the surface forces by means of a soil repellant finish leads to a decrease in that zeta potential and a lower propensity for soiling to occur. In studies of soil redeposition during the washing of shrink–resistant wool, Fité [51] found that decreasing the negative zeta potential (becoming more positive) resulted in an increase in soil redeposition.

Peper and Berch [44] summarized an ideal soil resistant finish as being one that is thin, hard, hydrophilic, and negatively charged. It should be also added that this finish needs to have a low polar surface energy.

9.1.5 Soil–Resistant Finishes for Fibers and Textiles

A wide range of chemical materials has been evaluated for use on fibers and fabrics to retard soiling. Although wearing apparel, especially items made with fiber blends and those with permanent–press treatment, is predisposed to soiling, it can be cleaned by efficient laundering. Carpets, however, are not cleaned as frequently and soil tends to build up with time. One benefit to carpet life and appearance is the use of soil–resistant additives. Many of the additives for both apparel and carpet fabrics are addressed in this section, but it should be noted that fabric manufacturers may use proprietary formulations and probably are not be included in these discussions. These finishes are only examples of the possible combinations.

9.1.5.1 Metal Oxides and Salts for Fabric Treatment

About forty years ago carpet manufacturers began adding submicron particles of insoluble silica (silicon dioxide) or aluminum salts in attempts to hide the soil embedded in the carpet [52]. Although these salts were attracted to the fiber surfaces by electrical charges, they were not as effective as first thought and were removed during cleaning. When the insoluble aluminum salt, aluminum phosphate ($AlPO_4$), with a particle size of about 70 millimicrons, was dispersed in water with trisodium phosphate ($Na_3PO_4 \cdot 12H_2O$) and sprayed onto the surface of an Axminster pile carpet, an improvement in soil resistance was observed [53].

Another aluminum salt, an ammonium orthophosphate with the composition $AlPO_4 \cdot NH_4H_2PO_4$ was found particularly effective in decreasing the level of soil on wool carpet [54]. If the amount of added salt is lower that 0.25%, little improvement was noted, but add–on levels between 0.5 and 1.5% were effective in diminishing the amount of soil without affecting the hand or whitening of the fabric.

Hydrated titanium oxide has also been recommended as an antisoiling treatment for carpets [55]. When titanium oxide monohydrate ($TiO_2 \cdot H_2O$) with a particle size below 0.1 microns is dispersed in an acidic aqueous medium at a pH below 5 it can be used as a carpet additive. The size of the titanium oxide particle is very critical in preparing the dispersion and in its success in soil reduction. The size should be between 0.02 and 0.075 microns for commercial applications. After a dispersion of 1% solids was applied to wool carpet, dried in hot air, and soiled in a rotating drum the samples were vacuumed. It was readily apparent that the treated samples were much cleaner that the untreated ones. Wool–viscose blends, viscose, and cotton carpets were also processed with similar results. Textiles have also been treated with tetraalkyl titanate and poly(methylhydrogensiloxane) [56]. A nylon carpet treated with this mixture and soiled showed a reflectance loss of 5 NBS units versus 23 NBS units for a carpet treated with an aqueous dispersion of TiO_2 hydrate.

Zirconium salts have also been used in finishes for imparting soil resistance to nylon carpets. Smith [57] formulated a finish composed of polyethylene glycol–aryl poly(oxyethylene) glycol mix and zirconium acetate (or stannic tetrachloride pentahydrate) that was shown to be effective in soil resistance of nylon carpet. Another carpet was prepared from nylon 66 and treated with 1% by weight aqueous finish composed of a 1:9 methyl benzyl phenol:ethylene oxide reaction products, a silicone product, acetic acid, and a solution of zirconium oxide and hydroxylamine sulfate. It had a brightness loss of 28% compared with a 59% loss for the untreated carpet. In another patent, Smith [58] showed that the soil resistance of nylon carpet treated with fluorocarbon finishes was enhanced by the addition of zirconium salts.

9.1.5.2 Carboxymethyl Cellulose Finishes for Soil Resistance

Carboxymethyl cellulose (CMC) is a cellulose ether prepared by reacting alkaline cellulose with sodium chloroacetate. The commercial product contains about 0.5 carboxymethyl ($-CH_2COOH$) groups on each C_6 glucose unit. The molecular weight can range from about 21,000 to 500,000. Usually sold as the alkali salt, this film forming, water–soluble polymer has many applications, including a sizing agent in woven fabric manufacturing.

Since cotton fibers have a rough surface that entraps soil particles, it has been shown [31] that the coating of cotton fabric smooths the fiber surface and reduces the adherence of soil. Permanent–press resins have been shown to substantially increase the soiling of cotton fabrics during laundering [59]. After a section of white cotton fabric was treated with dimethylol cyclic ethyleneurea, it was soiled with 220 Å carbon black and the degree of soiling measured by reflectance on a Hunter Reflectometer. When a 0.1% sodium salt of CMC add–on as an after treatment or in the soiling solution, the reflectance values were one–third less with the CMC than without that product.

During other experiments on the wet soiling of cotton, Berch and Peper [60] evaluated several different finishes, including CMC. Iron oxide and carbon black, used both as a dry powder and dispersed in oleic acid, and vacuum cleaner dirt were applied to cotton fabric, rinsed in distilled water and air dried. Instrumental measurement of reflectance was used to calculate K/S values. When the fabrics were treated with carboxymethyl cellulose prior to soiling, only light soiling of the fabric was measured for all types of soil. These results have been confirmed by Pinault [52] and Reeves et al. [3].

9.1.5.3 Acrylic and Vinyl Polymer Soil–Repellant Finishes

Since particles of soil normally have a negative electrical charge, it may be assumed that if a polymer surface had a similar charge, some soil repellence might be observed. A coating that induces that charge will also increase the hydrophilic character since more hydrogen bonding might occur. Acrylic polymers are good examples of that type of coating with poly(acrylic acid) being quite effective. The structure of poly(acrylic acid) is:

$$\left[-CH_2-CH-CH_2-CH- \right]$$
$$\qquad\quad COOH \quad\ COOH$$

The benefit of the anionic poly(acrylic acid) coating on cotton and cotton blends has also been reported by Pinault [52], and Berch and Peper [60].

However, it has been shown that poly(methyl methacrylate) with a structure of:

$$\left[-CH_2-\underset{\underset{OCH_3}{\overset{\overset{CH_3}{|}}{\underset{|}{C=O}}}{\overset{|}{C}}-CH_2-\underset{\underset{OCH_3}{\overset{\overset{CH_3}{|}}{\underset{|}{C=O}}}{\overset{|}{C}}-CH_2-\underset{\underset{OCH_3}{\overset{\overset{CH_3}{|}}{\underset{|}{C=O}}}{\overset{|}{C}}- \right]$$

has no antisoiling property, but actually increased soiling when applied to cotton or polyester/cotton blends [61].

In a comprehensive investigation into the consequences of changing the hydrophilic nature of acrylic acid polymers, Bille et al. [62] prepared a series of ethyl acrylate–acrylic acid and ethyl acrylate–methacrylic acid copolymers. When applied to cotton/polyester blends, the best antisoiling results were obtained when the copolymers contained at least 15% by weight acrylic acid or methacrylic acid, with acrylic acid polymers slightly better. Copolymers consisting of methacrylic acid blends, however, were more permanent than those with acrylic acid. Similar copolymers of C_{6-18} alkyl (meth) acrylates with acrylic acid dissolved in chlorinated hydrocarbons were claimed to be effective in soil reduction [63] on textiles. Byrne and Arthur [64] also concluded that poly(methacrylic acid) was helpful in removing oily soil from cotton fabric.

After polyester film, nylon film, and nylon fabric were treated with ionizing radiation from an electron accelerator, they were immersed in monomeric acrylic acid, N–methylol acrylamide, and sodium styrene sulfonate, the samples were dried, and the contact angles measured [65]. These treatments resulted in significant reductions in water contact angles on all samples. The lower contact angles on the polymer surfaces indicate an increased ability to dissipate static charges and increased ability to release oily stains from these surfaces in aqueous solutions. The increase in hydrophilicity and thus oil repellency after radiation induced grafting of acrylate to polyester fabric was also described by Vrakami and Okada [66] in other experimental studies of acrylic acid grafted onto polyester surfaces; they were characterized by contact angles, Fourier Transform Infrared Spectroscopy (FT–IR), and Electron Spectroscopy for Chemical Analysis (ESCA) [67]. Contact angles for the grafted polymer were measured in a water/octane system and showed an elevation with increasing acrylic acid content up to about 7% acrylic acid, after which there was no change. FT–IR and ESCA analyses revealed that the hydrophilic –COOH groups were not only on the surface but also in the interior regions. These analytical procedures also showed that the number of $-COOH$ displayed a linear correlation with the polar component of the surface energy, γ_s^p, thus agreeing with contact angle measurements. No effect was noted for the dispersion component of surface energy, γ_s^d. In a second paper these authors [68] treated the grafted fabric with squalane, triolein, and oleic acid as examples of oily soil. The removal of these soils during washing correlated directly with increasing acrylic acid content, and thus with increasing polar surface energy. It would then appear that low polar surface energy will repel soil particles and high polar surface energy will aid in their removal.

A variation in the acrylate polymer soil release is the use of acrylamide

$$CH_2=CH-CONH_2$$

or monomethylol acrylamide

$$CH_2=CH-COONH(CH_2OH)$$

which polymerize and may cross-link with the cotton substrate [69]. Both the wicking action and soil removal were enhanced. Marco [70] described a process for imparting both soil release and durable press characteristics to fabric that contain polyester fiber. This finishing composition contained an aminoplast textile resin, such as dihydroxy dimethylol ethylene urea, and acrylic compounds as n-methylol acrylamide and a copolymer of 70 parts ethyl acrylate and 30 parts of acrylic acid. The finish was cured with ionizing radiation with the potential varying from 100,000 to 500,000 electron volts.

Another formulation was prepared by the interpolymerization of vinyl acetate, ethylene, and n-methylol acrylamide in an aqueous dispersion system. This polymer was applied to woven fabric in the form of an aqueous latex emulsion [71]. This polymeric coating, resistant to soil, was found to be completely intact on the fabric surface after repeated laundering. Other terpolymers were made by polymerizing a mixture of (1) a vinyl monomer containing a curable methylol or epoxy group, (2) a vinyl monomer containing an anionic group, (3) and other polymerizable vinyl monomers such as vinyl esters of aliphatic acids [72]. This coating on polyester fabric displays antistatic properties and some degree of soil resistance.

9.1.5.4 Silicone Soil–Resistant Finishes

Surface energies of fiber surfaces could be lowered by using silicone finishes, but some studies have demonstrated that poly(dimethyl siloxane) can increase soiling since it has no hydrophilic properties. In two inquiries by Berch, Peper, and Drake [60][73] cotton fabrics were treated with three different silicones, two of which are strongly hydrophobic and the third, while also water repellant, is a softener. The two hydrophobic silicones, poly(dimethyl siloxane) and a silicone rubber, dramatically increased the soiling of the fabric. Although the silicone softener did not cause the cotton to be soiled to the degree of the first two, it was worse than untreated cotton. The silicones also retained considerably more soil during laundering than the fabric for which no finish has been applied.

Several patents have been issued that claim soil resistance for some modified silicones, but their commercial uses are limited. A polysiloxane, transformed by the addition of 1.4 to 2 monovalent hydrocarbon radicals per silicon atom, especially with hydroxyl end groups, was applied to a wide variety of fabrics. In all cases the fabrics had excellent water repellency and were resistant to grease spotting [74]. No claims were made for repelling soil. Another silicone finish was prepared from poly(dimethyl siloxane) that was cured on fabrics with heat [75]. This gum was also water and oil resistant, but no soiling data were reported.

Monomers of siloxanes containing fluoroalkoxyalkyl groups were synthesized and polymers prepared from these monomers applied to textile fabric to yield water-, oil-, and soil-resistant material [76]. One example of these monomers is 3-(heptafluoroisopropoxy) propyltrichlorosilane, which has a structure of:

$$
\begin{array}{c}
\text{CF}_3 \\
| \\
\text{F---C---O---CH}_2\text{---CH}_2\text{-CH}_2\text{---SiCl}_3 \\
| \\
\text{CF}_3
\end{array}
$$

This monomer was polymerized by adding the monomer to an excess of water and the polymeric mass was isolated. This polymer was insoluble in acetone and toluene but soluble in trichlorotrifluoroethane. A sample of this solution was added to toluene, preparing a 50% solids solution. Fabric swatches were added to the solution, rinsed, and dried. Superior oil and water repellency ratings were obtained. The same inventors suggested another monomer

$$
\begin{array}{c}
\text{CF}_3 \qquad\qquad\qquad\qquad\qquad \text{CF}_3 \\
| \qquad\qquad\qquad\qquad\qquad\qquad | \\
\text{F---C---O---CH}_2\text{-CH---CH}_2\text{---CH}_2\text{-O---C---F} \\
| \qquad\qquad\quad | \qquad\qquad\qquad | \\
\text{CF}_3 \qquad\quad \text{SiCl}_3 \qquad\qquad\quad \text{CF}_3
\end{array}
$$

for polymerization and addition to wool and cotton fabric. This product also yielded excellent oil and water resistance [77].

A somewhat different approach has been proposed by Shane and

Wieland [78]. No polymers were prepared but mixtures of silanes and fluorocarbons were added to woven cotton fabric and heated at 300°F for two to three minutes. One of the silane materials is octadecyl tris[1–(2 methyl)–aziridinyl] silane, which has an unusual structure:

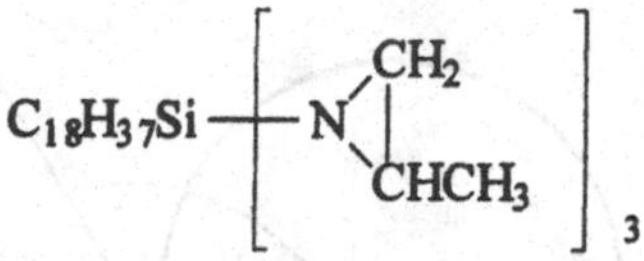

$$C_{18}H_{37}Si \left[N \begin{array}{c} CH_2 \\ \diagdown \\ CHCH_3 \end{array} \right]_3$$

When this silane was mixed with a fluorocarbon, such as the acrylate ester of N–alkyl N–alkanol perfluorooctane sulfonamide, applied to cotton textile fabric and cured, oil– and water–resistant material was the result.

9.1.5.5 Fluorochemical Compounds Affecting Textile Soiling

A wide variety of chemical compounds have been evaluated as additives to reduce the soiling of textiles, as addressed in the preceding sections. These materials were probably effective on some fabrics, but they were not acceptable for all fiber types nor were they successful in reducing all types of soil. Very few of these compounds are being used commercially today. Other soil–retarding chemicals that contain fluorine are more generally accepted, are valuable additives for treating almost all fabrics, and have soil–repelling abilities that are more noticeable. A large number of fluorochemical compounds have been synthesized and several adopted for commercial use, many of which will be discussed in this section. Of course, one must realize that the exact commercial fluorochemical materials in use today are probably not the ones depicted, but are variations of these model compounds.

For about 50 years researchers have recognized the phenomenon that fluorochemicals produce surfaces with low surface energies [79]. On such surfaces, a liquid will form a drop with a characteristic contact angle. The contact angle, Θ, the surface energy of solid surface, and the liquid surface tension are related through Young's equation [80]:

$$\gamma_{LA} \cos \Theta = \gamma_{SA} - \gamma_{SL} \tag{9–4}$$

where:

γ_{LA} = surface tension of the liquid in air,
γ_{SA} = interfacial energy at the solid–air junction, and
γ_{SL} = interfacial energy at the solid–liquid junction.

The equilibrium contact angle of a high surface tension liquid, such as water, on a low energy surface, like polytetrafluoroethylene, can be visualized as in Figure 9–1.

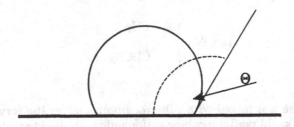

Figure 9–1. Equilibrium Contact Angle of a High
Surface Tension Liquid on a Low Energy Surface.

If the surface tension of the liquid contacting the low energy surface is lower than that of water, the contact angle decreases as the interfacial surface energy difference is reduced. When the liquid surface tension is less than the critical surface energy of the substrate, the liquid will spread.

Smooth, uniform surfaces are characterized by a single value of the contact angle, as in Figure 9–1. Typical textile surfaces are neither smooth nor homogeneous. On such surfaces, a drop is depicted as having both advancing and receding contact angles [81]. The advancing contact angle, Θ_A, is the maximum value that the contact angle may achieve as the drop spreads or is inflated in size. The receding contact angle, Θ_R, is defined as the minimum value that a drop assumes when the drop front is moved toward the center of the drop, either through deflation or tilting. These contact angles are displayed by Figure 9–2.

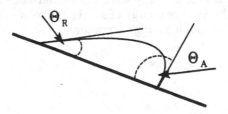

Figure 9–2. Diagram of the Advancing, Θ_A, and
Receding, Θ_R, Contact Angles on a Tilted Surface.

In the capillary system of yarn and textiles, the advancing contact angle affects spreading and wicking [82]. From these experiments, it follows that oil and oily soil repellency are directly related to the values of the advancing contact angles. The values of the receding contact angles, in contrast, help predict whether an oil drop will be removed during blotting or cleaning [81]. Studies by Smith and Sherman [30] showed that if the advancing contact angle is greater than 90°, fabric soiling is significantly reduced and fluorochemical finishes on fabrics are quite effective in attaining advancing contact angles that were in excess of 90°. Other investigations [83] into the effects of fiber content, fabric construction, and cleaning on fluorochemically treated fabric found that fiber type and fabric construction had little effect on soiling, but cleaning did reduce the effectiveness of soil repellency. Even after cleaning, the fluorochemical finishes maintained good soil resistance.

One might assume that the most effective soil-resistant fluorocarbon coating on a fiber might be one similar to poly(tetrafluoroethylene) in which all hydrogen atoms are replaced with fluorine. This assumption, however, is erroneous. Researchers in several different laboratories have established that there must be a balance between oleophobicity to repel oily soil and hydrophilicity to attract water, or other polar liquids, in the most effective fluorochemical compounds [84]–[86]. Many different compounds have been synthesized that combines both oily soil repelling fluorine groups and hydrophilic groups.

As early as 1957 several different fluorochemicals were applied to cotton fabric to obtain oil- and water-repellent surfaces [87]. Two different types were described, chromium coordination complexes of saturated perfluoromonocarboxylic acids and polymers of fluoroalkyl esters of acrylic acid. The coordination complexes had the general formula:

$$\left[CF_3(CF_2)_nCO_2Cr_2OH \right]^{4+}$$

and the fluoroacrylate polymers had a structure of:

$$\left[\begin{array}{c} -CH_2CH- \\ | \\ COOCH_2R \end{array} \right]_x$$

in which R is any perfluorinated alkyl group. When the chromium complex was made from perfluoro-octanoic or perfluoro-decanoic acid, about 1.5% complex solids in the treating solution were necessary to significantly inhibit oil penetration of cotton fabric. Two fluoroacrylate polymers were chosen as representatives of that class of compounds, poly(1,1-dihydrotrifluoroethyl acrylate), abbreviated PTEA, and poly(1,1-dihydroperfluorobutyl acrylate), also called Poly FBA. The penetration time of motor oil or peanut oil was measured and the results showed that Poly FBA was far more effective than PTEA. When the dry pickup of Poly FBA was only 2.7%, no penetration occurred during the test period, but even a dry pickup of 25% PTEA could not prevent oil from wetting the fabric in fewer than three hours. These experiments demonstrated that changes in chemical structure, from trifluoroethyl to perfluorobutyl, can drastically influence oil repellency.

During that same year a patent claiming effective soil resistance for fluorinated sulfoamido acrylate esters was issued [88]. Two of the typical monomers synthesized were the acrylate esters of N–propyl–N–ethanol perfluorooctanesulfonamide:

$$C_8F_{17}SO_2N \underset{CH_2CH_2OCOCH=CH_2}{\overset{C_3H_7}{<}}$$

and the methacrylate ester of N–ethyl–N–ethanol perfluorooctane-sulfonamide:

$$C_8F_{17}SO_2N \underset{CH_2CH_2OCOC=CH_2}{\overset{C_2H_5}{<}} \quad CH_3$$

These monomers were polymerized to yield fluorocarbon acrylate and methacrylate homopolymers. Coatings of the hydrophilic polymers were applied to textiles from solution or emulsion and provided firmly bonded surface coatings. Due to the orientation of the polymer molecules, the fluorocarbon "tails" provided an inert outer surface that was both hydrophobic and oleophobic. Drops of water and drops of oil deposited on that surface remained or ran off rather than spreading or wetting the surface. Some modifications of these compounds are vinyl–type esters [89][90], such as N–methyl–N–methyl perfluorooctanesulfonamide, with a typical structure of:

$$C_8F_{17}SO_2N \underset{CH_2-C}{\overset{CH_3}{<}} \overset{O}{\underset{O-CH=CH_2}{\overset{\|}{-}}}$$

Polymers prepared from these compounds and applied to surfaces had properties similar to other perfluorinated sulfonamide materials.

Some experiments designed to explore the comparison between the chromium coordination complex with perfluorooctanoic acid and a fluorine–containing acrylic polymer, poly(1,1–dihydroperfluorobutyl acrylate) were also reported [91]. Although the perfluoroacrylate polymers imparted excellent oil repellency to cotton fabric, lower levels of the chromium complex yielded superior results. This complex, with only 1.5% solids in the treatment bath, produced an oil–resistant fabric even after 18 weeks of exposure. Other studies [92] investigated the effect of the perfluoroalkyl chain length on oil repellency of cotton. The overall conclusions were that longer chain lengths improved the repellency results. The minimum length was at least four carbons, but seven to nine carbon atoms were preferable. Acrylate polymers with the following structure were the most satisfactory of the ones examined.

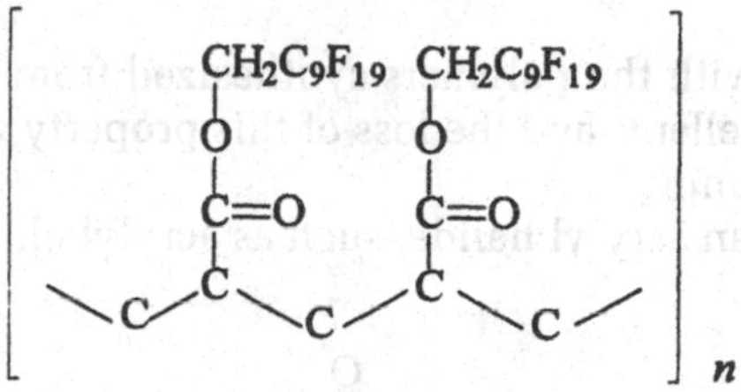

$$
\left[\begin{array}{ccc}
 & CH_2C_9F_{19} & CH_2C_9F_{19} \\
 & | & | \\
 & O & O \\
 & | & | \\
 & C=O & C=O \\
 & | & | \\
\diagdown C \diagup\!\!\!\!^{C}\diagdown_{C}\diagup\!\!\!\!^{C}\diagdown_{C}\diagup & &
\end{array} \right]_n
$$

Various fluorinated alcohols, with the carbon chain containing the fluorine atom ranging in length from C_3 to C_{11}, were synthesized and reacted with methyl methacrylate to prepare another fluoroacrylate monomer [93]. One example of this monomer is heptafluoropentyl methacrylate, which has a structure of:

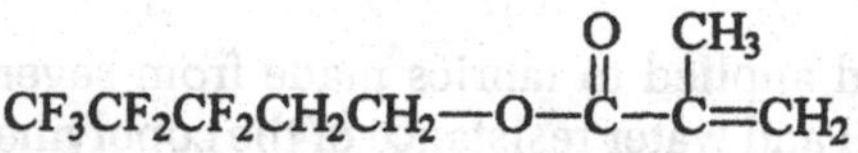

$$
CF_3CF_2CF_2CH_2CH_2-O-\overset{\overset{O}{\|}}{C}-\overset{\overset{CH_3}{|}}{C}=CH_2
$$

After emulsion polymerization at 65°C for six hours, an acrylate latex polymer was formed. This fluorinated methacrylate was blended with poly(n–octyl methacrylate) in several ratios and diluted to yield several concentrations of total solids. Samples of cotton poplin fabric were padded with these emulsions to obtain 100% wet pickup. In oil–repellency tests, the inventors found that at least 3% of the fluorinated polymer in the mixture and about 0.1% of the fluorinated polymer on the fabric were necessary for adequate oil resistance.

Kleiner [94] synthesized an acrylate monomer reported to be very effective in resisting water. It is poly(heptafluoroisopropyl acrylate) with a structure of:

$$\underset{\underset{CF_3}{|}}{\overset{\overset{CF_3}{|}}{F-C}}-O-\overset{\overset{O}{\|}}{C}-O-CH_2CH_2-O-\overset{\overset{O}{\|}}{C}-\overset{\overset{CH_3}{|}}{C}=CH_2$$

Another class of fluoroacrylate polymers is based on acrylamide. Typical patents describing these polymers have been filed by Kleiner [94] and by Langerak, Nelson, and Wright [95]. An example of the chemical structure of the methacrylamide monomers is:

$$C_9F_{19}-\overset{\overset{O}{\|}}{C}-\overset{\overset{CH_3}{|}}{N}-(CH_2)_2-\overset{\overset{H}{|}}{N}-\overset{\overset{O}{\|}}{C}-\overset{\overset{CH_3}{|}}{C}=CH_2$$

Wool fabrics treated with the polymers synthesized from these monomers were excellent soil repellents and the loss of this property were not affected by washing or dry cleaning.

Copolymers of an acrylyl halide, such as acrylyl chloride, which has the chemical structure

$$CH_2=CH-\overset{\overset{O}{\|}}{C}-Cl$$

and fluorochemical monomers similar to

$$C_8F_{17}SO_2\overset{\overset{CH_3}{|}}{N}C_{11}H_{22}OCOCH=CH_2$$

were synthesized and applied to fabrics made from several different fiber compositions [96]. Oil and water resistance of the copolymers were generally superior to the results reported for the fluorochemical homopolymers, which were described by Sherman, Smith, and Johannessen [84].

Very similar fluorocarbon vinyl ether polymers were outlined in a patent filed by Heine [97]. One of the preferred monomers was:

$$C_8F_{17}SO_2\overset{\overset{C_3H_7}{|}}{N}-C_2H_4-O-CH=CH_2$$

Polymers prepared from these monomers provided superior oil and water repellency on all types of wool fabrics and were very resistant to hydrolysis or loss of the fluoroalkyl group when in contact with fatty oils at elevated temperature. This property is particularly important when coating paper packaging materials that will be in contact with foodstuffs.

Another variation of the fluorochemical acrylate and vinyl polymers that have been recommended for use as repellents on textile materials are the acrylol perfluorohydroxamates [98]. One of the most useful monomers is methacrylyl perfluorooctanoylhydroxamate with a structure of:

$$
\underset{\text{H}}{\overset{\overset{\displaystyle O}{\|}}{C_7F_{15}-C-N}}-O-\overset{\overset{\displaystyle O}{\|}}{C}-\overset{\overset{\displaystyle CH_3}{|}}{C}=CH_2
$$

After polymerization and application to fabric surfaces, these fabrics were especially resistant to hydrolysis and the soil–repellent properties were maintained after repeated washing.

Fluorinated cationic compounds have also been manufactured as additives to impart soil resistance to textiles [99]. One example of these quaternary amines that was useful for cotton was described as [100]:

$$
\left[\underset{\overset{|}{C_2H_5}}{\overset{\overset{H}{|}}{C_4F_9-C_4H_8-N}}-CH_2-\overset{\overset{C_2H_5}{|}}{\underset{\displaystyle +}{N}}-C_2H_5 \right]^{+} \quad Cl^{-}
$$

After application to cotton denim fabric, it was highly resistant to oily soil and the effectiveness was retained after washing or dry cleaning. About 0.2% fluorine should be present on the treated fabric after curing to provide the desired soil repellency after laundering. This amount is readily furnished by solutions containing 1.0 to 2.5% of the quaternary amine used as the treating agent. Fluorochemical acrylate quaternary amines were also effective in the soil–resistant treatment for wool and cotton [101]. A general structure is:

$$
\left[\underset{}{\overset{\overset{\displaystyle O}{\|}}{C_7F_{15}-C}}-\overset{\overset{\displaystyle H}{|}}{N}-C_3H_6-\overset{\overset{\displaystyle CH_3}{|}}{\underset{\underset{\displaystyle CH_3}{|}}{N^+}}-CH_2CH_2-O-\overset{\overset{\displaystyle O}{\|}}{C}-CH=CH_2 \right] \quad Cl^{-}
$$

These monomers polymerize upon standing at room temperature and provide outstanding oil– and water–repellent properties to fabrics and other surfaces. About 0.9% polymer solids on cotton fabric will provide excellent resistance to both water and oil.

The hydrophilic nature of antisoiling fluorochemical compounds may be enhanced by the addition of propyl ethers, such as 1,1–dihydropentadeca–fluorooctyl glycidyl ether, with a structure represented by:

$$C_7F_{15}CH_2-O-CH_2-\overset{\overset{\displaystyle H}{|}}{\underset{\underset{\displaystyle O}{\diagdown\diagup}}{C}}-CH_2$$

This type of compound can be reacted directly with cotton at about 100°C for 30 hours [102]. This fabric had excellent oil– and water–repellent properties that were durable after boiling for 20 minutes in an aqueous, alkaline soap solution. Another perfluorinated ether to increase hydrophilicity

$$\underset{\underset{\displaystyle CF_3}{|}}{\overset{\overset{\displaystyle CF_3}{|}}{F-C}}-O-CH_2-\underset{\underset{\displaystyle H}{|}}{C}\overset{O}{\overbrace{}}\underset{\underset{\displaystyle H}{|}}{C}-CH_2-O-\underset{\underset{\displaystyle CF_3}{|}}{\overset{\overset{\displaystyle CF_3}{|}}{C}}-F$$

was reported by Pittman, Roitman, and Sharp [86]. This material was polymerized and applied to various fabrics, and then its soiling was measured. Although the critical surface tension was quite low, the soil release properties were not as good as expected, probably because the hydrophilicity is too low.

An additional soil–repellent compound that reacts directly with cotton can be an acid halide, as perfluorooctanoyl chloride:

$$C_8F_{17}-\overset{\overset{\displaystyle O}{\|}}{C}-Cl$$

as has been reported [103]. A sample of cotton fabric was placed in a flask with a small amount of dimethylformamide and the fluorinated acid chloride. After shaking for 4 minutes, the fabric was removed, washed, and dried. This cotton sample possessed excellent oil– and water–repellent properties that remained even after 1 hour Soxhlet extraction with carbon tetrachloride.

If ethyl perfluorooctanoate is allowed to react with ethylenimine (also known as azridine) and followed by limited polymerization, another oil–repellent agent is prepared [104][105]. The reactions are:

$$C_7F_{15}\overset{\overset{\textstyle O}{\|}}{C}-O-C_2H_5 \; + \; HN\underset{CH_2}{\overset{CH_2}{\diagdown\!\!\!\diagup}}$$

Ethyl perfluorooctanoate   Ethylenimine

$$C_7F_{15}\overset{\overset{\textstyle O}{\|}}{C}-N\underset{CH_2}{\overset{CH_2}{\diagdown\!\!\!\diagup}}$$

$$H-\!\!\left[\!\!\underset{\underset{C_7F_{15}-C=O}{|}}{N}\!-CH_2CH_2\right]_{w}\!\!\left[NH-CH_2CH_2\right]_{x}\!\!\left[N-CH_2CH_2\right]_{y}$$

$$\left[CH_2CH_2N\right]_{z}\!\!-H$$

$$C_7F_{15}-C=O$$

Cotton fabrics processed with this polymer were superior to many other fluorochemical finishes in both oil– and water–repellent aspects. Add–on percentages of as low as 0.2% were effective.

A soil–resistant compound composed of fluoroalkyl derivatives of monocyclic ureas was suggested by Gagliardi [106]. The structure of one of these compounds was:

$$CF_3(CF_2)_6CH_2-\overset{\overset{\textstyle O}{\|}}{C}-\overset{\overset{\textstyle H}{|}}{N}-CH_2-\overset{}{N}\underset{CH_2-CH_2}{\overset{\overset{\textstyle O}{\overset{\|}{C}}}{\diagup\diagdown}}N-CH_2-O-C_2H_5$$

A fluorinated melamine was also prepared by this inventor by reacting hexa(methoxymethyl) melamine and pentadecafluorooctyl alcohol in equimolar proportions while using a small amount of orthophosphoric acid as an ether interchange catalyst. The structure of this compound was:

$$H_3COH_2C$$ ... structure of triazine ...

$$\begin{array}{c} H_3COH_2C \\ H_3COH_2C \end{array} N-C \overset{N}{\underset{N}{\diagup}} C-N \begin{array}{c} CH_2OCH_3 \\ CH_2OCH_3 \end{array}$$

$$\underset{C}{\overset{}{}}$$

$$N-CH_2OCH_3$$

$$CH_2OCH_2(CF_2)_6CF_3$$

After either of these compounds was padded onto cotton fabric, effectual reduction in soiling was observed.

A novel polyfluorourea finish for cotton that provided both oil– and water–repellent features for cotton fabric was described by Connick and Ellzey [107]. This finish component, fluorinated ethoxymethyl urea (FEMU), was prepared by reacting 1,1–dihydroperfluorooctylamine with isocyanic acid according to the scheme:

$$C_7F_{15}CH_2NH_2 + HN{=}C{=}O \longrightarrow C_7F_{15}CH_2NH{-}\overset{O}{\overset{\|}{C}}{-}NH_2$$

$$\downarrow \quad \overset{-}{O}H, CH_2O, C_2H_5OH$$

$$(C_7F_{15}CH_2NH{-}\overset{O}{\overset{\|}{C}}{-}NH)_2CH_2 \overset{H^+}{\longleftarrow} C_7H_{15}CH_2NH{-}\overset{O}{\overset{\|}{C}}{-}NHCH_2OC_2H_5$$

FMBU   FEMU

FEMU, either in an ethanol solution or as an aqueous dispersion, was applied to cotton fabric with an acid catalyst, dried at 80°C for five minutes, and then cured at 155°C for three minutes. During the curing step the authors believed that FEMU was converted to FMBU and that no polymerization occurred. Oil repellency was greater than any of the commercial fluorocarbon finishes that were also evaluated in these experiments.

Many other fluorinated organic molecules have found application as materials to reduce soiling, including carbamates. One of these carbamates is [108]:

$$C_7F_{15}(CH_2)_3{-}O{-}(CH_2)_2{-}O{-}\overset{O}{\overset{\|}{C}}{-}N \begin{array}{c} CH_2OH \\ CH_2OH \end{array}$$

and was effective on repelling oil applied to cotton fabric. Another highly effective carbamate that was prepared for use on cotton was [109]:

$$
\begin{array}{cccc}
HN-CH_2-N-CH_2-N-CH_2-NH \\
| \quad\quad | \quad\quad | \quad\quad | \\
C=O \quad C=O \quad C=O \quad C=O \\
| \quad\quad | \quad\quad | \quad\quad | \\
O \quad\quad O \quad\quad O \quad\quad O \\
| \quad\quad | \quad\quad | \quad\quad | \\
CH_2 \quad CH_2 \quad CH_2 \quad CH_2 \\
| \quad\quad | \quad\quad | \quad\quad | \\
C_8F_{17} \quad C_8F_{17} \quad C_8F_{17} \quad C_8F_{17}
\end{array}
$$

A piece of cotton fabric was padded with a 5% solution of this carbamate in acetone, dried, and heated in an oven for three minutes at 150°C. The oil repellency of this fabric was excellent, even after 14 washings and ironings. After tumbling with dry soil, very little discoloration was apparent on that fabric.

Several fluorinated aminohydroxy compounds have been investigated as components to reduce soiling of textile fabrics. One of these that was effective on cotton, wool, and nylon was based on toluene–2,4–diisocyanate [110]. An example of these materials had a structure of:

$$
\begin{array}{l}
CH_3 \\
\quad\quad\quad\quad H \ O \ H \\
\quad\quad\quad\quad | \ \ || \ \ | \\
\quad\quad\quad\quad N-C-N-SO_3^- \ \overset{+}{N}H(C_2H_5)_3 \\
\\
\quad\quad N-C-O-CH_2(CF_2)H \\
\quad\quad | \ \ || \\
\quad\quad H \ \ O
\end{array}
$$

and as shown contains both hydrophilic and hydrophobic parts of the molecule. The compound was isolated, dissolved in dimethylformamide, and diluted to a 3% by weight solution with water. Cotton and polyester fabric was padded with the solution to provide a 100% wet pickup, air dried, and cured at 107°C for 1.5 minutes. These fabrics showed good oil repellency that was retained after several washings. Paraaminophenol was the starting reagent for another compound reported to give excellent resistence to the deposition of oily soil. After reaction with the butyl ester of perfluorooctanoic acid, methylolation was carried out by reaction with two moles of formaldehyde dissolved in methanol [111]. The structure of one example of these compounds is:

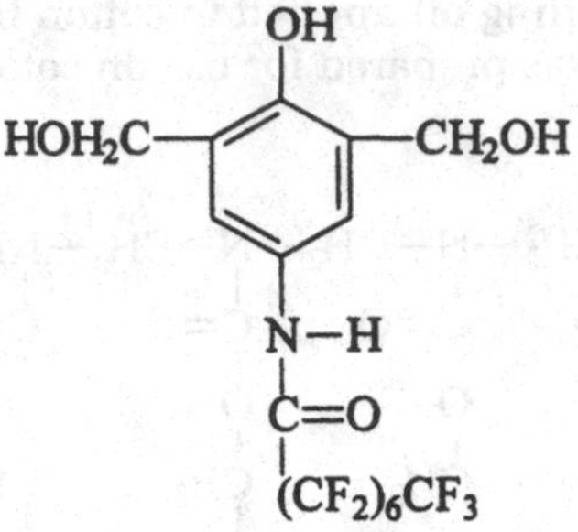

$$
\begin{array}{c}
\text{OH} \\
\text{HOH}_2\text{C}-\!\!\!\bigcirc\!\!\!-\text{CH}_2\text{OH} \\
\text{N}-\text{H} \\
\text{C}=\text{O} \\
(\text{CF}_2)_6\text{CF}_3
\end{array}
$$

A very important class of chemicals was designed by Guenthner and Lazerte [112] comprising fluorocarbon urethanes having the general structural formula:

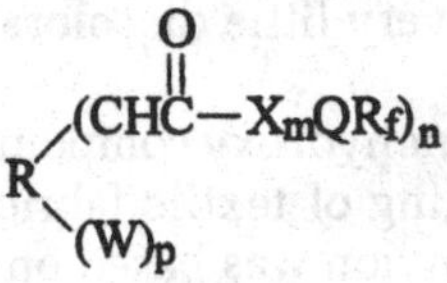

$$
R{\overset{\displaystyle(\text{CH}\overset{\text{O}}{\overset{\|}{\text{C}}}-\text{X}_m\text{QR}_f)_n}{\underset{(\text{W})_p}{}}}
$$

in which R is an organic radical having various structures, X a hydrophilic group such as a sulfonamide, R_f a monovalent fluorocarbon radical, Q a sulfonyl or carbonyl divalent radical, and W an isocyanate radical. As might be expected, a large variety of fluorochemical urethanes were prepared having this general structure. Three examples of the compounds tested that yielded extremely effective oil and water resistance were as follows:

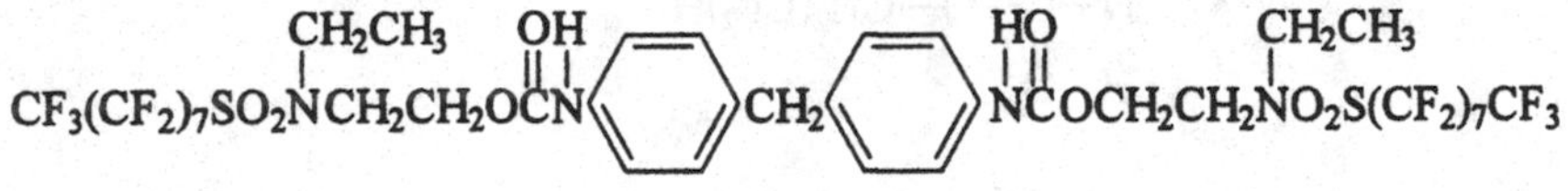

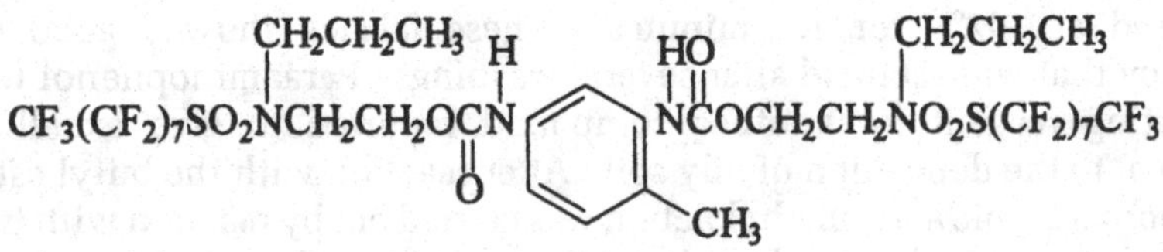

$$
\text{CF}_3(\text{CF}_2)_7\text{SO}_2\overset{\text{CH}_2\text{CH}_3}{\text{N}}\text{CH}_2\text{CH}_2\text{O}\overset{\text{OH}}{\overset{\|}{\text{C}}}\text{N}\!\!\!\bigcirc\!\!\!\text{CH}_2\!\!\!\bigcirc\!\!\!\overset{\text{HO}}{\text{N}}\overset{\|}{\text{C}}\text{OCH}_2\text{CH}_2\overset{\text{CH}_2\text{CH}_3}{\text{N}}\text{O}_2\text{S}(\text{CF}_2)_7\text{CF}_3
$$

$$
\text{CF}_3(\text{CF}_2)_7\text{SO}_2\overset{\text{CH}_2\text{CH}_2\text{CH}_3}{\text{N}}\text{CH}_2\text{CH}_2\text{O}\overset{\text{H}}{\underset{\|}{\overset{}{\text{C}}}}\text{N}\!\!\!\bigcirc\!\!\!\overset{\text{HO}}{\text{N}}\overset{\|}{\text{C}}\text{OCH}_2\text{CH}_2\overset{\text{CH}_2\text{CH}_2\text{CH}_3}{\text{N}}\text{O}_2\text{S}(\text{CF}_2)_7\text{CF}_3
$$
$$
\text{CH}_3
$$

$$CH_3CH_2CH_2CH_2N\overset{\overset{H}{|}}{C}\overset{\overset{O}{||}}{C}\overset{\overset{H}{|}}{N}\quad N\overset{\overset{HO}{|}}{\underset{||}{C}}OCH_2CH_2NO_2S(CF_2)_7CF_3$$

These compounds were applied to fabric surfaces by the usual methods for coatings and may be mixed with usual fabric finish components without any loss of soil–resistant properties. Oil repellence on wool gabardine, polyester–wool blends, and cotton was superior to most conventional fluorochemical treatments.

Derivatives of trimellitic acid were described as being effective in reducing the soiling of nylon carpets [113][114]. The structure of one example of these compounds is:

$$\overset{\overset{O}{||}}{C}-SC_2H_4C_9F_{19}$$

Other organic compounds that were disclosed as effective in the reduction of soiling of nylon are fluorochemically substituted guanamines [115]. A representative structure is:

$$F-\left[\overset{\overset{F}{|}}{\underset{\underset{CF_3}{|}}{C}}-CF_2O\right]_{10}-\overset{\overset{F}{|}}{\underset{\underset{CF_3}{|}}{C}}-CF_3$$

$$HOCH_2-\overset{}{\underset{\underset{H}{|}}{N}}\quad N\quad \overset{}{\underset{\underset{H}{|}}{N}}-CH_2OH$$

These guanamines may be heat polymerized on textiles to give dry soil–resistant finishes, or copolymerized with diisocyanates or epoxides to give flexible polymeric coatings.

Several unique combinations of fluoroalkyl groups and inorganic substances have been shown to be highly effective in reducing oil repellency of many different types of textile fabrics. One representative chemical series consists of the alkali metal or ammonium salts of fluoroalkyl phosphoric acid [116]. If the compound is ammonium bis(1H, 1H, 9H–hexadecafluorononyl) phosphate, which has a structure of:

$$\left[H(CF_2)_8CH_2O \right]_2 P{\Large\diagup}^{\textstyle O}_{\textstyle ONH_4}$$

is applied to cotton, wool, nylon, polyester, or acrylic fabric, significant improvement in oil repellency is observed. Equivalent oil repellency was found for another phosphate, bis(perfluorooctanamido–ethyl) phosphochloridate [117], which has a structure of:

$$\left[\begin{matrix} & O \\ & \parallel \\ C_7F_{15} & C\,N\,CH_2CH_2O \\ & | \\ & H \end{matrix} \right]_2 P{\Large\diagup}^{\textstyle O}_{\textstyle Cl}$$

Silicate and titanate esters of fluorinated aliphatic alcohols have also been described [118]. These materials were especially formulated to resist dry soil after application to cotton fabric. One example of this type of chemical compound is:

$$\begin{matrix} O(CF_2)_{10}CF_3 \\ | \\ CH_3O-Ti-O(CF_2)_{10}CF_3 \\ | \\ O(CF_2)_{10}CF_3 \end{matrix}$$

An alternate procedure for treating cotton to resist soiling involves two steps; the first step is to immerse the fabric in a solution of a salt such as aluminum acetate or zirconium nitrate and the second scheme is to treat the fabric with a fluoromonocarboxylic acid, such as perfluorooctanoic acid [119]. The structure of the acid is:

$$CF_3(CF_2)_6C{\Large\diagup}^{\textstyle O}_{\textstyle OH}$$

When fabrics were treated with aluminum acetate and the fluoroacid, repellency to dry soil, water, and oil was significantly improved.

9.1.5.6 Miscellaneous Chemicals Affecting Fabric Soiling

Treatment of polyester–cotton poplin fabric with a strong alkali solution (about 10% aqueous concentration of sodium hydroxide) was shown to be as effective as soil–release finishes in reducing the soiling by synthetic sebum [120]. The exact theoretical reason for this improvement is not clear, but the observed improvement in wash fastness suggests a modification of the fiber surface was achieved. This treatment also increased the hydrophilicity of the samples.

Although static electricity certainly does not account for all fabric soiling, treatment with antistats to reduce initial soiling and improve soil removal has been reported by several authors. Wilson [121] concluded that electrostatic effects may be responsible for approximately half the total soiling of women's nylon, polyester, and acetate slips. Cross–linked polyamine resin antistats actually increased soiling, but quaternary ammonium compounds caused a marked reduction in the ease of soil removal in all cases. Nonionic antistatic agents also increased the ease of soil removal, but they were also lost during washing. A commercial finish with a proprietary composition, Permalose–T®, reduced static and soiling of polyester fabrics [122] and showed a high resistance to domestic laundering. When a solid poly(alkylene ether) is intimately co–spun with melt–spun polyamide filaments, the poly(alkylene ether) is distributed in the form of elongated, microscopic particles oriented so that the long dimension is parallel to the major fiber axis [123]. These filaments had improved static and soil performance. Nylon fiber first treated with an antistatic agent and then with a fluorochemical to decrease the suface energy substantially reduced the dry soiling of carpets [124].

Some of the specific cationic antistatic agents that were effective in soil reduction are derivatives of imidazolines. One material suggested for use on nylon, polyester, and acrylic carpet was an imidazoline–urea reaction product [125]. A typical structure for these compounds is:

$$
\left[CH_3(CH_2)_{17}-C \overset{\displaystyle N-CH_2}{\underset{\displaystyle \underset{\displaystyle \underset{\displaystyle NH_2}{NH_3}}{H_2N-\overset{O}{\overset{\|}{C}}-NH-\overset{O}{\overset{\|}{C}}-NH_3}}{N-CH_2}} \right]^{+} \quad \overset{O^-}{\underset{O}{\overset{\displaystyle\|}{C}}}-(CH_2)_2CH_3
$$

This compound is removed during carpet cleaning and must be reapplied at levels of greater than 1% weight percent solids. A similar imidazoline–urea product was effective after pretreatment of the fiber with a salt, such as zirconium acetate [126]. A film–forming imidazoline with a structure of:

$$\left[CH_3CHH_2)_{11}-C{\overset{N-CH_2}{\underset{N-CH_2}{\overset{\diagup}{\diagdown}}}}\underset{H_3CH_2C \quad CH_2CH_3}{} \right]^{+} \quad R^{-}$$

in which R is the acid form of sodium carboxymethyl cellulose, sodium phosphated cellulose, or sodium sulfonated cellulose [127] was also prepared. The static on the fabric surface was decreased, as was the soiling.

9.2 STAINING AND STAIN–RESISTANT FINISHES

While the soiling of fabrics may occur by spontaneous exposure to atmospheric contaminants, staining is almost always the result of an accidental spill, mistakes made by pets, or some other unforeseen occurrence. Although one might spill a liquid on wearing apparel, most staining problems occur on carpets, especially ones made from polyamides such as wool or nylon. Most of the common household staining materials are acidic and are attracted to the basic dye sites on the polyamides. Silk, also consisting of a natural polyamide, will stain with acid dyes, but the use of silk in carpet construction is rare. Other carpet fibers, polyolefin, polyester, or acrylic, are not readily dyed by acid dyes, but may be affected by other stains unless treated with fluorochemical antisoiling compounds to enhance repellency. Olefin polymers also have low surface energies since they do not possess a polar component, but only embody the dispersive type of surface energy [128]. Polyester and acrylic fibers are usually dyed with disperse dyes since they do not have basic dye sites.

All types of carpet fibers can be partially protected from staining if they are coated with an antisoiling fluorochemical agent. Since the fluorochemical surface is not easily wetted, if the stain is retained on the carpet surface it can be quickly blotted with an absorbent material and removed. Frequently, however, if the staining substance is dropped from waist height, typical of many spills, the hydrodynamic force allows the substance to penetrate the carpet tuft and wick into the yarn bundle. This occurrence instigated a massive research effort by fiber producers and carpet manufacturers to develop a product to resist such staining.

9.2.1 Types of Staining Materials

During the mid 1980s a study was conducted by the National Family Opinion Research organization concerning problems that consumers experienced with carpets [129]. This study examined the responses and evaluated them with respect to frequency of occurrence and the severity of the concern. After a complete analysis of all results, the most prevalent complaint about carpets was both frequent and severe problems with spots and stains. The sponsor of the study, the Fibers Division of the Monsanto Company, was able to classify staining materials into four groups, primarily based on staining characteristics. The first group, as listed in Table 9–1, is generally coarse particulates or non–acid–dye colorants in a water, oil, or paste base. They can be completely removed from fluorochemically treated carpet following recommended cleaning procedures.

Table 9–1

Group I Common Household Stains

Beer	Ink, water soluble
Beet juice	Latex paint
Blood	Lipstick
Catsup	Nail polish
Chocolate	Orange juice
Cooking oil/soil	Prune juice
Cranberry juice	Red clay soil
Dark carbonated cola	Rouge
Dirty motor oil	Rust
Furniture stain	Topsoil
Grease	Water colors
Ink, ball point	

(Ref. [129]. Reprinted by permission of the Monsanto Company)

The staining materials listed in Group II are finer sized particulates or natural or synthetic colorants in waste, oil, or paste bases. Carpets stained by Group II substances can be cleaned to a faint but discernable stain. Two items in this group, coffee and red wine, should be especially considered. Green coffee beans contain about 7.3% sucrose but during roasting this sugar content decreases to 0.3% [130]. The color of brewed coffee results from the degradation products of the sugar. Caramel color bodies are disperse dyes and can stain many carpet fibers, especially if the coffee contains cream and sugar. Although this is a common stain, it can be lightened by treatment with dilute acetic acid [131]. The dyes in red wine are probably anthocyanine based since these dyes are responsible for most of the red, pink, violet, and blue colors of fruits, flowers, and leaves [132]. The large variation in color is determined by the number of hydroxyl groups, their degree of methylation, the acidity of the biological environment, and several other factors relating to the attachment of monosaccharides. The structure of a typical example, cyanidine, is:

Materials causing stains from Group II are shown in Table 9–2.

Table 9–2

Group II Common Household Stains

Asphalt	Mustard
Coffee, cream & sugar	Oil paint
Crayon	Red wine
Egg	Shoe polish
Grape juice	Tea

(Ref. [129]. Reprinted by permission of the Monsanto Company)

The substances found in the Group III list of stains were considered the ones most often encountered in a home and were some of the ones to cause the most severe problems on polyamide carpets. These stains all contain acid dyes as the principal color body. Children's drink mixes, such as Kool–Aid®, colored gelatins, and fruit flavored sodas generally contain FD & C Red No. 40, or C. I. Food Red 17, as the dye to give the products their red color. The structure of this dye is [133]:

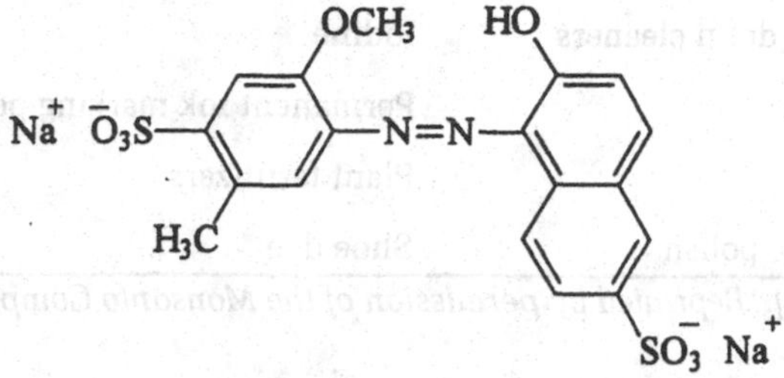

The sulfonic acid groups are attracted to the amine end groups on polyamides as well as to the amide within the chain. Many pet foods are also dyed red to simulate the color of meat and after eating this food pets will occasionally regurgitate their food, thus staining the carpet. Other pet effluents, such as urine and feces, are also frequent staining materials for all types of carpets, but they are especially effective in discoloring nylon. Table 9–3 lists these staining materials.

Table 9–3
Group III Common Household Stains

Fruit–flavored sodas	Liqueurs
Pet vomit	Liquid fruit punch
Cough syrup	Liquid medicines
Children's drink mixes	Mouthwashes
Gelatins	Pet body effluents

(Ref. [129]. Reprinted by permission of the Monsanto Company)

A fourth type of stainant can contain compounds that are bleaches or oxidizing agents. These materials can penetrate both fluorochemically treated and stainblocked carpet fibers and degrade the dye present in the carpet, usually leaving a light–colored spot. Fortunately for the consumer, these materials represent only a small percentage of the problem stains in the normal home [129]. These are presented by Table 9–4.

Table 9–4

Group IV Common Household Stains

Acid toilet bowl cleaners	Hair dye
Acne medication	Insecticides
Alkaline drain cleaners	Iodine
Bleaches	Permanent ink marking pens
Soot	Plant fertilizers
Furniture polish	Shoe dye

(Ref. [129]. Reprinted by permission of the Monsanto Company)

9.2.2 Laboratory Evaluation of Carpet Staining

Since the staining of carpets is a major complaint of the user, the laboratory measurement of the degree of staining is necessary in the process of developing an acceptable stainblocking substance. One of the earliest approaches [134] was simply to immerse a small sample of nylon carpet into an aqueous solution of 0.054 grams/liter of FD & C Red Dye No. 40, leave for one hour, rinse with tap water, and observe whether the sample is dyed or not. This procedure was soon modified [135]. In this test an aqueous solution of 0.08 grams/liter of FD & C Red Dye No. 40 and citric acid at a concentration of 0.4 grams/liter was prepared. This solution is equivalent to a cherry flavored commercial soft drink mix containing the red dye and has an optical density of 3.5 and a pH of 3.0. Several spots on the carpet were made by pouring 10 ml of the solution into a section of plastic tubing about 3.8 cm in diameter and 5.0 cm in length that was placed upright on the carpet sample. The solution was allowed to stand on the carpet for one, two, four, six, and eight hours. The carpets were then washed with water to remove as much of the dye as possible and the spots visually graded using a scale of one (white) to eight (red) which roughly correspond to the International Gray Scale [136] in which a scale value of one is completely stained (Gray Scale black) and a scale value of five was unstained, or white. Since many spills occur from waist height, the dimensions of the plastic tubing were finalized to a diameter of about 4.5 cm and a height of about one meter [129]. Because many carpets are coated with a fluorochemical soil repellent and do not wet easily, the red liquid is frequently worked into the carpet by hand to simulate the worst type of spill that might occur. A similar test has been reported [137] in which 30 ml of an aqueous solution of 0.056 grams/liter of FD & C Red 40, adjusted to a pH of 2.8 with citric acid, is poured from a height of 12 inches. The stains stood for 24 hours and were removed with a fine spray. Staining was graded using a visual scale ranging from zero to eight.

Another technique for the estimation of staining consists of preparing a solution of FD & C Red Dye No. 40 at a concentration of 0.1 grams/liter [138]. Thirty milliliters of that solution were placed in a flat aluminum pan measuring about 3" x 4". A small section of polyamide carpet was fully immersed with the tufts down into the staining solution for one hour. The sample was removed, washed with tap water, and dried in an oven for 15 minutes at 100°C. In a modification [139] the time was changed to 30 minutes or 24 hours. Staining was evaluated with a Minolta® tristimulus color analyzer in the L*a*b Difference Mode with the target sample set for the unstained carpet, with "a" the value on the red–green axis [135]. These samples were also determined visually using a 1–5 scale in which 1 equals heavy staining and 5 no staining. Another variation was reported [140] in which the 1–5 scale was reversed in which 1 was unstained and 5 heavily stained. Instead of carpet samples, nylon sleeves have also been used as the substrate for staining with cherry Kool-Aid® or coffee [141] and ranked on a 0–8 scale in which zero represents no staining.

A completely different approach is to measure the optical density of the dye solution before and after dyeing nylon fiber [134]. The optical density of a weighed amount of dye solution was measured on a Cary® 15 Spectrophotometer using a ½ cm cell with light absorption being measured at 520 millimicrons wave length. Light absorption was recorded as T_0. A sample of fiber was weighed and placed into a stoppered container containing a sufficient amount of cherry flavored soft drink mix to provide a weight ratio of drink to fiber of 40:1. The stoppered container was shaken for one hour while maintaining a temperature of 80°C. After removing the fiber from the container, the optical density was again measured. The reading was recorded as T_1. Test results were expressed as a percentage of light penetration, calculated as:

$$\% = \frac{T_0 - T_1}{T_0} \times 100 \tag{9-5}$$

Low percentage values indicate a yarn resistant to staining.

Stain resistance of carpet is important when the carpet is new, but it is also imperative that this resistance be retained as the carpet ages. Although a mechanical device has been suggested [28] that provides wear and soiling of test carpets, many laboratories employ contract walkers to insure that realistic traffic evaluation is accomplished [129]. Stain testing on carpet before and after traffic is the normal procedure.

9.2.3 Chemicals Used as Stain–Resistant Additives

Chemicals intended for use as stain–resistant additives evolved from materials adopted as aftertreatments of dyed fabrics to improve their wet fastness. In a review of aftertreatments for improving the fastness of dyes, Cook [142] traced the development of stain resistance from about 1850 until 1980. Metal salts and natural tannins were some of the first chemicals found to fix dyes on cotton fabrics. Tanning agents, intended to protect leather from decomposition by various processes, can be classified into three classes of polyphenolic compounds. These are (1) hydrolyzable pyrogallol, (2) hydrolyzable ellagitannins, and (3) condensed or catechol tannins. Natural tanning agents were used on silk and wool with limited success. Although they were evaluated on nylon, they were valuable only for crocking and wet fastness. Their high cost and toxicity severely limited their use [142]. A series of synthetic tanning agents, called syntans, was synthesized to reduce the cost and toxic hazards and have found extensive use as stain reducing agents. One class of these syntans are the water–soluble novolac compounds that have a linear chainlike structure:

In this structure R_1 and R_2 are hydrogen or an alkyl group.

One of the early attempts to resist the acid dying of nylon was the addition of the sodium salt of sulfonated isophthalic acid

to the polymer chain of nylon 66. This polymer is normally prepared by reacting adipic acid with hexamethylene diamine, but when some of the isophthalic acid derivatives are substituted for some of the adipic acid, the dyeing properties are significantly altered. Huffman [143] synthesized a polymer that could be easily dyed with basic dyes but Magat [144] found that the polymer also resisted acid dyes. This polymer was used for several years in the production of differential dyeing nylon. One example of commercial products prepared with this polymer was womens' pantyhose for which the panty portion was knitted using the modified polymer. When the hosiery was dyed with acid dyes, the leg portion was dyed brown, or the desired color, and the panty portion remained white.

Two major reviews of the chemical syntans that have been developed during recent years were published by Cooke and Weigmann [145] and by Kamath *et al.* [128]. The historical development of current stainblocker technology will be followed in subsequent discussions.

Hydroxy compounds similar to the novolacs and tannic acid complexes were evaluated in reducing the adsorption of anionic dyes by nylon 6 by Cook and Majisharifi [146]. These authors synthesized several aromatic hydroxy compounds, including 2,2'–dihydroxy–1,1–dinapthyl-methane, with the structure:

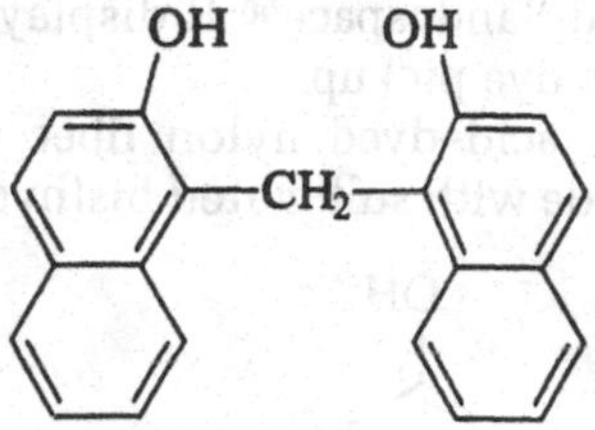

They also studied the effects of Cibatex® PA that is believed to have the structure:

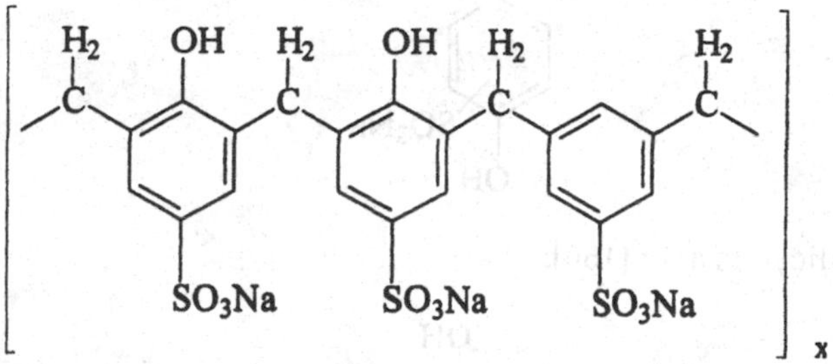

Nylon fabric was treated with three anionic dyes, C. I. Acid Red 102, C. I. Acid Red 41, and C. I. Red 27, but only Cibatex® PA exhibited any significant reduction in dye adsorption, and it was only moderately effective.

Shore [147][148] explored natural tannins as aftertreatments on both nylon 6 and nylon 66 and dyed with C. I. Acid Blue 25. He found that the tannins significantly reduced the rate of dye diffusion of the dye into nylon 66, but had much less effect on nylon 6. Wet fastness was also improved.

Since an increase in the number of amine end groups will increase the acid dyeability [149], it follows that if one could decrease the number of amine groups, the fibers would be more resistant to acid dyes. Acetic and maleic acids both react with amine end groups and reduce the number of amine ends from 64 for untreated to 27 for maleic and 10 for acetic acid reaction products. Another chemical that was quite effective is Sandospace® R with a structure of:

Fiber reactants of nylon and Sandospace® R displayed excellent acid dye resistance and good cationic dye pickup.

A leveling agent for acid–dyed nylon fiber was prepared by the condensation of formaldehyde with sulfonated bis(hydroxyphenyl) sulfone:

and 5–sulfosalicyclic acid [150]:

This condensation product was very effective as a dye leveling agent in a one or two bath dyeing and fixation system, but it was not suggested to be an agent that will resist acid dyes.

In 1981 Guthrie and Cook [151] reported on their studies on the adsorption into nylon 6 of Cibatex® PA and a sulfonated novolac that they had synthesized. The synthesized novolac was believed to have the structure:

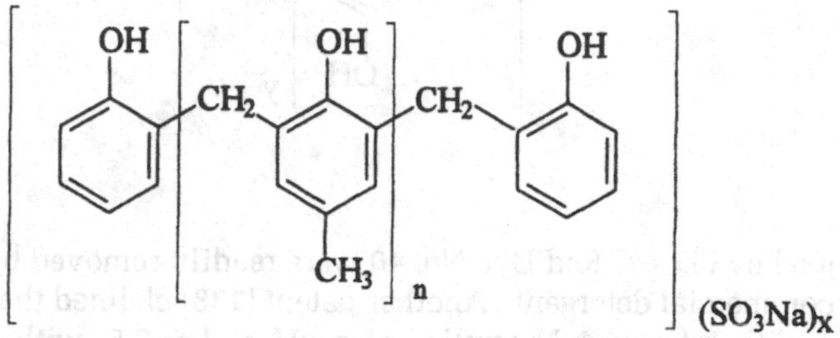

The authors measured the adsorption of each product by nylon 6 fiber from aqueous solutions at a pH of 4 at 80°C, and calculated the apparent diffusion coefficient of Cibatex® PA to be 10.5×10^{-15} m²/s (meters squared per second) and the synthesized condensate to be 4.5×10^{-15} m²/s. They estimated that these values are about ten times smaller that those found for anionic dyes. The shapes for the curves for equilibrium adsorption isotherms were about the same. Guthrie [152] then assayed the diffusion of C. I. Acid Red 18 through nylon 6 and concluded that the syntan Cibatex® PA did not suppress dye diffusion significantly. In another paper Guthrie and Cook [153] concluded that the initial deposition of the syntan is in the outer region of the fiber and acts as only a partial barrier to dye diffusion into nylon 6. The syntan then diffuses slowly into the fiber and gradually loses some of its retarding influence. This confirms the earlier conclusions of Dawson and Todd [154] on the ring dyeing of certain fibers.

In 1985 Ucci and Blyth [134] received a patent for a process to provide stain resistance for polyamide carpets by using a condensation product of phenol sulfonic acid and dihydroxyphenylsulphone when mixed with an alkali metal silicate. The condensation products were sold commercially as Intratex® N or Erinoal® PA and the probable chemical composition is:

Stains developed by FD & C Red Dye No. 40 were readily removed by using water with a commercial detergent. Another patent [138] claimed that nylon carpets made with Intratex® N applied at a pH of 1.5–2.5, with the pH adjusted with sulfamic acid, was more resistant to yellowing from atmospheric NO_2 than carpets treated at higher pH.

Another class of stain resistant compounds was proposed that same year in a patent by Matsuo *et al.* [155] consisting of several polyfluoroalkyl compounds. One specific example is:

$$R_fCH_2CH_2OCONH(CH_2)_6N \begin{matrix} CONH(CH_2)_6NHCOOC_{18}H_{37} \\ CONH(CH_2)_6NHCOOC_{18}H_{37} \end{matrix}$$

in which R_f is a polyfluoroalkyl group containing 1 to 20 carbon atoms. These compounds were resistant to oily materials and partially resistant to blue acid dyes.

Improved stain–resistant nylon carpet was manufactured by changing the syntan condensation product to one consisting of p–phenolsulfonic acid, diphenylsulfone, and 4,4–diphenylsulfone–2–sulfonic acid [133]. When the ratio of sulfonated phenylsulfone to phenylsulfone is at least 20:1, more Red Dye No. 40 was removed during washing than seen with previously described stain resistant agents. Harris and Hangey [137] suggested that a generalized formula for a syntan of this type is:

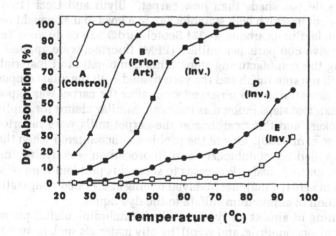

Although most of the chemicals previously described in this section exhibited some reduction in the diffusion of acid dyes or aided the removal of dyes from carpet samples, Blyth and Ucci [156] discovered a process that prevented acid dyes from staining nylon carpet. When used with a fluorochemical soil repellent, a condensation product similar to the one described by Harris and Hangey [137] and heat treated under certain conditions blocked the absorption of FD & C Red Dye No. 40, as contained in Kool–Aid® soft drink mix. Figure 9–1 presents data for dye absorption at various temperatures.

Figure 9–1. Effect of Dyeing Temperature on Dye Absorption.
*(Reference [156]. Reprinted from U. S. Patent 4,680,212
by permission of the Monsanto Company)*

Dye absorption values in the above graph were obtained by measuring the light absorption value of a 0.054 grams/liter aqueous solution of Red Dye No. 40 (the concentration of the dye in cherry Kool–Aid®), adding nylon test fiber to a quantity of the solution, shaking for three hours, removing the fiber and remeasuring the dye absorption of the solution. Plot A, the control, represents the absorption of non–heat–set nylon carpet staple and the value at 25°C shows that much of the dye was absorbed and above that temperature the dye was all absorbed. The data described by Plot B show the absorption of fiber that had been treated with Erinoal® NW, a sulfonated phenolic material used in previous experiments. If yarn similar to the one in Plot B was treated and then heat–set, improved dye resistance was obtained. The condensation products described in the patent were used to treat nylon yarn and the dye absorption measured for both non–heat–set (D) and heat–set (E) yarn. No dye was absorbed by either (D) and (E) at room temperature and with the heat–set yarn, no absorption was observed below 45°C. Resistance to both staining and soiling by oily materials can be provided when the sulfonated novolac resin was blended with a nonionic fluorocarbon [157].

Not only phenolic sulfonated compounds were considered to be of value in stain resistance, but also nylon treated with aliphatic sulfonic acids, especially ones with branched aliphatic chains, were evaluated [158]. The best results in resisting staining by acid dyes were obtained when the carbon chain consisted of from 14 to 17 carbons.

During the commercial development of stain–resistant polyamide carpet, it became apparent that carpet subjected to heavy human foot traffic stained to a deeper shade than new carpet. Blyth and Ucci [159] were granted a patent in which both stainblocker, at a level of 0.32 weight percent, and commercial fluorocarbons, as 3M Scotchgard® 358 or du Pont Teflon® at levels to give 800 parts per million (PPM) fluorine, were applied to the fiber during the manufacturing step. This combination of materials was blended with the spin finish and the yarn treated with the total composition. Essentially no staining was observed even after the carpet was exposed to 30,000 human foot steps (known as traffics). Similar claims for application of stainblockers and fluorocarbons at the carpet mills were reported in a paper by Jose *et al.* [160]. One of the problems encountered when the fiber producer applied the stainblocker and fluorocarbon was dyeing the final carpet to the desired shade. Processes to solve this problem were related in several patents [161]–[163] that involved dyeing at elevated temperatures or by the addition of ammonium sulfate to the dye liquor.

Staining of almost all types of fibers (including olefin, polyamide, polyester, polyacrylonitrile, and wool) by oily materials such as suntan oils, may be resisted by a procedure described by Pfeifer [164]. In the composition described in this patent, a fluoroaliphatic acrylate or methacrylate monomer, with a structural formula similar to

$$CH_3$$
$$C_7F_{15}CH_2CH_2OOCC=CH_2$$

was polymerized and the resulting polymer mixed with a rewetting agent, such as sodium dioctyl sulfosuccinate in ratios of 0.6% by weight fluoroacrylate and 0.6% by weight rewetting agent. Oily liquids were easily removed by normal laundering. No claims were made that this composition would block acid dyes.

An improvement in blocking of Red Dye No. 40 on nylon carpet by partially sulfonated novolac resins and fluorocarbons and also excellent oil and water repellency was obtained if about 1% magnesium chloride was included in the mixture [140]. The fluorocarbon was 3M FX–364®.

Soon after the commercial introduction of nylon carpets on which stainblockers and fluorocarbons had been applied, a distinct yellowing was observed and the dye shades were not as clear as needed [145]. A study of this phenomenon was made by the South Carolina Section of the American Association of Textile Chemists (AATCC) [165]. In their experiments, treated and untreated carpet yarn made from nylon 6 and nylon 66 were exposed to a xenon arc light source for up to 60 hours and the yellowing noted. They concluded that direct exposure to the light was necessary for yellowing to occur and that nylon 6 yellowed more than nylon 66. The exact cause of yellowing was unclear, but the presence of oxides of nitrogen was suspected to be a catalyst for that reaction. It is also well known that other chemical compounds yellow when exposed to NO_2. Carpet manufacturers also reported those homes in highly industrial areas, with elevated levels of atmospheric nitrogen oxides, experienced yellowing of stain–resistant carpets. A variation of sulfonated novolac resins proposed for improving stain resistance and reducing the yellowing of nylon carpet when exposed to atmospheric NO_2 was described by Olson, Chang, and Muggli [166]. The novolac was similar to Intratex® N and identified by 3M as FX–369®. This compound was mixed in an aqueous bath with a methacrylic acid–based copolymer and applied to nylon carpet where the mixture provided good stain resistance and lower yellowing.

Another composition that achieved similar results was a mixture of modified polymeric sulfonated phenol–formaldehyde condensates with hydrolyzed polymers of maleic anhydride and an ethylenically unsaturated aromatic material [167]. A significant improvement in manufacturing stainblockers that also resist yellowing was achieved by Moss, Sargent, and Williams [168]. A composition was prepared by polymerizing an α–substituted acrylic acid or ester in the presence of a sulfonated aromatic formaldehyde condensation product. Thus a mixture was no longer required but a single polymeric entity could be used.

As can be seen in the generalized stainblocker structure on page 427, the aromatic groups are phenolic so the hydroxyl groups are available for reaction with some other chemical. Two patents were issued to Liss and Beck [169][170] in which the inventors reacted the pendent hydroxyls with acetic anhydride to form acylated derivatives or with dimethyl sulfate, sodium chloroacetate, or ethyl chloroformate to prepare ether substituted stainblockers. These products, after application to nylon carpet, do not yellow upon exposure to light or oxides of nitrogen. Another approach is to prepare a copolymer from a hydrolyzed ethylenically unsaturated aromatic compound and maleic anhydride, with no sulfonated material present [139], but some mercaptans were added to the polymerization mix. Carpets manufactured with these polymers also were resistant to Red Dye No. 40 and did not yellow. Improved durability to shampooing of the maleic anhydride polymeric treatment was obtained when they were blended with water–dispersible epoxy resins [171]. A possible structure of the epoxy resin is:

$$CH_2-CHCH_2O(CH_2CHCH_2O)_xCH_2CH-CH_2$$

Coatings made from polymerized itaconic acid were also useful [172].

Hangey *et al.* [173] found that nylon carpets could be dyed, and then treated with an aromatic sulfonated condensate stainblocker and a fluorocarbon, followed by addition of a thiocyanate, such as ammonium thiocyanate, and magnesium sulfate, and thus yellowing was reduced. The stainblocking ability of the carpet was not affected.

A new stainblocker composition was claimed by Fitzgerald [174]. Bis–hydroxyphenylsulfone (BHPS) was polymerized with formaldehyde and a mercapto carboxylic acid to create the product:

in which X is:

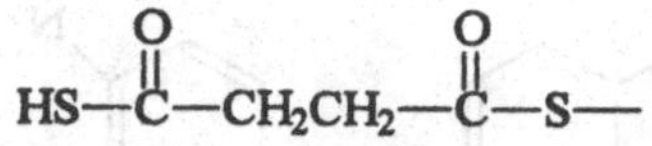

$$HS-\overset{\overset{\displaystyle O}{\|}}{C}-CH_2CH_2-\overset{\overset{\displaystyle O}{\|}}{C}-S-$$

This composition produced excellent stain resistance to Red Dye No. 40 and may be applied to carpet that has already been installed.

Stain resistance that is more durable to shampooing that the usual acrylic–acid–aromatic sulfonate condensation products have also been developed. Two of these [175] were partially sulfonated, partially phosphated resins with different structures. Resol A was:

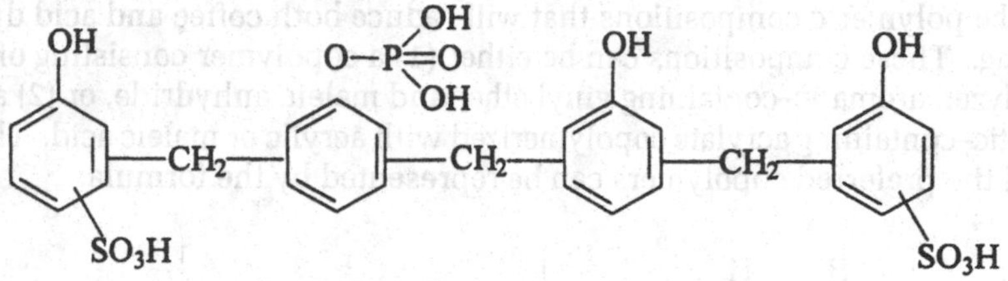

and Resol B was:

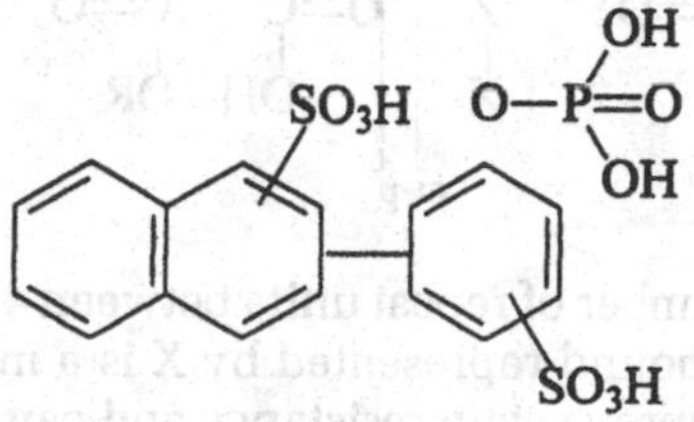

The acrylic–acid–Resol blends on nylon carpet yielded excellent resistance to Red Dye No. 40 and to yellowing by nitrogen oxides that remained after cleaning. A permanent stain–resistant textile fiber was made by treating a polyamide, as nylon or wool, or a cellulosic fiber with a compound suitable for nucleophilic displacement reactions. Chemicals that meet these requirements are triazine, pyrimidine, or quinazoline. The treated fiber can then be coated with a novolac–methacrylic acid copolymer. The stainblocker is then covalently bound to the fiber that exhibited no staining by many materials, including chlorine bleach and Cherry Kool–Aid® [176].

Haarer and Höcker [177] synthesized several compounds and evaluated the resistance of wool to Acid Red 27. The resistance effectiveness increased with increasing molecular weight of the stainblocker and increasing the acidity also gave a better effect. The structure of one of their typical products is:

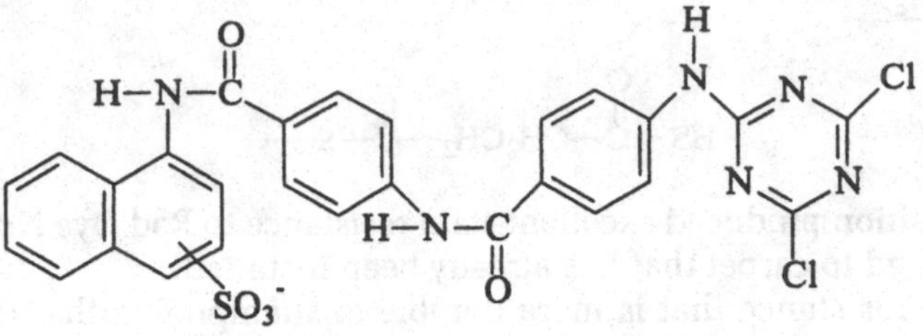

In addition to Red Dye No. 40, coffee is another composition that can stain polyamide carpets. Two patents have been issued [178][179] that describe polymeric compositions that will reduce both coffee and acid dye staining. These compositions can be either (1) a copolymer consisting of a hydrolyzed aromatic–containing vinyl ether and maleic anhydride, or (2) an aromatic–containing acrylate copolymerized with acrylic or maleic acid. The first of the preferred copolymers can be represented by the formula:

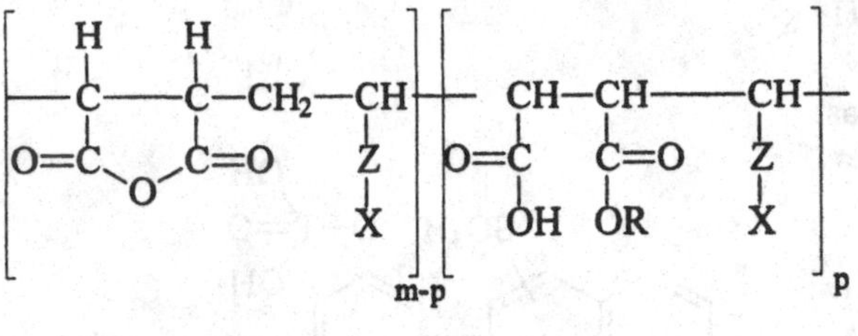

In this structure m is a number of repeat units between 4 and 100 and p is 0.5 to 0.7 times m. The compound represented by X is a moiety of an aromatic compound effective to improve stain resistance and can be either a napthyl– or phenyl–based material. Z is preferably $-O-CH_2-CH_2-O-$ and R is an alkyl group.

In the second type of copolymer, the acrylate, the structure is:

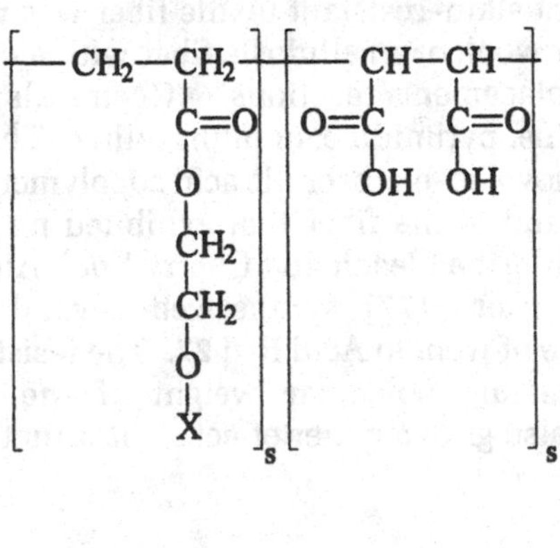

where the number of repeat units, s, is 2 to 50 and X is a phenyl–based stain-resisting compound. Protection against coffee staining was considered to be excellent.

9.2.4 Yarn Processing and Procedures for Stainblocker Application

After the carpet yarn has been spun, drawn, and crimped or bulked, the crimp is set into the yarn by one of three processes. In one of these the yarn is set in dry heat at 195–205°C, on Suessen equipment; in another the yarn is subjected to steam at 134–140°C on Superba machinery; and in the third the yarn is autoclaved with steam and pressure. These three processes cause changes in the internal structure of the fiber in which autoclaved and Superba treated yarn have a more open structure than yarn that has been Suessen heat set. Heat–set yarn is tufted into a carpet backing fabric and the carpet is processed into the final product.

Stainblocker may be applied in two distinctly different methods. In one, it may be added to the fiber finish during spinning and heat set on the yarn [163]. In this process the stainblocker is not applied as uniformly as in the second procedure, application in the carpet mill. Figure 9–2 presents a generalized flow diagram for carpet processing, including the addition of stainblockers and fluorocarbons.

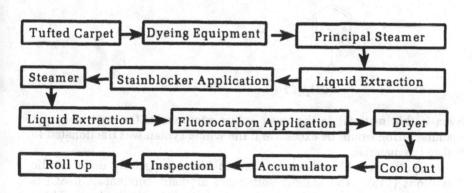

Figure 9–2. General Flow Chart for Dyeing and Finishing in Carpet Mills.

After the roll up in the preceding chart, the carpet goes through a process where a latex emulsion is used as an adhesive to attach a secondary backing to the product and a final inspection is made.

9.2.5 Mechanism of Stainblocking

The most widely used syntans for blocking stains on polyamide carpets are the sulfonated aromatic–formaldehyde condensation products. It seems reasonable that the condensation products are not pure compounds but mixtures having various structures and molecular sizes. Kelson and Holt [180] separated the components of commercial stainblockers and found that many of these components have limited stainblocking activity. In another laboratory [181], high pressure liquid chromatography was used to separate the fractions of a stainblocker and found that there were 67% soluble monomers, 20% substituted monomers, and only about 11% active material. The investigators concluded that the most active ingredient in the syntan that they studied has a structure of:

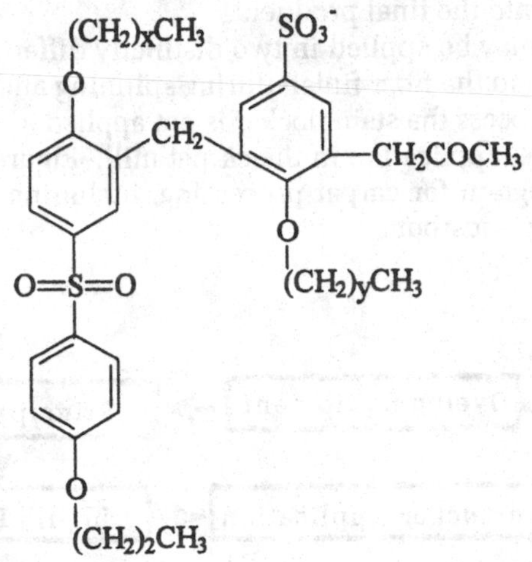

Even if this material is the most active component, the cost of carpet manufacturing would be excessive if the whole syntan was fractionated to isolate the material.

Syntans readily adsorb into nylon fibers, especially under acidic conditions [145]. The anionic sulfonated aromatic condensate (SAC) is adsorbed in a manner similar to the adsorption of anionic dyes. These molecules are generally bound to nylon through electrovalent bonds between the sulfonate anion and the protonated amine end groups; however, covalent linkages though methylol groups cannot be completely ruled out [128].

The depth of molecular penetration into the fiber is dependent upon several factors, including the molecular structure, the size of the molecule, and the yarn treatment prior to tufting. If we assume that the syntans have similar structures, then molecular size and yarn pretreatment are the controlling factors. In some important experiments by Hangey [182], Size Exclusion Chromatography was the tool used to establish that there was an optimum hydrodynamic volume for yarn penetration and stain resistance. The fractions that eluted were isolated, yarn treated with the fractions, and stain resistance determined. Although the exact molecular size of the fractions was not calculated, those that eluted between 6.3 and 6.5 milliliters yielded the best stain resistance. The size was not so small that they migrated too far into the fiber to be ineffective, nor too large that they required excessive steaming to achieve penetration. The optimum size on the molecule is such that some penetration occurs, yielding a yarn that is ring dyed with the stainblocker. Harris and Hangey [137] studied the penetration of a SAC by obtaining an ultraviolet microscope analog image of fiber cross-sections and observed that the SAC was found in the outermost regions of the fiber. The depth of penetration is shown in Figure 9–2.

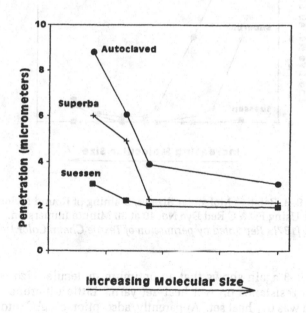

Figure 9–2. Penetration of SAC into Nylon Fiber with Molecular
Size After Various Yarn Pretreatments.
(Ref. [137]. Reprinted by permission of Textile Chem. Color.)

This chart clearly shows that autoclaved and Superba heat–set yarn allows further penetration than Suessen heat set, most likely because the wet heat treatments tend to open the fiber morphology and allow the SAC to adsorb more freely. The differences are much less pronounced with larger molecular size.

The depth of penetration is directly related to the stain resistance of the fiber. Figure 9–3 shows the stain resistance to Red Dye No. 40 of these same yarns, in which low values represent superior resistance.

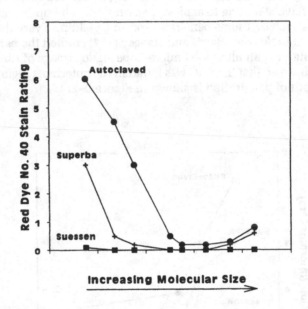

Figure 9–3. Relative Molecular Size vs. Staining of Round Nylon Fiber Using FD & C Red Dye No. 40 at 30 Minute Immersion. *(Ref. [137]. Reprinted by permission of Textile Chem. Color.)*

Figure 9–3 again shows that an optimum molecular size is best for superior stain resistance on wet heat–set yarn. Little difference is noted when the yarn was dry heat set. Apparently, adsorption of SAC into the fiber to a depth of about 2 micrometers yields a fiber that is most resistant to acid dyes.

The electrical double layer at the fiber surface is the most commonly cited mechanism to describe the efficiency of stainblocking by sulfonated aromatic condensates. According to this mechanism, the negatively charged syntan adsorbed on the fiber surface establishes an electrical double layer with a negative zeta potential. This prevents the adsorption of anionic dyes on the surface of the fiber, which is necessary before diffusion into the fiber. Formation of the electrical double layer has some importance on the exhaustion of syntans. Because of the larger number of charges, high molecular weight syntans manifest greater repulsion from the fiber surface and therefore they are deposited to a lower extent than the low molecular weight syntans.

The effect of pH on stain resistance can be explained based on clustering of ionomers within the fiber. At a pH greater than 7.5 the carboxyl groups on the fiber surface are ionized and thus repel the applied negatively charged stainblocker molecules. This results in a lower pick–up and the formation of a discontinuous stainblocker sheath on the fiber surface, resulting in poor resistance to staining. At low pH, about 4, the carboxyl groups are all protonated. The absence of a charge results in the deposition of a more continuous stainblocker layer that is responsible for preventing the diffusion of staining molecules [128].

REFERENCES

1. Rees, W. H. (1954). The Soiling of Textile Materials, *J. Textile Inst.*, **45**: P612
2. Getchell, N. F. (1955). Cotton Quality Study. III. Resistance to Soiling, *Textile Res. J.*, **25**:150
3. Reeves, W. A., Beninate, J. V., Perkins, R. M., and Drake, G. L. (1968). Soiling and Soil Removal Studies on Cotton and Polyester Fabrics, *Am. Dyestuff Reptr.*, **57**: 35
4. Compton, J. and Hart, W. J. (1954). The Theory of Soil–Fiber Complex Formation and Stability, *Textile Res. J.*, **24**: 263
5. Benisek, L. (1972). Service Soiling of Wool, Man–Made Fiber, and Blended Carpets, *Textile Res. J.*, **42**: 490
6. Florio, P. A. and Mersereau, E. P. (1955). Control of Appearance Changes due to Soiling. The Mechanism, Measurement, and Reduction of Soiling Changes in Carpet During Use, *Textile Res. J.*, **25**: 641
7. Chemspec® (1996). Understanding Carpet Soil, *http://enterprise.–newscomm.net/sani/csoil.htm*, Chemspec Inc., Baltimore, MD
8. ASTM Method D 2401–87 (1991). Standard Test Method for Service Change of Appearance of Pile Floor Coverings, *Annual Book of ASTM Standards, Volume 07.01*, ASTM, Philadelphia, pp. 654–660
9. Venkatesh, G. M., Dweltz, N. E., Madan, G. L., and Alurkar, R. H. (1974). Study of Soiling of Textiles and Development of Anti–Soiling ans Soil Release Finishes: A Review, *Textile Res. J.*, **44**:352
10. Ranney, M. W. (1970). *Soil Resistant Textiles*, Noyes Data Corporation, Park Ridge, NJ
11. Fort, T., Billica, H. R., and Sloan, C. K. (1966). Studies of Soiling and Detergency. Part I: Observations of Naturally Soiled Textile Fibers, *Textile*

Res. J., **36**: 7

12. Bowers, C. A. and Chantry, G. (1969). Factors Controlling the Soiling of White Polyester Cotton Fabrics. Part I: Laboratory Studies, *Textile Res. J.*, **39**: 1

13. Compton, J. and Hart, W. J. (1953). A Study of Soiling and Soil Retention in Textile Fibers. Grease–Carbon Black Soil–Cotton Fiber Systems, *Textile Res. J.*, **23**: 158

14. Kennedy, J. M. and Stout, E. E. (1968). A Method for Soiling White One–Fiber Test Fabrics, *Amer. Dyestuff Rept.*, **57** (1): 2

15. Brown, C. B., Thompson, S. H., and Stewart, G. (1968). Oil Take Up and Removal from Polyester, Polyester/Cotton Blend and Other Fabrics, *Textile Res. J.*, **38**: 735

16. Fort, T., Billica, H. R., and Grindstaff, T. H. (1966). Studies of Soiling and Detergency. Part II: Detergency Experiments with Model Fatty Soils, *Textile Res. J.*, **37**: 99

17. Park, E. K. C. and Obendorf, S. K. (1994). Chemical Changes in Unsaturated Oils upon Aging and Subsequent Effects on Fabric Yellowing and Soil Removal, *J. Amer. Oil Chemists' Soc.*, **71**: 17

18. Tripp, V. W., Moore, A. T., Porter, B. R., and Rollins, M. L. (1958). The Surface of Cotton Fibers. Part IV: Distribution of Dry Soil, *Textile Res. J.*, **28**: 447

19. Hoffman, J. H. (1968). Some Physical Aspects of Oil Release–A Microscopic Look, *Amer. Dyestuff Rept.*, **57**: P992

20. Shimauchi, S. and Mizushima, H. (1968). Soil Redeposition of Polyester Fiber and Its Test Methods, *Amer. Dyestuff Rept.*, **57**: P462

21. Wagg, R. E. and Britt, C. J. (1962). Detergency Studies Using a Radioactive Tracer, *J. Textile Inst.*, **53**: T205

22. Komeda, Y., Mino, J., Imamura, T., and Tokiwa, F. (1970). Studies of Detergency. Part II: Contribution of Influencing Factors–Analysis of Variance, *Textile Res. J.*, **40**: 733

23. Grindstaff, T. H., Patterson, H. T., and Billica, H. R. (1967). Studies of Soiling and Detergency. Part III: Detergency Experiments with Particulate Carbon Soil, *Textile Res. J.*, **37**: 564

24. Hart, W. J. and Compton, J. (1953). A Study of Soiling and Soil Retention in Textile Fibers. The Effect of Yarn and Fabric Structure in Soil Retention, *Textile Res. J.*, **23**: 418

25. Kubelka, P. and Munk, F. (1931). Ein Beitrag zur Optik der Farbanstricke, *Z. Tech. Physik*, Leipzig, Germany, **12**: 593

26. Weatherburn, A. S. and Bayley, C. H. (1955). The Soiling Characteristics of Textile Fibers. Part I., *Textile Res, J.*, **25**: 549

27. Berch, J., Peper, H., Ross, J., and Drake, G. L. (1967). Laboratory Method for Measuring the Dry Soilability of Fabrics, *Amer. Dyestuff Rept.*, **56** (6): 27

28. Lamb, G. E. R. (1992). A New Carpet Soiling Test, *Textile Res. J.*, **62**: 325

29. Luo, R. and Rhodes, P. (1996). AATCC Colour Science Glossary, *http:// ziggy.derby.ac.uk/colour/info/glossary/n/HunterLabDiff.html*

30. Smith, S. and Sherman, P. (1969). Textile Characteristics Affecting the Soil Release During Laundering. Part I: A Review and Theoretical Consideration

of the effects of Fiber Surface Energy and Fabric Construction on Soil Release, *Textile Res. J.*, **39**: 441

31. Porter, B. L., Peacock, C. L., Tripp, V. W., and Rollins, M. L. (1957). The Surface of Cotton Fiber. Part III: Effects of Modification on Soil Resistance, *Textile Res. J.*, **27**: 833

32. Beninate, J. V., Kelly, E. L., Drake, G. L., and Reeves, W. A. (1966). Soiling and Soil Removal Studies of Some Modified Crosslinked Cottons, *Amer. Dyestuff Rept.*, **55** (2): 25

33. Beninate, J. V., Kelly, E, L., and Drake, G. L. (1963). Influence of Fiber Swell–ability in Selected Fabrics on Wet Soiling and Ease of Soil Removal, *Amer. Dyestuff Rept.*, **52**: 752

34. Weatherburn, A. S. and Bayley, C. H. (1957). The Soiling Characteristics of Textile Fibers. Part II: The Influence of Fiber Geometry on Soil Retention, *Textile Res. J.*, **27**: 199

35. Weatherburn, A. S. and Bayley, C. H. (1957). The Soiling Characteristics of Textile Fibers. Part III: The Effect of Twist on Soil Retention, *Textile Res. J.*, **27**: 358

36. Katsumi, M. and Tsuji, T. (1969). Study of Soil Removal from Grafted Polypropylene, *Textile Res. J.*, **39**: 627

37. Fowkes, F. M. (1962). Determination of Interfacial Tensions, Contact Angles, and Dispersion Forces in Surfaces by Assuming Additivity of Intermolecular Interactions in Surfaces, *J. Phys. Chem.*, **66**: 382

38. Morton, G. P. (1969). Capillary Rise Related to the Soil Release Performance of Durable Press Fabrics, *Textile Chem. Colorist*, **1**: 202

39. Miller, B., Coe, A. B., and Ramachandran, P. N. (1967). Liquid Rise Between Filaments in a V–Configuration, *Textile Res. J.*, **37**: 919

40. Saito, M., Otani, M., and Yabe, A. (1985). Work of Adhesion of Oily Dirt and Correlation with Washability, *Textile Res. J.*, **55**: 157

41. Otani, M., Saito, M., and Yabe, A. (1985). Surface Energy Analysis of the Detergency Process–Surface Tension Components of Binary Mixtures of Organic Liquids and Aqueous Solutions of Surfactants, *Textile Res. J.*, **55**: 582

42. Berch, J., Peper, H., and Drake, G. L. (1965). Wet Soiling of Cotton. Part IV: Surface Energies of Cotton Finishing Chemicals, *Textile Res. J.*, **35**: 252

43. Zisman, W. A. (1964). Relation of Equilibrium Contact Angle to Liquid and Solid Constitution, *Contact Angle, Wettability, and Adhesion, ACS Advances in Chemistry Series, Vol. 43*, American Chemical Society, Washington, pp. 1–51

44. Peper, H. and Berch, J. (1965). Relation Between Surface Properties of Cotton Finishes and Wet Soiling, *Amer. Dyestuff Rept.*, **54**: P863

45. Fowkes, F. M. (1962). Determination of Interfacial Tensions, Contact Angles, and Dispersion Forces in Surfaces by Assuming Additivity of Intermolecular Interactions in Surfaces, *J. Phys. Chem.*, **66**: 382

46. Wu, S. (1971). Calculation of Interfacial Tension in Polymer Systems, *J. Poly. Sci.*, Part C, **34**: 19

47. Saito, M. and Yabe, A. (1983). Dispersion and Polar Force Components of Surface Tension of Some Polymer Films, *Textile Res. J.*, **53**: 54

48. Saito, M. and Yabe, A. (1984). Dispersion and Polar Force Components of Surface Tensions of Oily Soil, *Textile Res. J.*, **54:** 18

49. Harris, J. C. (1958). Electrical Forces Affecting Soil ans Substrate in the Detergency Process–Zeta Potential, *Textile Res. J.*, **28:** 912

50. Jacobasch, H. L. (1970). Ueber die physikalisch–chemischen Grundlagen des Anschumutzungs– und Haftverhaltens von Faserstoffen, *Textilveredlung*, Basel, Switzerland, **5:** 385

51. Fité, F. J. C. (1992). Soil Deposition and the Electrokinetic Behavior of Shrink–resist Wool Fabrics during Washing with Surfactant Mixtures at Different pH Levels, *J. Text. Inst.*, **83:** 69

52. Pinault, R. W. (1968). What's Ahead for Soil Release, *Textile World*, **118** (5): 112

53. Florio, P. A., assignor to Mohasco Industries, Inc. (1957).Textile Fabric Rendered Soil Resistant with Aluminum Phosphate, *U.S.* Patent 2,786,787

54. Lawler, E. B., Vartanian, R. D., and Roth, P. B. , assignors to the American Cyanamide Company, New York (1959). Process for Preparing Aluminum Phosphate Dispersion and Process of Treating Pile Fabric with the Resulting Dispersion, *U.S. Patent 2,909, 451*

55. Cooke, T. F. and Pierce, E. S., assignors to the American Cyanamide Company (1957). Titania Monohydrate Soil–Retarding Treatment for Textiles, *U.S. Patent 2,788,295*

56. Furuike, S. and Yamamoto, K. (1974). Soiling–Resistant Textile Finishes, *Japanese Patent 74 08, 840*

57. Smith, A. I., assignor to the Monsanto Company (1971). Finishes for Improving the Resistance to Soiling of Melt–Spun Polyamide and Polyester Fibers, *U.S. Patent 3,620,823*

58. Smith, A. I., assignor to the Monsanto Company (1980). Soil Resistant Yarns and Aqueous Finishes for Yarns, *European Patent 16,658*

59. Mazzeno, L. W., Kullman, M. H., Reinhardt, R. M., Moore, H. B., and Reid, J. D. (1958). Wet–Soiling Studies on Resin–Treated Cotton Fabrics, *Amer. Dyestuff Rept.*, **47:** P299

60. Berch, J. and Peper, H. (1963). Wet Soiling of Cotton. Part I: The Effect of Finishes on Soiling, *Textile Res. J.*, **33:** 137

61. Tsuzuki, R. and Yabuuchi, N. (1968). Oleophilic Stain Release, *Amer. Dyestuff Rept.*, **57:** P472

62. Bille, H. E., Albrecht, E., and Schmidt, G. A. (1969). Finishing for Durable Press and Soil Release, *Textile Chem. Colorist*, **1:** 600

63. Arkens, C. T., Kottke, R. H., and Moser, V. J., assignors to Rohm and Haas Company (1970). Dirt–Repelling Coatings for Textiles, *German Patent 2,007,925*

64. Byrne, G. A. and Arthur, J. C. (1971). Soil–Release Properties of Durable–Press Cotton–Polymethacrylic Acid Copolymer, *Textile Res. J.*, **41:** 271

65. Hoffman, A. S. and Berbeco, G. R. (1970). Hydrophilic Polymer Surfaces Via Radiation–Chemical Treatments, *Textile Res. J.*, **40:** 975

66. Vrakami, M. and Okada, T. (1970). Radiation Induced Graft Copolymerization to Polyester. VII. Grafting of 2–Hydroxyethyl Methacrylate onto Polyester

Fabric as a Stain Release Treatment. *Nihon Genshiryujo Kenkyujo Nenpo*, Tokai–mura, Japan, **JAERI:** 5026, 63

67. Kawase, T., Uchita, M., Fujii, T., and Minagawa, M. (1991). Acrylic Acid Grafted Polyester Surface: Surface Free Energies, FT–IR(ATR), and ESCA Characterization, *Textile Res. J.*, **61:** 146

68. Kawase, T., Uchita, M., Fujii, T., and Minagawa, M. (1992). Effects of Grafting with Acrylic Acid on Removal of Oily Soil from Polyester Fabric, *Textile Res. J.*, **62:** 663

69. Marsh, J. T. (1968). A Finishing Problem: Soil Release, *Textile Manufacturer*, **94:** 465

70. Marco, F. W., assignor to Deering Milliken Research Corporation (1968). Soil Release of Polyester–Containing Textiles, *U.S. Patent 3,377,249*

71. Lindermann, M. K. and Volpe, R. P., assignors to the Air Reduction Company, Inc. (1967). Vinyl Acetate–Ethylene–N–Methylol Acrylamide Coatings for Woven Fabrics, *U.S. Patent 3,345,318*

72. Collins, R. J. and Thompson, R. G., assignors to E. I. du Pont de Nemours and Company. (1963). Antistatic and Antisoiling Agents and the Treatment of Synthetic Linear Textiles Therewith, *U.S. Patent 3,090,704*

73. Berch, J., Peper, H., and Drake, G. L. (1964). Wet Soiling of Cotton. Part II: Effect of Finishes on the Removal of Soil from Cotton Fabrics, *Textile Res. J.*, **34:** 29

74. Dennett, F. L., assignor to Dow Corning Corporation (1957). Organosilicone Treatment of Fabrics, *U.S. Patent 2,807,601*

75. Law, P. A., assignor to Dow Corning Corporation (1965). Oleophobic Finishes for Textiles, *U.S. Patent 3,179,534*

76. Pittman, A. G. and Wasley, W. L., assignors to the U. S. Secretary of Agriculture (1967). Polysiloxanes Containing Fluoroalkoxyalkyl Groups for Treatment of Fabrics, *U.S. Patent 3,331,813*

77. Pittman, A. G. and Wasley, W. L., assignors to the U. S. Secretary of Agriculture (1969). Bis(fluoroalkoxy)alkyl Siloxanes, *U.S. Patent 3,420,793*

78. Shane, N. C. and Wieland, N. G. (1967). Water–and–Oil–Repellant Finishes for Fabric, *U.S. Patent 3,336,157*

79. Fox, H. W. and Zisman, W. A. (1950). The Spreading of Liquids on Low Energy Surfaces. I. Polytetrafluoroethylene, *J. Colloid Sci.*, **5:** 514

80. Young, T. (1855). *Miscellaneous Works*, (G. Peacock, ed.), Murry, London, England, p. 418

81. Read, R. E. and Culling, G. C. (1967). Fluorochemical Textile Finishing Oil Repellency and Stain Resistance, *Amer. Dyestuff Rept.*, **56:** P881

82. Kamath, Y. K., Hornby, S. B., Weigmann, H. –D., and Wilde, M. F. (1994). Wicking of Spin Finishes and Related Liquids into Continuous Filament Yarn, *Textile Res. J.*, **64:** 33

83. Crews, P. F., Rich, W., and Kachman, S. D. (1995). Effect of Fiber Content, Fabric Construction, and Cleaning on the Performance of Fluorochemically–Finished Fabrics, *Textile Chem. Colorist*, **27** (11): 21

84. Sherman, P. O., Smith, S., and Johannessen, B. (1969). Textile Characteristics Affecting the Release of Soil During Laundering. Part II: Fluorochemical

Soil–Release Textile Finishes, *Textile Res. J.*, **39**: 449

85.   Smith, S. and Sherman, P. O. (1969). The Physical Chemistry of Stain Release, *Textile Chem. Colorist*, **1** (2): 20

86.   Pittman, A. G., Roitman, J. N., and Sharp, D. (1971). Hydrophilicity in Fluorochemical Stain Release Polymers, *Textile Chem. Colorist*, 3(7): 49

87.   Phillips, F. J., Segal, L., and Loeb, L. (1957). The Application of Fluoro–chemicals to Cotton Fabrics to Obtain Oil and Water Repellent Surfaces, *Textile Res. J.*, **27**: 369

88.   Ahlbrecht, A. H., Brown, H. A., and Smith, S., assignors to the 3M Company (1957), Fluorocarbon Acrylate and Methacrylate Esters and Their Polymers, *U.S. Patent 2,803,615*

89.   Ahlbrecht, A. H. and Brown, H. A., assignors to the 3M Company (1958). Fluorocarbon Vinyl–Type Esters and Polymers, *U.S. Patent 2,841,573*

90.   Johnson, R. E. and Raynolds, S., assignors to E. I. du Pont de Nemours & Company (1966). Polymeric water and Oil Repellents for Textiles, Paper, Leather, etc., *U.S. Patent 3,256,230*

91.   Berni, R. J., Benerito, R. R., and Philips, F. J. (1960). Durability of Oleobhobicity of Cotton Fabrics Imparted by Fluorochemicals, *Textile Res. J.*, **30**: 576

92.   Grajeck, E. J. and Petersen, W. H. (1962). Oil and Water Repellent Fluoro–chemical Finishes for Cotton, *Textile Res. J.*, **32**: 320

93.   Fasick, R. W. and Raynolds, S., assignors to E. I. du Pont de Nemours & Company (1966). Fluorine Containing Esters and Polymers, *U.S. Patent 3, 282,905*

94.   Kleiner, E. K., assignor to Geigy Chemical Company (1969). Poly(N–Perfluoro–alkanolaminoacrylamides) and Methacrylamides Useful as Oil–and–Water–Repellent Finishes for Textiles, *U.S. Patent 3,428,709*

95.   Langerak, E. O., Nelson, J. A., and Wright, E. J., assignors to E. I. du Pont de Nemours & Company (1966). Copolymers of N–Methylolacrylamide and Fluoroacrylate Esters and Their Application as Water and Oil Repellents on Textiles, *U.S. Patent 3,248,260*

96.   Smith, S. and Sherman, P. O., assignors to the 3M Company (1967). Vinyl Polymers Containing Perfluorocarbon Groups and Acyl Halide Groups, *U.S. Patent 3,330,812*

97.   Heine, R. F., assignor to the 3M Company (1963). Aqueous Dispersions of Fluorocarbon Vinyl Ethers and Their Polymers for Sizing Fabrics and Coating Paper and Leather, *U.S. Patent 3,078,245*

98.   Pacini, P. L., assignor to Geigy Chemical Corporation (1968). Acrylol Per-fluorohydroxamates, *U.S. Patent 3,412,142*

99.   Gagliardi, D. D., assignor to Colgate–Palmolive Company (1967). Soilproofing Quaternary Ammonium Derivatives of Highly Fluorinated Carboxylic Acids, *U.S. Patent 3,350,218*

100. Koshar, R. J. and Brown, H. A., assignors to the 3M Company (1964). Quaternized Halomethylamides, *U.S. Patent 3,147,065*

101. Guenthner, R. A., assignor to the 3M Company (1965). Polymers Containing A Fluorinated Quaternary Salt of an Amine Substituted Acrylic Acid Ester,

U.S. Patent 3,207,730

102. Berni, R. J., McKelvey, J. B., and Benerito, R. R., assignors to the U. S. Secretary of Agriculture (1963). Perfluoroalkoxy Substituted Propyl Ethers of Cellulose Textile Fibers, U.S. Patent 3,079,214

103. Berni, R. J. and Fagley, T. F., assignors to the U. S. Secretary of Agriculture (1961). Perfluoroalkanoyl Esters of Cellulose for Fibers Repellent to Oil and Water, U.S. Patent 2,992,881

104. Moreau, J. P. and Drake, G. L. (1968). Soiling and Soil Release Properties of Some Fluorochemical Finishes, Amer. Dyestuff Rept., 57(9): 13

105. Dorset, D. C. M. (1970). Fluorochemicals: Development and Uses in Oil- and Water–Repellent Finishes, Text. Manufacturer, 96: 112

106. Gagliardi, D. D., assignor to the Colgate–Palmolive Company (1968). Soil Proofing Textiles With Fiber Reactive Fluoroalkyl Derivatives of Amino–Aldehyde Compounds, U.S. Patent 3,362,782

107. Connick, W. J. and Ellzey, S. E. (1970). A Novel Polyfluorourea Finish for Cotton, Textile Res. J., 40: 185

108. Bloechl, W., assignor to FMC Corporation (1971). Water- and Oil–Repelling Polyfluorocarbamates, French Patent 2,034,379

109. Nelson, J. A., Miller, T. G., and Smeltz, K. C., assignors to E. I. du Pont de Nemours and Company (1960). Water- and Oil–Repellency Agents for Textiles, U.S. Patent 2,958,613

110. Green, L. Q., assignor to E. I. du Pont de Nemours and Company (1959). Preparation of Fluorine Compounds Useful as Oil Repellents for Textiles, U. S. Patent 2,917,409

111. Gagliardi, D. D., assignor to Colgate–Palmolive Company (1967). Methylol–4–perfluorooctanamidophenol Fabric Repellents, U.S. Patent 3,352, 625

112. Guenthner, R. A. and Lazerte, J. D., assignors to the 3M Company (1968). Fluorocarbon–Urethane Compounds for Imparting Oil Repellency and Stain Resistance, U.S. Patent 3,398,182

113. Downing, A. P. and Powell, R. L., assignors to Imperial Chemical Industries, Ltd. (1974). Additives for Producing Dirt–Releasing or Dirt–Repelling Properties in Textiles, German Patent 2,424,447

114. Stanley, R., assignor to the Penwalt Corporation (1975). Polyoxyalkylene Fluoroalkyltrimellitates, U.S. Patent 3,994,951

115. Bartlett, P. L., assignor to E. I. du Pont de Nemours and Company (1972). Homopolymers of Substituted Guanamines, U.S. Patent 3,687,900

116. Cohen, W. V., assignor to E. I. du Pont de Nemours and Company (1963). Fluoroalkyl Phosphate Oil Repellent Compounds, U.S. Patent 3,096,207

117. Mackenzie, A. K., assignor to E. I. du Pont de Nemours and Company (1965). Polyfluoro Alkamidoalkyl Phosphates, U.S. Patent 3,188,340

118. Gagliardi, D. D., assignor to Colgate–Palmolive Company (1967). Silicate and Titanate Orthesters for Fabrics, U.S. Patent 3,342,630

119. Segal, L., Loeb, L., Tripp, V. W., and Clayton, R. L., assignors to the U. S. Secretary of Agriculture (1962). Oil and Water Resistant Cellulosic Fabrics, U.S. Patent 3,031,335

120. Liljemark, N. T. and Åsnes, H. (1971). Soil–Release Properties Imparted to Polyester/Cotton Fabrics by Alkali Treatment, *Textile Res. J.*, **41**: 732

121. Wilson, D. (1962). The Effect of Anti–Static Agents on the Soiling of Garments and on the Ease of Soil Removal During Washing, *J. Textile Inst.*, **53**: T1

122. Garrett, D. A. and Hartley, P. N. (1966). New Finishes for Terylene® and Terylene® Blend Fabrics, *J. Soc. Dyers Colourists*, **82**: 252

123. Magat, E. E. and Sharkey, W. H., assignors to E. I. du pont de Nemours and Company (1969). Static Resistant Filament, *U.S. Patent 3,475,898*

124. Bierbruer, C. J., Goebel, K. D., and Landucci, D. P. (1979). Fluorocarbon Soil Retardants for Carpets, *Am. Dyestuff Rept.*, **68**(6): 19

125. Menin, B., assignor to E. F. Houghton & Company (1964). Soil Resistant Carpet Fibers, *U.S. Patent 3,159,502*

126. MacDonald, F. J. and Cook, A. A., assignors to the Arkansas Company (1966). Imidazoline–Ethylene Urea Derivatives, Metal Salt, *U.S. Patent 3,293,178*

127. Press, J. J. (1996). Antistatic Agent for Textiles, *U.S. Patent 3,277,079*

128. Kamath, Y. K., Huang, X. X., Ruetsch, S. B., and Weigmann, H. –D. (1996). Fibers (Stainproofing) in *Polymeric Materials Encyclopedia, Volume 4*, (Joseph C. Salamone, ed.), CRC Press, Boca Raton, FL, pp. 2333–2341

129. Martin, D. H. and Floyd, D. P. (1987). Stain Resistant Nylon Carpets, presented at the 57th Annual Meeting of the Textile Research Institute, Charlotte, NC, April 22, 1987

130. Moores, R. G. and Stefanucci, A. (1964). Coffee, *Kirk–Othmer Encyclopedia of Chemical Technology, Second Edition, Volume 4*, John Wiley & Sons, New York, pp. 748–763

131. Slade, P. E., unpublished data

132. Zollinger, H. (1991). *Color Chemistry, Syntheses, Properties and Applications of Organic Dyes and Pigments, Second and Revised Edition*, VCH, New York, pp. 56–59

133. Blyth, R. C. and Ucci, P. A., assignors to the Monsanto Company (1986). Stain–Resistant Nylon Carpets Impregnated with Condensation Product of Formaldehyde with Mixture of Diphenylsulfone and Phenolsulfonic Acid, *U. S. Patent 4,592,940*

134. Ucci, P. A. and Blyth, R. C., assignors to the Monsanto Company (1985). Process for Conveniently Providing Stain–Resistant Polyamide Carpets, *U.S. Patent 4,501,591*

135. Ucci, P. A., assignor to the Monsanto Company (1986). Stain Resistant Carpet with Impervious Backing, *U.S. Patent 4,579,762*

136. AATCC Evaluation Procedure 1 (1995). Gray Scale for Color Change, *AATCC Technical Manual*, American Association of Textile Chemists and Colorists, Research Triangle Park, NC, p. 111

137. Harris, P. W. and Hangey, D. A., (1989). Stain Resist Chemistry for Nylon 6 Carpet, *Textile Chem. Colorist*, **21**(11): 25

138. Greschier, I., Malone, C. P., and Zinnato, A. P., assignors to E. I. Du Pont de Nemours and Company (1988). Method for Producing Stain Resistant Polyamide Fibers, *U.S. Patent 4,780,099*

139. Fitzgerald, P. H., Rao, N. S., Visod, Y. V., Henry, G. K., and Prowse, K. S., assignors to E. I. du Pont de Nemours and Company (1991). Stain–Resistant Aromatic/maleic Anhydride Polymers, U.S. Patent 5,001,004

140. Payet, G. L. and Chang, J. C., assignors to the 3M Company (1989). Treating Fibrous Polyamide Articles, U.S. Patent 4,875,901

141. Green, G. D., Munk, S. A., and Barnes, D. K., assignors to Allied–Signal, Inc. (1990). Stain–Resistant Polymers Derived from the Itaconic Acid Useful as Coatings for Fibers, U.S. Patent 4,925,906

142. Cook, C. C. (1982). Aftertreatments for Improving the Fastness of Dyes on Textile Fibres, Rev. Prog. Coloration, 12: 73

143. Huffman, W. A. H., assignor to the Monsanto Company (1962). Polycarboxamide Fibers, U.S. Patent 3,039,990

144. Magat, E., assignor to E. I. du Pont de Nemours and Company (1965). Polycarbonamides of Improved Dye Affinity Having the Benzene Sulfonic Acid Salt Moiety as an Integral Part of the Polymer Chain, U.S. Patent 3,184, 436

145. Cooke, T. F. and Weigmann, H.–D. (1990). Stain Blockers for Nylon Fibers, Rev. Prog. Coloration, 20: 10

146. Cook, C. C. and Majisharifi, M. (1977). The Effects of Some Aromatic Hydroxy Compounds on the Adsorption of Anionic Dyes by Nylon 6, Textile Res. J., 47: 244

147. Shore, J. (1971). Aftertreatment of Anionic Dyes on Nylon Fibres. I. Effect of Backtanning on Dyed Nylon, J. Soc. Dyer. Colour., 87(1): 3

148. Shore, J. (1971). Aftertreatment of Anionic Dyes on Nylon Fibres. II. Effect of Postsetting on Dyed and Backtanned Nylon, J. Soc. Dyer. Colour., 87(2): 37

149. Anton, A. (1981). Polyamide Fiber Reactive Chemical Treatments for Differential Dyeability, Text. Chem. Colorist, 13(2): 31

150. Tomita, M. and Tokitaka, M. (1980). Dihydroxy–Diphenylsulphone and Salicylic Acid Derivatives in the Aftertreatment of Dyed Nylon, J. Soc. Dyer. Colour., 96(6): 297

151. Guthrie, J. and Cook, C. C. (1981). The Adsorption of a Synthetic Tanning Agent and a Sulphonated Phenol–Formaldehyde Novolac by Nylon-6, Polymer, 22: 1939

152. Guthrie, J. (1982). Effect of a Synthetic Tanning Agent on the Diffusion Properties of an Acid Dye in Nylon 6, J. Appl. Polymer Sci., 27: 2567

153. Guthrie, J. and Cook, C. C. (1982). The Effect of a Synthetic Tanning Agent on the Uptake of Acid Dyes by Nylon 6, J. Soc. Dyer. Colour., 98(1): 6

154. Dawson, T. L. and Todd, J. C. (1979). Dye Diffusion — The Key to Efficient Coloration, J. Soc. Dyer. Colour., 95(12): 417

155. Matsuo, M., Itoh, K., Hayashi, T., and Oda, Y., assignors to Asahi Glass Company (1985). Stainproofing Agent and Process for its Preparation, U.S. Patent 4,504,401

156. Blyth, R. C. and Ucci, P. A., assignors to the Monsanto Company (1987). Stain Resistant Nylon Fibers, U.S. Patent 4,680,212

157. Kirjanov, A. S. and Hoecklin, D., assignors to Crompton & Knowles Corporation (1989). Stainblocker and Fluorocarbon Oil Repellents, U.S.

Patent 4,857,392

158. Munk, S. A. and Malloy, T. P., assignors to Allied Corporation (1987). Imparting Stain Resistance to Certain Fibers, *U.S. Patent 4,699,812*

159. Blyth, R. C. and Ucci, P. A., assignors to the Monsanto Company (1989). Stain Resistant Nylon Carpet, *U.S. Patent 4,839,212*

160. Jose, D. J., Lewis, B. J., Materniak, J. M., Rivet, E., Shellenbarger, R. M., Vinod, Y., and Williams, E. D. (1988). Stain Resistant Carpets, *Can. Text. J.*, **105**(11): 34

161. Blyth, R. C. and Ucci, P. A., assignors to the Monsanto Company (1989). Stain–Resistant Nylon Fibers, *U.S. Patent 4,879,180*

162. Blyth, R. C. and Ucci, P. A., assignors to the Monsanto Company (1990). Process for Dyeing Stain Resistant Nylon Carpets, *U.S. Patent 4,892,558*

163. Chao, N. P. C., assignor to the Monsanto Company (1991). Continuous Processes for Acid Dyeing of Stain Resistant Nylon Carpets, *U.S. Patent 5, 030,246*

164. Pfeifer, C. R., assignor to BASF Corporation (1989). Stain Resistant Composition for Synthetic Organic Polymer Fibers and Methods of Use: Fluorocarbon Polymer, *U.S. Patent 4,861,501*

165. South Carolina Section, AATCC (1989). Chemical Analysis of Yellowing of Stain Resist Finishes for Carpets, *Text. Chem. Colorist*, **21**(4): 21

166. Olson, M. H., Chang, J. C., and Muggli, I. A., assignors to the 3M Company (1989). Process for Providing Polyamide Materials with Stain Resistance with Sulfonated Novolac Resin and Polymethacrylic Acid, *U.S. Patent 4,822,373*

167. Fitzgerald, P. H., Rao, N. S., Vinod, Y, and Alender, J. R., assignors to E. I. du Pont de Nemours and Company (1989). Stain–Resistant Agents for Textiles, *U.S. Patent 4,883,839*

168. Moss, T. H., Sargent, R. R., and Williams, M. S., assignors to Peach State Labs (1990). Stain Resistant Polymeric Composition, *U.S. Patent 4,940,757*

169. Liss, T. A. and Beck, L. H., assignors to E. I. du Pont de Nemours and Company (1990). Stain Resistant Polymers and Textiles, *U.S. Patent 4,963, 409*

170. Liss, T. A. and Beck, L. H., assignors to E. I. du Pont de Nemours and Company (1990). Stain Resistant Polymers & Textiles, *U.S. Patent 4,965,325*

171. Pechhold, E., assignor to E. I. du Pont de Nemours and Company (1994). Process Providing Durable Stain–Resistance by Use of Maleic Anhydride Polymers, *U.S. Patent 5,358,769*

172. Green, G. D., Munk, S. A., and Barnes, D. K., assignors to Allied–Signal, Inc. (1991). Stain–Resistant Polymers Derived from Itaconic Acid Useful as Coatings for Fibers, *U.S. Patent 5,006,408*

173. Hangey, D. A., Friedberger, M. P., Archie, W. A., and Spitz, R. N., assignors to Allied–Signal, Inc. (1992). Methods and Compositions to Enhance Stain Resistance of Dyed Nylon Carpet Fibers: Thiocyanate to Reduce Yellowing, *U.S. Patent 5,110,317*

174. Fitzgerald, P. H., assignor to E. I. du Pont de Nemours and Company (1993). Phenolic Stain–Resists, *U.S. Patent 5,229,483*

175. Elgarhy, Y. M. and Knowlton, B. R., assignors to Trichromatic Carpet, Inc. (1995). Polyamide Materials with Durable Stain Resistance, *U.S. Patent 5,457,259*

176. Sargent, R. R. and Williams, M. S., assignors to Peach State Labs (1994). Permanently Stain Resistant Textile Fibers, *U.S. Patent 5,316,850*

177. Haarer, J. and Höcker, H. (1994). New Reactive Auxiliaries for Dye–Resist Treatment of Wool, *Textile Res. J.*, **64**: 480

178. Calcaterra, L. T., Koljack, M. P., Farishta, Q., Koehler, M. G., Bedwell, W. B., Hangey, D. A., and Green, G. D., assignors to Allied–Signal, Inc. (1992). Method to Impart Stain Resistance to Polyamide Textile Substrates, *U.S. Patent 5,118,551*

179. Calcaterra, L. T., Koljack, M. P., Farishta, Q., Koehler, M. G., Bedwell, W. B., Hangey, D. A., and Green, G. D., assignors to Allied–Signal, Inc. (1994). Method to Impart Coffee Stain Resistance to Polyamide Textile Substrates, *U. S. Patent 5,359,010*

180. Kelson, J. S. and Holt, L. A. (1992). Chromatographic Separation of the Components of Commercial Syntans and Assessment of Their Effectiveness as Stain–Blocking Agents, *J. Soc. Dyers Colour.*, **108**: 327

181. Bauers, M. Keown, R. W., and Malone, C. P. (1993). Separating and Identifying the Active Ingredient of a Stain Resistant Compound, *Textile Res. J.*, **63**: 540

182. Hangey, D. A., assignor to Allied Signal Inc. (1992). Molecular Size of Hydrodynamic Volume of Sulfonated Aromatic Condensates Used to Impart Stain Resistance to Polyamide Carpets, *U.S. Patent 5,131,909*

CHAPTER 10

ENVIRONMENTAL DEGRADATION OF FINISH MATERIALS

Four possible routes exist for finish materials to enter the ecosystem, especially the aquatic segment of that system. These are (1) through an unintentional spill or discharge during the synthetic fiber manufacturing process, (2) during removal of finishes at a textile processing plant, (3) from discarded yarn or fabric in land fills, or (4) from carpet cleaning and subsequent disposal of the wastewater.

No manufacturing process is completely foolproof. Errors are made in piping, employees can make mistakes in materials handling, and sometimes finishes can enter waste streams through cleaning of equipment. Although most synthetic fiber producers have waste treatment facilities that can handle any discharge, occasionally some accident happens that will allow finish materials to find their way into a stream or the ground water.

Perhaps the one most predominate route for finishes to escape into aquatic systems is at a fabric processing plant. Many types of yarn are dyed, either as the yarn itself or as a finishing step in fabric preparation. Finishes can be removed in a pre–scouring rinse, in a dyeing process, or in a washing procedure after dyeing. Although many textile mills have adequate on–site wastewater treatment, some do not. The effluent is allowed to flow directly into a stream or to a community water treatment facility. This has been a problem in the past and certainly will be one in the future.

The third route for finishes to reach the environment is becoming more significant. Used carpet and other fabrics are often placed in landfills and rain can easily remove finish components and allow them to reach the ground water supply. Recycling carpet will certainly become a most desirable step, and probably a future requirement. Although the fourth process of finish release, carpet cleaning, is not usually considered significant, it can discharge large quantities of textile finish into domestic waste disposal systems.

The finish chemists should be aware that their creation will, eventually, become a part of the aquatic ecosystem and they should plan for that reality. Components should be chosen with care and all efforts must be made to prevent accidental discharges.

10.1 WATER RESOURCES

Solar radiation is used by many biological and physical processes on the earth. A small percentage is employed by growing plants during photosynthesis but a large portion, approximately 25%, is used to evaporate water and initiates what is known as the hydrologic cycle. Water not only evaporates from any wet surface, but plants also release water to the atmosphere by transpiration. The evaporated water returns to the earth's surface as some form of precipitation, but not necessarily back to the location from which it originally came. Water is distributed on the earth in the ratios listed in Table 9–1 [1].

Table 9–1

Distribution of Water Resources

Location	Water Volume, 10^{15} gallons	Percent of Total
World oceans	362,000	94.20
Glaciers	6, 350	1.65
Lakes	60.5	0.016
Soil moisture	21.6	0.006
Atmospheric vapor	3.7	0.001
River water	0.32	0.0001
Total ground water	15, 800	4.13
Ground water less than 5 miles down	1, 160	0.28

The average annual rainfall on the conterminous United States is about 30 inches, which is equivalent to a daily precipitation of 4,200 billion gallons of water. About 70% of this, or 3,000 bgd (billion gallons per day), is lost to evaporation, leaving 1,200 bgd as the daily runoff. Of this amount, at most only 315 bgd can be used as a sustained, dependable water supply [2]. This water is used for irrigation, industry, cooling water, and domestic requirements. The rest is either required for navigation, hydroelectric power, and fish and wildlife, or could not economically be developed into a dependable supply. Most of this 315 bgd of water is consumed in agriculture (84%) with only about 6% used by public water supplies and only 4% by industry [3]. Although only this small amount of water is used by industry, it is extremely important to avoid any pollution that would contaminate the water supply for other end uses.

10.2 WATER POLLUTION

Much of the water withdrawn for use has already been used in one application or another, and some of it will be reused again before it finally returns to the oceans. With each use of the water, various forms of pollution contribute to a degradation of its quality. Sometimes the degradation of water quality is only temporary—natural self–purification is sufficient to eventually restore the water quality—but sometimes the pollutant is one that does not degrade naturally. In other cases the sheer volume of pollution is sufficient to overload the self–purification mechanisms. These two circumstances can cause the water quality to be more permanently damaged.

There are several ways to categorize water pollutants, with the following list proposed by the Environmental Protection Agency [3].

1. Oxygen–demanding wastes
2. Disease–causing agents
3. Synthetic organic compounds
4. Plant nutrients
5. Inorganic chemicals and mineral substances
6. Sediments
7. Radioactive substances
8. Thermal discharges

An expanded discussion of this list follows in the next section.

    1.   Oxygen–demanding wastes are biodegradable organic compounds contained in domestic sewage or some industrial effluents. Although these wastes are decomposed by bacteria, the oxygen dissolved in the water system (DO) utilized in the bacterial

       degradation (aerobic decomposition) can be depleted. If the amount of oxygen drops to a low enough level, another decomposition mechanism, anaerobic decomposition, proceeds.

2. Disease–causing agents are various pathogenic microorganisms that usually enter the water from human sewage. Contact with these organisms can be made through drinking water or other activities in which water contact occurs.

3. Synthetic organic compounds include soaps, detergents, and other household products. Agricultural chemicals such as pesticides, herbicides, and various organic chemicals from industrial effluent waste streams can also cause pollution. Some of these products are toxic to wildlife and may also be harmful to humans.

4. Plant nutrients are principally derived from agricultural runoff from fertilized farm land, but may also have as their source most sewage treatment plants and from industrial waste.

5. Inorganic chemicals and mineral substances originate from mines, acids and bases, and heavy metals such as mercury and cadmium. These can be extremely toxic to both wildlife and humans.

6. Sediments are particles of soils, sands, and minerals washed from the land. Sedimentary material can fill reservoirs, harbors, and rivers thus causing hazards to navigation and marine life.

7. Radioactive substances can enter water systems from mining and the processing of radioactive ores, possibly from nuclear power plants and some medical facilities.

8. Thermal discharges from steam–electric and nuclear power plants can raise the water temperature by as much as 20°F, resulting in local changes in the water system.

10.3 DISSOLVED OXYGEN

The amount of dissolved oxygen (DO) in water is an important parameter of water quality. Fish, for example, require some minimum amounts of DO depending on their species, stage of development, level of activity, and the water temperature. In one study, McKee and Wolf [4] recommended that the DO remain above 5 ppm for at least 16 hours of the day, and during the remaining eight hours it should not drop below 3 ppm. Some more desirable species, such as trout, require more oxygen than other species.

       There are four processes that affect the amount of oxygen in the water: reaeration, photosynthesis, respiration, and the oxidation of organic matter. *Reaeration* is the process in which oxygen enters the water through the contact the water surface makes with the atmosphere. The solubility of oxygen in fresh water, at a pressure of one atmosphere, decreases with temperature, as shown in Figure 10–1 [5].

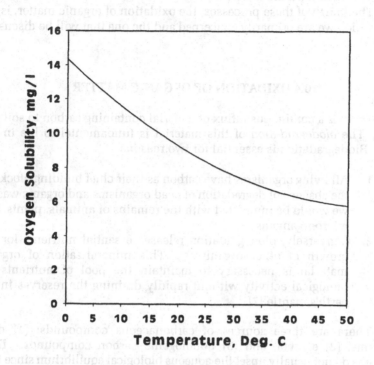

Figure 10–1. Effect of Temperature on Dissolved Oxygen.
(Ref. [5]. From Standard Methods for the Examination of Water and Wastewater, 19*[th] Edition. Copyright 1995 by the American Public Health Association, the American Water Works Association, and the Water Environment Federation. Used with permission.)*

When the actual amount of oxygen is less than the saturation value given in the figure, atmospheric oxygen absorbs into the water at a rate that is proportional to the deficit. By increasing the surface area of the water in contact with the atmosphere the transfer of oxygen to the water is increased, thus bubbling air into a system or spraying the water into the air will dramatically increase the rate of absorption.

During the process of *photosynthesis*, solar energy is used to generate oxygen during the daylight hours. In photosynthesis carbon dioxide and water, catalyzed by light and chlorophyll, generate oxygen and some carbohydrate or organic matter. Many aspects of photosynthesis are known, but some of the details of energy transfer and photooxidation–reduction are still unknown [6]. *Respiration* is defined as a procedure that continually removes oxygen from the water.

The fourth of these processes, the oxidation of organic matter, is the one in which we are primarily concerned and the one that will be discussed in detail.

10.4 OXIDATION OF ORGANIC MATTER

There is a continuous influx of material containing carbon to soil and water. The *biodegradation* of this material is fundamental to life in our world. Biodegradation is essential for two reasons:

1. All living organisms have carbon as their chief building block. In the absence of degradation of dead organisms and organic wastes, we would be inundated with the remains of animals, plants, and microorganisms.
2. Conversely, biodegradation releases essential nutrients for the growth of other organisms. This mineralization of organic material is necessary to maintain the pool of nutrients for biological activity without rapidly draining the reserves in the earth's mantle [7].

There are three sources of carbonaceous compounds: (1) dead organisms, (2) excreta, and (3) nonbiogenic carbon compounds. Dead organisms do not usually upset the aqueous biological equilibrium since they are normally decomposed in the soil. A major source of material that disrupts this equilibrium comes from animal and human excreta in the natural waters. In urban areas, sewage treatment plants provide microflora to degrade the carbon compounds that the native microflora in the natural waters cannot handle because of the great quantities.

In rural areas, however, the amount of animal waste far exceeds the amount of human excreta in the urban areas. A major factor contributing to the problem of disposal of animal wastes is the urbanization of farm animals. Livestock in the United States produce more than one billion tons of solid waste and almost ½ billion tons of liquid waste annually. This amount of organic matter is equivalent to that produced by two billion people. A typical feedlot for cattle contains about 10,000 animals and is equivalent in organic load to a city of 45,000 people.

The third group of carbonaceous compounds being disposed into the soil and water is nonbiogenic. These include fuel, oils, pesticides, and industrial wastes, and many of these are *recalcitrant*, meaning that degrading them biologically is difficult. The amount of these waste products is very large, with more than 120 million people served by sewers. These people produced more than 5,000 billion gallons of wastewater with a combined biochemical oxygen demand (BOD) of more than 3,500,000 tons. Table 10–2 lists the sources of nonbiogenic wastes in the Unites States in 1963 [7].

Table 10–2

Sources of Domestic and Industrial Organic Pollutants

Source	Waste Water Volume (billion gal/yr)	Strength (B.O.D.) (million lb/yr)
Domestic wastes	5300	7300
Chemical industries	3700	9700
Pulp and paper wastes	1900	5900
Textile mills	140	890
Food processing	690	4300
Petroleum and coal	1300	500
Rubber and plastics	160	40

(Reference [7]. Reprinted by permission of Prentice–Hall, Inc.)

This wide range of organic compounds continually enters the aquatic environment and presents many challenges to the microbial systems. There are equally complex groups of microorganisms present in the ecosystem producing the enzymes that degrade these compounds. If there is a supply of dissolved oxygen in the system, the decomposition proceeds through an *aerobic* mechanism. If the rate of oxygen consumption is more rapid than reaeration and photosynthesis can increase the rate of oxygen production, then the water becomes *anaerobic*. The consequences of anaerobic decomposition included a rapid decline in fish population and the production of volatile materials that have very unpleasant odors. The rate of substrate decomposition under aerobic conditions is measured by the BOD of the water.

A general first–order rate equation that describes the degradation of organic materials under these aerobic conditions is [7]:

$$y = L(1 - e^{kt}) \qquad (10-1)$$

in which y is the oxygen demand in time t, L is the first stage BOD, and k is a reaction rate constant. It should be noted that L represents only the first stage BOD while two other stages of oxygen usage can occur. In this first stage the microflora uses the available nonrecalcitrant organic compound in the water. During growth they assimilate the organic matter and excrete reduced inorganic chemicals. At the end of the first stage all available organic compounds have disappeared and the medium is rich in ammonium compounds. In this second stage, bacteria that oxidize ammonium ions to nitrate proliferate, using CO_2 as a carbon source. The third stage of oxygen utilization involves action of protozoa and other microorganisms on the large bacterial population developed in the first two stages. The five-day BOD test, which will be discussed later, measures the first stage microflora activity. The later two stages are ignored. When scientists speak of a BOD of 50 mg/liter or of the removal of 99% of the BOD in a waste by some treatment process, they refer only to the immediately available organic matter [7].

The oxidation of organic wastes in streams can be calculated by the Streeter–Phelps equation [8]. This equation is based on the ability of microorganisms in a stream to degrade a specific concentration of wastes by observing the rate of oxygen utilization and the rate of reaeration from the atmosphere and algal photosynthesis. When K_1 is defined as the deoxygenation rate for a stream flowing from point a to point b, it can be calculated as:

$$K_1 = \frac{1}{T_b - T_b} \log \frac{L_a}{L_b} \qquad (10\text{--}2)$$

in which T is the time of sampling at each point, and L the ultimate first stage BOD at these points. The reaeration rate for stream conditions at the same points is K_2, which is:

$$K_2 = \frac{(K_1)(L_{avg})}{D_{avg}} - \frac{D_b - D_a}{(2.3)(D_{avg})(t)} \qquad (10\text{--}3)$$

In Equation 10–3, L_{avg} is the average value of the BOD from point a to point b, D_{avg} is the average value of the dissolved oxygen deficit, D is the dissolved oxygen deficit at the stated points, and t is the time for the stream to flow from a to b.

These deoxygenation and reaeration rates are combined into the Streeter–Phelps equation:

$$D = \frac{K_1 L_0}{K_2 - K_1}(10^{-K_1 t} - 10^{-K_2 t}) + D_0(10^{-K_2 t}) \qquad (10-4)$$

in which D is the dissolved oxygen deficit at any time of flow t, L_0 is the initial BOD load in the stream, D_0 the initial dissolved oxygen deficit in the stream, and t is the time of flow.

The absence of oxygen does not completely inhibit the microbial decomposition of organic matter. Under *aerobic* conditions, oxygen is the ultimate electron donor for energy transfer in the microbial cell. Under *anaerobic* conditions, the organic compounds themselves serve as the electron donors; however, the energy yield in the anaerobic process is lower than the aerobic one. This results in a slower rate of decomposition and incomplete degradation of the carbonaceous materials.

The addition of large quantities of organic matter to a natural water source results in a high rate of biochemical oxidation by the microflora that are present and ultimately in a deficiency in dissolved oxygen. This aerobic decomposition has carbon dioxide and water as their final products. The shift to anaerobic metabolism yields organic acids, with the most common being lactic, butyric, and acetic. A broad range of anaerobic microorganisms is responsible for these transformations [7].

The formation of flammable methane gas is one of the most important anaerobic transformations. This reaction requires the complete absence of oxygen; however, these conditions are seldom met in oxygen–deficient waters. Biochemical methane synthesis is common in swamps and deep sediments. Rumen bacteria in cattle produce methane and are frequently used to study this reaction.

Methane–producing bacteria can be used in the secondary treatment of domestic sewage. Sludge is digested in an anaerobic reactor, kept free of oxygen, and contains a culture of methane–forming bacteria, with methane gas as the principal product. Methane produced in this process is frequently sufficient to provide power for the sewage treatment plant.

The production of methane is a two–step process that requires two separate microorganisms. In the first stage, an organic acid or alcohol is formed from complex organic substrates. This is not a specific process but many different anaerobic microflora can be responsible for this reaction. The second step is the specific reduction of the acid or alcohol to methane by a methane bacterium.

Formation of acetic acid is an essential step in the formation of methane. The carboxyl group is converted to CO_2 and the alkane portion to CH_4. The carbon dioxide then reacts with hydrogen ions formed in the oxidation of the organic compound to acetic acid yielding more methane.

The methane–forming bacteria belong to a separate family, the *methanobacteriaceae*. They are gram–negative and are strict anaerobes [7].

10.5 PREDICTION OF WASTE DEGRADATION

There are several laboratory tests that are used to show aerobic decomposition of organic matter in wastewater. Biodegradation is an important factor determining the behavior of these chemicals in the environment. Data on biodegradation are required by legislation governing the manufacture and use of chemicals in many countries, although there is no general agreement on standard methods for the measurement of biodegradation nor on the evaluation of the results among those countries.

Methods for characterization of biodegradation include nonspecific measurements, such as theoretical oxygen demand (ThOD), chemical oxygen demand (COD), biochemical oxygen demand (BOD), and dissolved organic carbon (DOC). Specific methods measure the disappearance of the parent compound, and in more detailed studies, the concentration of the degradation products. These specific methods include a variety of separation techniques such as thin–layer, gas, and high pressure liquid chromatography. In some studies mass spectrometry has been used as a tool for characterization [9].

10.5.1 Nonspecific Methods

10.5.1.1 Theoretical Oxygen Demand (ThOD)

The calculated amount of oxygen required to oxidize hydrocarbons to carbon dioxide and water is known as the theoretical oxygen demand (ThOD) and is usually expressed as grams of oxygen/gram of material [10]. For example, when a hydrocarbon is oxidized to carbon dioxide and water:

$$C_nH_{2m} + \left(n + \frac{m}{2}\right)O_2 \longrightarrow n\ CO_2 + m\ H_2O$$

The ThOD of the hydrocarbon, C_nH_{2m}, equals:

$$\frac{32\left(n + \dfrac{m}{2}\right)grams}{(12n + 2m)\,grams} = \frac{8\,(\,2n + m\,)}{6n + m} \tag{10–5}$$

When the organic molecule contains elements other than C, H, or O (such as N, S, P, etc.) the ThOD depends on the final oxidation state of that element.

As an example, the ThOD of a hydrocarbon can be calculated. In butane, with the formula $CH_3CH_2CH_2CH_3$, the number of carbons, n, equals 4, the number of hydrogens, $2m$, is 10, and the chemical has a molecular weight of 58. It is oxidized by $(n + m/2)$ molecules of oxygen, or 6.5. The ThOD is then:

$$\frac{8(2*4+5)}{6*4+5} = \frac{104}{29} = 3.586g\, O_2/g\, C_4H_9 \qquad (10\text{--}6)$$

This ratio can obviously be restated as:

$$\frac{Mol.\, Wt.\, 6.5\, O_2}{Mol.\, Wt.\, C_4H_{10}} = \frac{208}{58} = 3.586\, gO_2/gC_4H_{10} \qquad (10\text{--}7)$$

10.5.1.2 Chemical Oxygen Demand (COD)

If the structure is unknown, the total amount of organic material in a waste stream, or an aqueous solution, that is readily oxidized can be measured by the COD. The usual procedure is to reflux the material with potassium dichromate in sulfuric acid in the presence of silver sulfate as a catalyst, as proposed by the American Public Health Association (APHA), Method 5225 C [5]. In this procedure, 10 ml of sample, or an aliquot diluted to 10 ml, is placed in a flask fitted with a ground glass joint. Six milliliters of a 0.0167 molar potassium dichromate solution is added to the sample together with several glass beads. At this time, slowly add 14 ml of concentrated sulfuric acid to which a small amount of silver sulfate has been dissolved and mix well. Mercuric sulfate may also be included to complex any chloride ions. Connect the flask to a condenser and reflux for two hours. After cooling, dilute with distilled water and titrate the excess dichromate with standard ferrous ammonium sulfate using a ferroin indicator. A blank containing 10 ml of distilled water is carried through the same method. The COD, in mg/l, is calculated as:

$$COD, mg/l = \frac{(A - B) \times M \times 8000}{ml\, sample} \qquad (10\text{--}8)$$

where:

A = ml ferrous ammonium sulfate used for the blank,
B = ml ferrous ammonium sulfate used for the sample, and
M = molarity of the ferrous ammonium sulfate.

10.5.1.3 Biochemical Oxygen Demand (BOD)

Biochemical oxygen demand is usually exerted by dissolved and colloidal organic matter, and imposes a load on the biological units of the waste treatment plants. Oxygen must be provided so that bacteria can grow and oxidize the organic matter. An added BOD load, caused by an increase in organic waste, therefore requires more biological activity, more oxygen, and more biological–unit capacity for its treatment. All dissolved or colloidal organic matter, however, does not oxidize at the same rate, with the same ease, or to the same degree. Sugars, for example, are more readily oxidized then starches, proteins, or fats. The rate of decomposition for industrial organic matter may be faster or slower than that for sewage organic matter [11].

Biochemical degradation of organic materials in treatment plants or in streams can be empirically assayed in the laboratory by measuring the reduction in dissolved oxygen as the microflora assimilates and oxidizes the material. The most common procedure is to incubate the sample in the dark for five days followed by a measurement of the oxygen content. This BOD procedure, like the COD procedure, is also proposed by the American Public Health Association in method 5210 B [5]. Seed bacteria used in the laboratory test are normally purchased from a scientific supply source.

In the prescribed procedure, all water used in the test must be distilled or treated by a special deionizing system. It should be aerated thoroughly to achieve maximum oxygen saturation. One liter of the water or wastewater sample is aerated for about 45 minutes, followed by the addition of the microbial seed solution and various amounts of dilution water containing a nutrient buffer to fill the container completely. The initial dissolved oxygen (DO) is then measured with a DO meter. Stopper without trapping air bubbles and place in a dark incubator at $20 \pm 1°C$ for five days. After that time, remove the sample and again measure the DO. The biochemical oxygen demand is calculated as:

$$BOD\, mg/l = \frac{(D_1 - D_2) - (B_1 - B_2)f}{P} \qquad (10\text{--}9)$$

where:

D_1 = initial sample DO,
D_2 = DO of the sample after incubation,
B_1 = initial DO of the control,
B_2 = DO of the control after incubation,
f = ratio of seed in the sample to seed in the control, and
P = decimal fraction of the sample used.

The aerobic decomposition of organic matter in streams is dependant on the complex native microflora capable of degrading this mixture of ever–changing substrates. Two groups of microorganisms are responsible for the degradation:

1. Microorganisms suspended on colloidal particles,
2. Sessile microorganisms.

The vast majority of microorganisms in natural waters are not freely suspended. They are attached to detritus, clay, and other colloids in the water. This portion of the microflora lives on benthic algae and on the sediment. This group includes *Sphaerotilus*, a filamentous aerobic bacterium that attaches to surfaces in rivers containing high concentrations of organic matter. Steamers are formed from chains of cells covered in a sheath [7].

Usually, BOD values are lower than ThOD or COD values since some organic materials, known as *recalcitrant*, are not easily metabolized by bacteria. The BOD test, as a useful measure of potential substrate degradation and oxygen utilization, has several pitfalls:

1. It ignores the contribution of recalcitrant organic compounds.
2. It assumes first–order kinetics. The reaction is much more complex and may not follow first–order kinetics.
3. It ignores the mineralization of the organic substrate and the potential for algal photosynthesis and therefore further BOD at a later time or place [7].

10.5.1.4 River Die–Away Test

A variation of the biochemical oxygen demand procedure is the river die–away test. This technique has been used extensively to measure the biodegradability of both nonionic and anionic surfactants and shows the rates of breakdown and the degree of ultimate disappearance [12]. In this method a specific quantity of the detergent test substance is added to a sample of river water contained in a glass jar and allowing the mixture to incubate at room temperature. Degradation of the detergent in this system is measured by some analytical procedure, especially by a cobaltothiocyanate colorimetric technique that assesses the degradation of polyoxyethylene surfactants. Disappearance of anionic surfactants is measured by a methylene blue method [12].

10.5.2 Specific Methods

These methods are specific either in respect to a particular compound, or to its physical properties, or biological activity. Methods for analyzing for a selected compound may not be unequivocally specific, but are selective and may be subject to interference by other compounds. The best example for methods using a physical property is the analytical procedure for surfactants, when they are based on properties such as solubility, complex formation, or surface activity of the parent compound [13]. The determination of biological activity such as the inhibition of acetylcholinesterase by organophosphorus is an example of the third method noted above; however, techniques based on biological activity are rare in biodegradation studies [9].

1. Thin–Layer Chromatography (TLC)

Thin–layer chromatography is a frequently used separation technique in biodegradation experiments. The reasons for popularity of TLC are its simplicity, low cost, and high separation efficiency. Because of the latter, it is usually possible to see at a glance the parent compound and many of its metabolites on a developed TLC plate. It is also possible to estimate quantitatively how much material was left behind at the start. This is very useful to know and more difficult to estimate in other techniques. Complex mixtures of products can be separated by two–dimensional TLC without introducing much complication into the procedure. The commercial availability of high–quality, coated TLC plates increase the convenience of the technique [9].

2. High Pressure Liquid Chromatography (HPLC)

HPLC is a versatile and fast technique for analytical or preparative separation and quantitative determination of a variety of chemicals. The equipment is much more expensive than that for TLC, but it can be bought in pieces and added on as required. The essential equipment consists of a pump, a column, and a detector (frequently a UV detector). Some excellent commercial HPLC systems are available; however, they are quite expensive.

One advantage of HPLC in biodegradation studies is that an analysis can be performed on a mixture of compounds with differing polarities in one run. This is usually impossible in other separation methods. In HPLC, these separations are achieved on reversed–phase columns. These are columns containing silica particles chemically bonded to a hydrocarbon, usually C_{18} aliphatic chains. The commonly used solvent systems are mixtures of either methanol or acetonitrile and water. The solvent is more polar than the stationary phase, therefore the term reversed–phase. By decreasing the water content of the mobile phase, compounds are eluted from their column in order of decreasing polarity, as opposed to the order of elution in classic liquid/solid chromatography. This is of a considerable advantage in the separation and quantitation or isolation of metabolites since these are as a rule more polar then the parent compounds.

3. Gas Chromatography (GC)

Gas chromatography is a routine separation and quantitative determination technique with a large variety of instrumentation available commercially. GC offers a larger choice of detectors than HPLC and generally of higher sensitivity. As a rule, derivatives of more polar compounds must be prepared before GC analysis. Common examples of such derivatives are esters of carboxylic acids and trimethylsilyl ethers in cases of alcohols or phenols.

4. Gel Permeation Chromatography (GPC)

This technique separates compounds primarily based on molecular weight, with elution proceeding generally in the order of decreasing molecular weight, although adsorption on gel particles may play a role as well. In biodegradation studies, the primary importance of GPC is in the separation of compounds of interest from coextractives. This technique has not been widely used in biodegradation studies.

5. Structure Determination After Separation

After the complex mixtures have been separated by one of the above procedures, the individual compounds may be characterized by ultraviolet and visible spectrophotometry, fluorescence spectrophotometry, infrared spectrophotometry, mass spectrometry, and nuclear magnetic resonance.

10.6 DEGRADATION OF SPECIFIC CHEMICALS

Much of the organic matter that is degraded either in community waste treatment facilities or in natural water systems is easily oxidized by the bacteria that reside in those aqueous environments. Unfortunately, many chemicals, either natural or synthetic, are finding their way into soil and water. Some of these are strongly resistant to biodegradation, and are known, as previously noted, as recalcitrant organic compounds.

The dramatic increase in technological development in the United States during the past two or three decades has put an ever–increasing strain on the water that receives these recalcitrant compounds. These include complex organic chemicals produced as by–products in the synthetic chemical industry, and in the plastics industry, together with petroleum hydrocarbons.

The agricultural industry produces many tons of recalcitrant pesticides that are partially responsible for improved crop yields. Paradoxically, some of the most effective pesticides are also responsible for a health hazard to human and animal life and present a serious danger to the aquatic environment.

Fortunately, most of the components used as components in spin finishes are not classified as recalcitrant. Most of these compounds are degraded but can, at times, overload the community treatment facilities and the manufacturing and processing locations must provide at least some pre–discharge treatment before discharge. The degradation of the various types of finish components is discussed below.

10.6.1 Mineral Oils and Other Petroleum Products

Crude oil is a mixture of aliphatic and aromatic hydrocarbons; however, the mineral oils used in textile lubricants and coning oils are generally aliphatic. Straight–chain aliphatic hydrocarbons are normally degraded in natural water by a large number of bacteria and yeasts [14]. Branched chain alkanes are more resistant and may remain in the aqueous environment for a longer period. The biodegradation of petroleum hydrocarbons in a natural habitat follows these empirical rules:

1. Alkanes are more rapidly degraded than aromatic hydrocarbons.
2. Within the alkanes, straight chains are more susceptible than branched chains.
3. Chains of lengths between 10 and 18 are most rapidly oxidized. Methane, ethane, and propane are only attacked by highly specialized organisms. Waxes containing more than 30 carbons are quite insoluble and, therefore, recalcitrant.
4. Within the aromatic series, both alkyl–substituted benzenes and polycyclic compounds are more readily degraded than benzene [15].

An adequate supply of nitrogen and phosphorus is essential for hydrocarbon degradation. These elements are present in very low concentrations in the deep ocean and degradation of oil spilled at sea is limited by nitrogen and phosphorus deficiency. Decomposition is also strongly affected by the water temperature. At temperatures below 5°C, very little degradation occurs. The optimal temperature for hydrocarbon decomposition is approximately 25°C.

10.6.2 Esters

Organic compounds classified as esters are readily degraded by the common enzymes present in most aqueous environments. The fats that we eat are mostly triglyceride esters and the enzymes in our digestive tract convert the fats to glycerine and the fatty acids. The aliphatic fatty acids are subsequently degraded by two possible biochemical mechanisms [16][17]: β oxidation and methyl oxidation. In β oxidation two carbons at a time are converted to acetyl groups that are used for energy or synthesis reactions by the cell [12]. The steps in β oxidation, where HSCoA is coenzyme A, are:

$$HSCoA + RCH_2CH_2CH_2CH_2\overset{\overset{\textstyle O}{\|}}{C}OH$$

$$\downarrow$$

$$RCH_2CH_2CH_2CH_2\overset{\overset{\textstyle O}{\|}}{C}SCoA \ + \ H_2O$$

$$\downarrow$$

$$RCH_2CH_2CH=CH\overset{\overset{\textstyle O}{\|}}{C}SCoA \ + \ (2H)$$

$$H_2O \quad \downarrow$$

$$RCH_2CH_2\overset{\overset{\textstyle O}{\|}}{C}H-CH_2\overset{\overset{\textstyle O}{\|}}{C}SCoA$$

$$\downarrow$$

$$RCH_2CH_2\overset{\overset{\textstyle O}{\|}}{C}CH_2\overset{\overset{\textstyle O}{\|}}{C}SCoA \ + \ H$$

$$HSCoA \quad \downarrow$$

$$RCH_2CH_2\overset{\overset{\textstyle O}{\|}}{C}SCoA \ + \ CH_3\overset{\overset{\textstyle O}{\|}}{C}SCoA$$

The stages in methyl oxidation to carboxyl are:

$$RCH_2CH_2CH_3$$

O_3 ↓

$$RCH_2CH_2CH_2OOH$$

↓

$$RCH_2CH_2CH_2OH$$

↓

$$RCH_2CH_2\overset{\displaystyle O}{\overset{\|}{C}}H$$

↓

$$RCH_2CH_2\overset{\displaystyle O}{\overset{\|}{C}}OH$$

The various stages in the decomposition of an aromatic acid, after hydrolysis of the ester, are described by Schick [12] for benzoic acid.

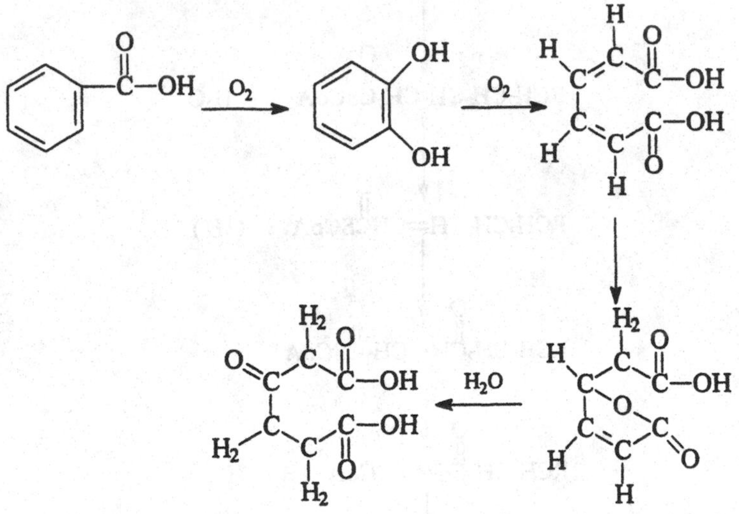

10.6.3 Nonionic Surfactants

The most plausible mechanism for the biodegradation of nonionic surfactants follows two routes—oxidation of the hydrophobic groups as outlined in Section 10.6.2 above and in the hydrolysis of the ethylene oxide chain. This hydrolysis probably occurs through an autoxidation reaction that is shown as follows [18][19]:

$$—O—CH_2CH_2—O—CH_2CH_2—O—$$

$$\downarrow \quad R \cdot \ \ or \ \ O_2$$

$$—O—CH_2CH_2—O—CHCH_2—O—$$

$$\downarrow \quad O_2 \ and/or \ RH$$

$$—O—CH_2CH_2—O—\underset{\underset{OOH}{|}}{C}HCH_2-O—$$

The hydroperoxide moiety then decomposes and the adjacent C–C bond is broken as:

$$—O—CH_2CH_2-O—\underset{\underset{OOH}{|}}{C}HCH_2-O—$$

$$\downarrow \quad Heat$$

$$—O—CH_2CH_2—O—\underset{\underset{O \cdot}{|}}{C}HCH_2-O— \ + \ \cdot OH$$

$$\downarrow$$

$$—O—CH_2CH_2-O—\underset{\underset{H}{|}}{C}=O \ \ + \ \cdot CH_2—O— \ \longrightarrow \ CH_2O$$

The unzipping of the poly(ethylene oxide) chain continues with the evolution of formaldehyde and the formation of other free radicals. If the chain is poly(propylene oxide), the material evolved is acetaldehyde.

In the early stages of biodegradation of polyoxyethylene surfactants four molecular species are present [12]:

1. With intact hydrophobe
    1.1   Intact original molecule
    1.2   Degraded ethylene oxide chain
2. With carboxylated hydrophobe
    2.1   Intact ethylene oxide chain
    2.2   Degraded ethylene oxide chain

Using various isolation procedures, Osburn and Benedict [20] have proved the validity of the carboxylation and hydrolysis mechanisms in the biodegradation of polyoxyethylene surfactants. These authors propose stepwise degradation of the ether chain by a bacterial– or enzyme–induced hydrolysis with one unit of ethylene oxide being converted to a mole of ethylene glycol and the chain always ending with the terminal hydroxyl group. This mechanism does not greatly differ from that suggested by the preceding equations if the aldehyde ending in the latter reaction is reduced to the alcohol.

1. Polyoxyethylene Alkylphenols

Alkylphenol ethoxylates are excellent scouring agents, emulsifiers, and detergents. It has been reported [12] that when compared with polyoxyethylene alcohols, the presence of a phenoxy group retards microbial degradation in wastewater systems. Some reports [21][22] also suggest that alkylphenol ethoxylates will be disallowed for washing and cleaning agents in Europe by the year 2000. Evaluations in the United States have shown that the biodegradation of these compounds proceeds more rapidly that those studies [23]–[26].

The degradation of the ethylene oxide chain was studied by Frazee *et al.* [27] in river die–away tests on a branched hydrophobe chain ethoxylated nonylphenol. The branched chain material was chosen to retard the loss of nonylphenol ethoxylated with 10 moles of EO by degradation of the alkyl chain and thus allow time for changes in the ethylene oxide chain to occur. Appreciable degradation of the ethylene oxide chain was displayed between the second and third weeks of the test and a compound that approximated nonylphenol with 4 moles of EO resulted. This same material was observed after 34 days. The ether chain of POE (4) nonylphenol may be further degraded but at a much slower rate.

The effect of chain length and degree of branching of the hydrophobe on the biodegradability of polyoxyethylene alkylphenols was investigated by Huddleston and Allred [28]. The data from a river die–away test showed that the degradation of polyoxyethylene straight–chain alkylphenols increases with the increasing length of the alkyl chain. Branched–chain alkylphenol ethoxylated compounds are much more resistant to biodegradation than the corresponding straight–chain molecules. The data obtained in this study is shown in Table 10–3.

Table 10–3
Biodegradation of Polyoxyethylene Alkylphenols in Mississippi River Water, Percent Remaining

Hydrophobe	EO %	Incubation days									
		0	1	2	4	6	8	12	15	22	26
Straight–chain tetradecylphenol	70	10^2	10^2	76	30	7	7	7	7	7	7
Straight–chain dodecylphenol	74	10^2	10^2	69	19	10	9	9	9	9	9
Straight–chain decylphenol	66	10^2	80	46	24	20	18	11	10	10	4
Straight–chain nonylphenol	65	10^2	80	55	48	47	43	40	35	35	35
Straight–chain octylphenol	59	10^2	91	72	70	70	68	65	64	64	46
Branched–chain nonylphenol	65	10^2	82	80	80	80	80	80	75	72	46

(Reference [12])

An experiment to study the dependance of biodegradation of nonylphenols on the length of the ethylene oxide chain has also been described [27]. In a river die–away test, the experiments showed that ether chains longer than 10 EO units are very resistant to degradation. Straight–chain hydrophobe structures containing more than 10 oxyethylene units per mole of hydrophobe degrade only by the carboxylation of the alkyl group. Branched–chain structures that have more than 10 moles of EO are essentially nonbiodegradable.

Other studies [12] report on the position of attachment to the hydrophobe on the biodegradation of ethoxylated alkylphenol compounds. These experiments also show that *secondary* attachment of the alkyl chain is more recalcitrant than *primary* attachment, again showing problems with chain branching.

2. Polyoxyethylene Alcohols

The emphasis on biodegradability has brought to the forefront the polyoxyethylene alcohols as active components in household products and as finish components. Frazee *et al.* [27] claim that degradation of straight-chain ethoxylated alcohols are so rapid that degradation of the ethylene oxide chain need not be considered an important factor. However, Pitter and Trauc [29] and Ruschenberg [30] have shown that the rate of degradation of the ethylene oxide chain is inversely proportional to its length. Blankenship and Piccolini [31] examined the variation in rates of degradation of polyoxyethylene *n*–dodecanols ($C_{12}EO_n$) with ethylene oxide content. The pertinent results from a river die–away test are illustrated in Figure 10–2.

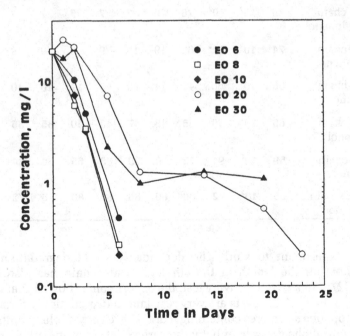

Figure 10–2. Variation in Rate of Degradation of Ethoxylated
n–Dodecanols With Ethylene Oxide Content.
*(Reference [31]. Reprinted by permission of
Soap/Cosmetic/Chemical Specialties)*

Surface tension measurements have also been used as an analytical technique. Almost no differences between the rates of degradation of the lower homologs in the series EO_6, EO_8, and EO_{10} were observed. In line with the findings reported for polyoxyethylene alkylphenols the rates of the higher homologs in the series, EO_{20} and EO_{30}, are much lower than those of the lower homologs [12].

Several investigations have dealt with the effect of the structure of the hydrophobic group and the position of attachment of the polar group on the biodegradability of ethoxylated alcohols [29][32][33]. For example, Huddleston and Allred [28] have studied the degradation of polyoxyethylene primary alcohols. The pertinent results from a river die–away test are given in Table 10–4. Colorimetry was used as the analytical technique.

Table 10–4

Biodegradation of Primary Polyoxyethylene Alcohols in Mississippi River Water, Percent Remaining

Hydrophobe	EO %	Incubation, days									
		0	1	2	3	4	5	6	8	10	13
1–Decanol	65	10^2	89	11	1	1	0				
1–Hexadecanol	63	10^2	80	13	1	1	0				
1–Dodecanol	67	10^2	95	23	0	0	0				
1–Octanol	62	10^2	91	64	—	34	17	2	0	0	
1–Octadecanol	63	10^2	93	56	—	11	5	2	2	1	0

Hydrophobe	EO %	0	1	2	4	6	8	12	15	22	26
Branched–chain tridecyl alcohol	76	10^2	10^2	64	55	55	51	49	47	19	14

(Reference [12])

The polyoxyethylene alcohols made from straight–chain alcohols are very biodegradable, with rates approaching those of soap and straight–chain alkyl sulfates. The presence of alkyl chain branching significantly retards microbial degradation. In similar experiments, Blankenship and Piccolini [31] also showed that the ethoxylated straight–chain alcohols degrade in river water at rates comparable to soft anionics. The analytical method used was surface tension. In these studies a series of polyoxyethylene fatty alcohols with differing hydrocarbon chain lengths, ranging from C_{10} to C_{16}, containing equal fractional weights of ethylene oxide was degraded. Within experimental accuracy all compounds seem to degrade at about the same rate. These investigators did find, however, that there was a variation in degradation rate depending on the position of the secondary attachment of the ethylene oxide chain. A plot of degradation rates of this type of dodecanols, with the same ethylene oxide content, is shown by Figure 10-3.

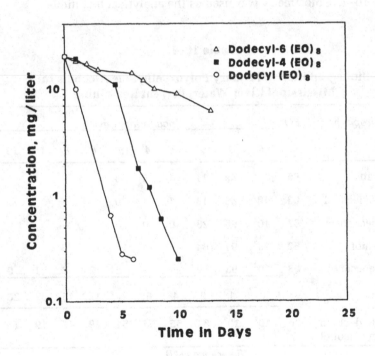

Figure 10-3. Secondary Attachment Position and Degradation Rate.
(Reference [31]. Reprinted by permission of
Soap/Cosmetic/Chemical Specialties)

The radical change in rate, which occur if the point of attachment is moved from the end of the chain toward the center, is worth noting. The change from dodecyl-4 EO_8, which is reasonably fast compared with the primary compound, to dodecyl-6 EO_8, which is rather slow, is especially noteworthy.

The methods of synthesis of ethoxylated alcohols seem to have an effect on the rate of biochemical decomposition of the surfactants. If the base alcohol is made by the oxo [34] process, as shown in the equations:

$$RCH=CH_2 + CO + H_2 \longrightarrow RCH_2CH_2CHO$$

$$RCH_2CH_2CHO + H_2 \longrightarrow RCH_2CH_2CH_2OH$$

the degradation rate may be different from those made by Ziegler [35] process. The first step in this synthesis is the reaction of ethylene with triethyl aluminum under controlled conditions of about 100°C and at 800 to 1000 psi to obtain a high molecular weight trialkyl aluminum product.

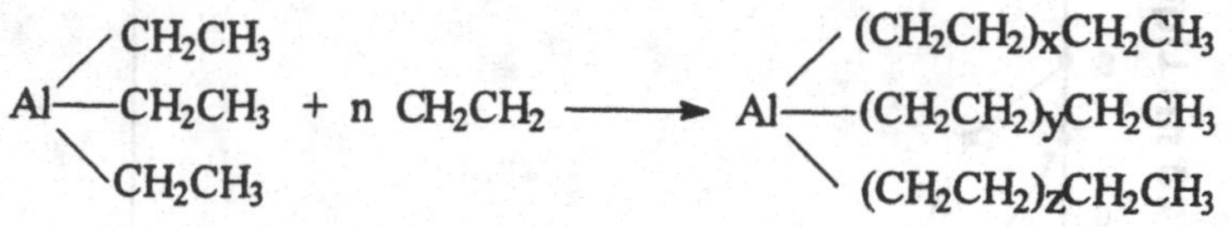

The alkyl aluminum is then oxidized with air to form an alcoholate:

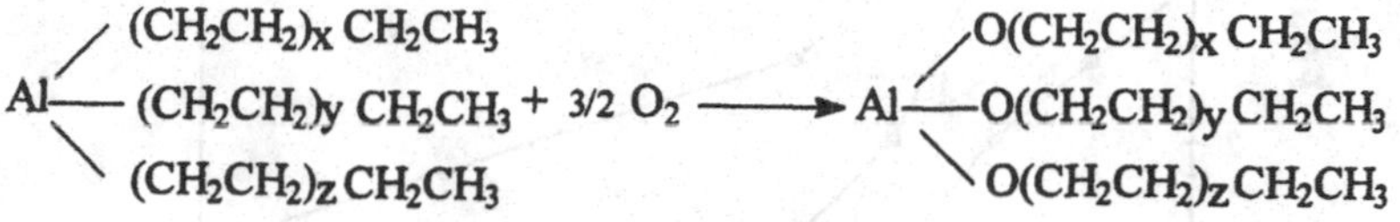

The aluminum alcoholate is then hydrolyzed with dilute sulfuric acid to yield a straight–chain primary alcohol. A Poisson mathematical relationship governs the distribution of the straight–chain alcohols made via this route.

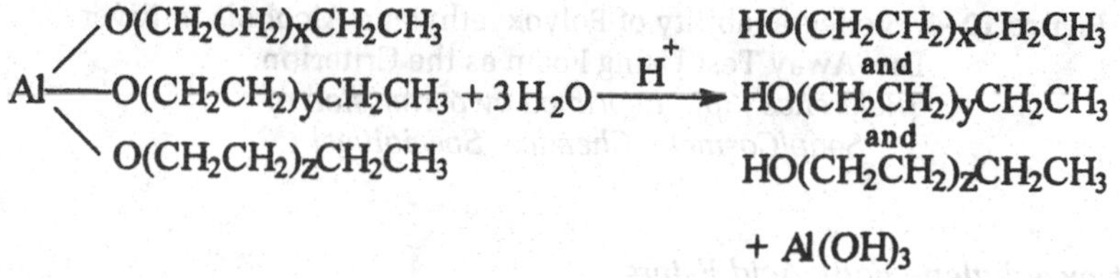

In the study of the biodegradation of ethoxylated alcohol made by different techniques, Myerly et al. [36] used secondary alcohols, Ziegler alcohols, and oxo alcohols in their experiments. The secondary alcohol was a mixture of C_{11} to C_{15} homologs, the Ziegler primary alcohol a 35:65 mixture of C_{12}/C_{14}, and the oxo alcohol a 50:50 mixture of C_{12}/C_{14}. The course of degradation in a river die–away test was followed by foam measurements and colorimetry. The data from the foam experiments are displayed in Figure 10–4. Although the colorimetry data are not shown, the data follow a similar trend. The rate of degradation of the oxo alcohol was slower than the Ziegler or secondary alcohol.

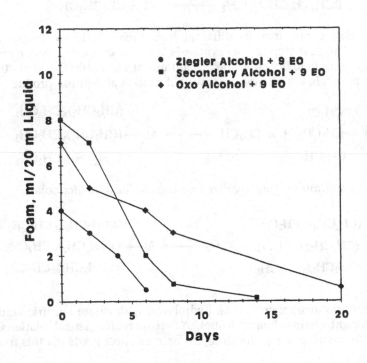

Figure 10–4. Biodegradability of Polyoxyethylene Alcohols in River
Die–Away Test Using Foam as the Criterion.
*(Reference [36]. Reprinted by permission of
Soap/Cosmetic/Chemical Specialties)*

3. Polyoxyethylene Fatty Acid Esters

Weil and Stirton [37] investigated the degradation of a nonionic surfactant derived from ethoxylated coconut fatty esters and compared them to three other types of nonionic surfactants. The concentration was determined by measuring the surface tension during a river die–away test. The results from these experiments are shown in Figure 10–5. Degradation of the ester type surfactant proceeds immediately and rapidly, whereas the ether type appears to go through a short induction period before the rapid degradation. In contrast, degradation of the nonylphenol type of nonionic surfactant was much slower.

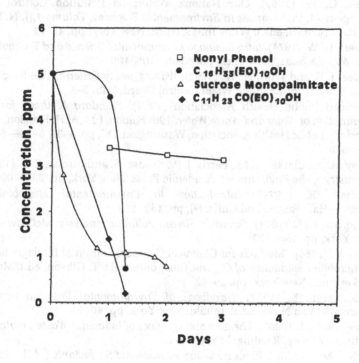

Figure 10-5. Biodegradation of Nonionic Surfactants.
*(Reference [37]. Reprinted by permission of the
American Oil Chemists' Society)*

All of the above data show that almost all finish components are degraded in the aqueous environment, except for some high molecular weight waxes. Branched aliphatic chains degrade more slowly than linear chains and aromatic ethoxylates might be less recalcitrant than once believed. Esters, ethoxylated alcohols, and polyethylene glycol esters degrade readily in almost all systems.

REFERENCES

1. Lvovitch, M. I. (1972). World Water Balance, *World Water Balance, Proceedings of the Reading Symposium, Volume 2*, UNESCO, Geneva, Switzerland, pp. 401–415

2. Kruse, C. W. (1969). Our Nation's Water: Its Pollution Control and Management, in *Advances in Environmental Sciences, Volume 1*,(J. N. Pitts and R. L. Metcalf, eds.). Wiley Interscience, New York, pp. 41–71

3. Masters, G. W. (1974). *Introduction to Environmental Science and Technology*, John Wiley & Sons, New York, pp. 85–86, 105–106

4. McKee, J. E. and Wolf, H. W. (1963). *Water Quality Criteria*, The Resource Agency of California, State Water Control Board, Pub. 3–A

5. American Public Health Association (1995). *Standard Methods for the Examination of Water and Waste Water, 19th Edition*, (M. A. H. Franson, ed.), American Public Health Association, Washington, DC, pp. 4–90, 5–14— 5–15, 5–3—5–6

6. Bailey, R. A., Clarke, H. M., Ferris, J. P., Krause, S, and Strong, R. L. (1978). *Chemistry of the Environment*, Academic Press, New York, pp. 280–300

7. Mitchell, R. (1974). *Introduction to Environmental Microbiology*, Prentice–Hall, Englewood Cliffs, NJ, pp. 132–150

8. Nemerow, N. L. (1974). *Scientific Stream Pollution Analysis*, McGraw–Hill, New York, pp. 289–290

9. Zitco, V. (1984). Methods for Chemical Characterization of Biodegradation in *Microbial Degradation of Organic Compounds*, (D. T. Gibson, ed.), Marcel Dekker, Inc., New York, pp. 29–42

10. Verschueren, K. (1977). *Handbook of Environmental Data on Organic Chemicals*, Van Nostrand Reinhold, New York, pp. 40–41

11. Nemerow, N. L. (1963). *Theories and Practices of Industrial Waste Treatment*, Addison–Wesley, Reading, MA, pp. 8-9

12. Schick, M. J. (1967). Biodegradation in *Nonionic Surfactants* (M. J. Schick ed.), Marcel Dekker, Inc., New York, pp. 971–996

13. Hummel, D. O. (1962). *Identification and Analysis of Surface–Active Agents, Volume 1*, Interscience, New York, pp. 23–34

14. Britton, L. N. (1984). Microbial Degradation of Aliphatic Hydrocarbons in *Microbial Degradation of Organic Compounds* (D. T. Gibson, ed.), Marcel Dekker, Inc., New York, pp. 89–129

15. Gibson, D. T. and Subramanian, V. (1984). Microbial Degradation of Aromatic Hydrocarbons in *Microbial Degradation of Organic Compounds* (D. T. Gibson, ed.), Marcel Dekker, Inc., New York, pp. 181–252

16. Swisher, R. D. (1963). The Chemistry of Surfactant Biodegradation, *J. Am. Oil Chemists' Soc.*, **40**: 648

17. Swisher, R. D. (1963). Intermediates in ABS (Alkylbenzenesulfonates), *Soap Chem. Specialties*, **39**: 58

18. Streitwiser, A. and Heathcock, C. H. (1985). *Introduction to Organic Chemistry*, Macmillian, New York, pp. 510–511

19. Slade, P. E., unpublished data

20. Osburn, Q. W., and Benedict, J. H. (1966). Polyethoxylated Alkyl Phenols: Relation of Structure to Biodegradation Mechanism, *J. Am. Oil Chemists' Soc.*, **43**: 141

21. Balekjian, J., Hoechst–Celanese Corporation (1992). *Waste–Water Regulations Affecting the Textile & Fiber Industries in Europe*, Presentation to Monsanto Company, August 14, 1992

22. Oslo and Paris Conventions for the Prevention of Marine Pollution (1992). *PARACOM Recommendations on Nonylphenol–Ethoxylates*, Oslo, Norway, June 22–26, 1992

23. Melnikoff, A., letter to Mr. Stig Borgvang (1993). *Comments on the PARACOM Recommendations on Nonylphenol Ethoxylates*, Union Carbide Chemicals and Plastics (Europe) S. A., January 11, 1993

24. Alkylphenol & Ethoxylates Panel (1994). *Alkylphenol Ethoxylates in the Environment: An Overview*, Chemical Manufacturers Association, Washington, DC, August, 1994

25. Naylor, C. G. (1992). Environmental Fate of Alkylphenol Ethoxylates, *Soap/Cosmetics/Chemical Specialties*, **68**(8): 27

26. Naylor, C. G., Mieure, J. P., Adams, W. J., Weeks, J. A., Castaldi, F. J., Ogle, L. D., and Romano, R. R. (1992). Alkylphenol Ethoxylates in the Environment, *J. Am. Oil Chemists' Soc.*, **69**: 695

27. Frazee, C. D., Osburn, Q. W., and Crisler, R. O. (1964). Application of Infrared Spectroscopy to Surfactant Degradation Studies, *J. Am. Oil Chemists' Soc.*, **41**: 808

28. Huddleston, R. L. and Allred, R. C. (1967). Effect of Structure on Biodegradation on Nonionic Surfactants, *Chem. Phys. Appl. Surface Active Subst, Proc. 4th International Congress Surface Activity, Brussels, Belgium, 1964, Volume 7,* 871

29. Pitter, P. and Trauc, J. (1964). Surface Active Agents in Waste Waters. IV. Biological Degradation of Nonionic Agents in Laboratory Models of Aeration Tanks, *Sb. Vysoke Skoly Chem.–Technol. Praze Technol. Vody*, Prague, Czech Republic, **7**: 201

30. Ruschenberg, E. (1963). Structural Elements of Detergents and Their Influence on Biochemical Degradation, *Vom Wasser*, Weinheim, Germany, **30**: 232

31. Blankenship, F. A. and Piccolini, V. M. (1963). Biodegradation of Nonionics, *Soap Chem. Specialties*, **39**(12) : 75

32. Steinle, E. C., Myerly, R. C., and Vath, C. A. (1964). Surfactants Containing Ethylene Oxide. Relationship of Structure to Biodegradability, *J. Am. Oil Chemists' Soc.*, **41**: 804

33. Vath, C. A. (1964). A Sanitary Engineer's Approach to Biodegradation of Nonionics, *Soap Chem. Specialties*, **40**(2): 56

34. Bent, R. L. (1963). Alcohols in *Kirk–Othmer Encyclopedia of Chemical Technology, Second Edition, Volume 1*, John Wiley & Sons, New York, pp. 531–541

35. Zeigler, K., and Gellert, H. (1955). Polymerization of Ethylene, *U.S. Patent 2,699,457*

36. Myerly, R. C., Rector, J. M., Steinle, E. C., Vath, C. A., and Zika, H. T. (1964). Secondary Alcohol Ethoxylates as Degradable Detergent Materials, *Soap Chem. Specialties*, **40**(5): 78

37. Weil, J. K. and Stirton, A. J. (1964). Biodegradation of Some Tallow–Based Surface–Active Agents in River Water, *J. Am. Oil Chemists' Soc.*, **41**: 355

AUTHOR INDEX

SUBJECT INDEX

Printed in the United States
by Baker & Taylor Publisher Services

Printed in the United States
by Baker & Taylor Publisher Services